NORTH-HOLLAND
PERSONAL LIBRARY

DIFFRACTION PHYSICS

DIFFRACTION PHYSICS

JOHN M. COWLEY

Galvin Professor of Physics
Arizona State University

NORTH-HOLLAND
AMSTERDAM · OXFORD · NEW YORK · TOKYO

ISBN: 0 444 86925 5

Published by:
North-Holland Physics Publishing
a division of
Elsevier Science Publishers B.V.
P.O. Box 103
1000 AC Amsterdam
The Netherlands

Sole distributors for the U.S.A. and Canada:
Elsevier Science Publishing Company, Inc.
52 Vanderbilt Avenue
New York, N.Y. 10017
U.S.A.

First edition 1975
Second, revised edition 1981
Paperback edition 1984
Reprinted 1986

Library of Congress Cataloging in Publication Data

Cowley, J. M. (John Maxwell), 1923-
Diffraction physics.

Bibliography: p.
Includes index.
1. Diffraction. I. Title.
QC415. C75 1981 535'.4 81-9599
ISBN 0 444 86121 1 AACR2
0 444 86925 5 (Paperback)

PRINTED IN THE NETHERLANDS

PREFACE TO THE FIRST EDITION

This book has evolved, over the last ten years, from lecture notes for classes in physical optics, diffraction physics and electron microscopy given to advanced undergraduate and graduate students. It reflects my own particular interests in electron diffraction and diffraction from disordered or imperfect crystals and employs an approach which is particularly suited to the treatment of these topics. This approach, using the Fourier transform from the beginning instead of as an extension of a Fourier series treatment, is not only more satisfying from a conceptual and theoretical basis but it provides the possibility for a unified treatment of all the different branches of diffraction physics, employing electrons, X-rays or neutrons.

Because this approach has been adopted, the generation of the familiar ideas of diffraction of radiation by solids is slow and detailed. Bragg's Law does not appear until Chapter 6 and then only as a corollary to the Ewald sphere construction. This may create difficulties for readers or students unfamiliar with the more conventional approaches. Hence this book is probably more appropriate for those who have had one elementary course in diffraction or who are already engaged in some area of diffraction work.

Much of the content relating to electron diffraction has been generated in collaboration with A.F. Moodie, of the Division of Chemical Physics, C.S.I.R.O., Australia, who has for many years been my mentor in matters of diffraction theory. My heartfelt gratitude goes to him and to all my colleagues in the C.S.I.R.O., the University of Melbourne and Arizona State University whose valuable and friendly collaboration has made this work possible and who have allowed me to use their results and their pictures to illustrate the text. I wish to thank Drs. Kato, Borrmann and Beauvillain for permission to reproduce Figs. 9.10, 14.4 and 14.5 respectively.

PREFACE TO THE SECOND EDITION

In preparing the second edition of this book, I have been greatly assisted by my students and others who have pointed out a number of errors in the first edition. Some valuable comments came from Professor Z.G. Pinsker and colleagues who prepared the Russian language edition. I am very grateful to these people and also to Dr. Peter Goodman for discussions on some of the additional sections which have been included.

Most of the changes which have been made to the first edition come within the final few chapters which deal with applications of the diffraction methods since it is in these areas that progress has been most rapid. In fact, the expansion of the field has been so great that it is impossible in a book of this size to provide any coherent account of the many interesting recent research developments. However, it is hoped that the mention of some of these topics, together with the appropriate references, will lead the reader towards more complete information on topics of a more specialized interest.

CONTENTS

Section I

PHYSICAL OPTICS

CHAPTER 1

Fresnel and Fraunhofer diffraction

1.1. Introduction

In attempting a unified treatment of the field of X-ray, neutron and electron diffraction, including electron microscopy and other imaging techniques based on diffraction, we are faced with the task of bringing together a number of theoretical treatments developed for convenience in handling particular topics. Often the "popular" treatment used by experimentalists is not just a simplification of the more rigorous methods of the theorist. It may have a different basis which is more conceptually pleasing but less tractable mathematically or is easier to visualize as in the geometric optics approximation for imaging theory.

What we hope to do is to provide the experimentalist with a more coherent account of the basic theoretical ideas of this field, using arguments which may be followed with a certain minimum of mathematics. For both theorist and experimentalist we hope to provide an appreciation of how the physical observations using different radiations and different types of sample may be knit together with a common thread of theory so that ideas and understanding in one area may be related and transfered to other areas.

For most people using X-ray diffraction for crystal structure analysis, diffraction is a three-dimensional extension of simple Fraunhofer diffraction theory applied to the idealized case of infinite periodic objects with sharply defined diffracted beam directions and a lattice of weighted points in reciprocal space. The basic mathematical tool is the Fourier series. With greater sophistication, the concept of a Fourier transform is introduced to deal with cases of finite or imperfect crystals in the same single-scattering, kinematic approximation.

This is far removed from the theoretical treatment of X-ray diffraction by Ewald [1916] or von Laue [1931] or even from the simpler, intuitive approach of Darwin [1914]. Yet these are the sources which must be relied upon for discussion of the limitations of the assumption of the simple practical theory. Also, for the understanding and interpretation of important new techniques such as X-ray topography and X-ray interferometry and the older methods using Kossel lines, there is the same basic diffraction theory but the simplifications for practical experimental use are developed along different lines.

Neutron diffraction was developed first by nuclear physicists with a jargon of differential scattering cross sections rather than atomic scattering amplitudes. Then it diversified, with structure analysis people transferring the X-ray diffraction ideas and solid-state physicists describing their experiments in terms of $\boldsymbol{k}$ vectors, Brillouin zones and so on. The added complication of inelastic scattering studies for time-dependent processes, including phonons and magnons, brought mainly an elaboration of this solid-state physics approach rather than an extension of the Fourier transform methods.

The development of electron diffraction paralleled that of X-ray diffraction in that the simple kinematic, Fraunhofer diffraction, approximation was used where possible and the dynamical theory of Bethe [1928], analogous to von Laue's X-ray theory was used where necessary. The difference from the X-ray case came from the greater importance of dynamical scattering effects and the greater complication of the dynamical effects which did occur. Consequently there was a need for relatively simple approximations for practical use, and several were developed to deal with different types of experiment.

An additional complication to the electron diffraction scene arose from the derivation of atomic scattering factors from scattering theory, the domain of nuclear and atomic physicists, with the attendant jargon. A legacy of this origin remains in that those engaged in electron diffraction from gases consider atomic scattering factors (in a unit of Å) as a function of the variable $s = 4\pi\lambda^{-1} \sin\theta$ rather than the variable $\lambda^{-1} \sin\theta$ used in X-ray diffraction or electron diffraction of the solid state, or the reciprocal lattice distance $u = 2\lambda^{-1} \sin\theta$.

The theory of image formation in the electron microscope naturally follows that of light microscopy, but this must be combined with the theory of the scattering or diffraction of electrons in the object. These two aspects of electron microscopy are brought together naturally by the unifying concept of the Fourier transform which has a basic significance for electron diffraction as for X-ray diffraction in the simple kinematical approximation and has also formed the basis for recent developments in imaging theory, introduced by Duffieux [1946].

The fact that the basic dynamical theories of X-ray diffraction and electron diffraction have been developed in terms of differential equations rather than in terms of integrals such as the Fourier transform then suggests a major cleavage of the necessary theory into two parts, formally related but very different in practice. However this gap has been bridged, at least for electron diffraction, by development of dynamical diffraction theories formulated in terms of integrals, with the Fourier transform playing a major part. A further major component can be recognized as related to the elementary physics text book treatment of Fresnel diffraction.

With this in mind, we start our discussion of diffraction with a reminder to the reader of the elements of physical optics and introduce in this way the description of diffraction, scattering and imaging in terms of the Fourier transform integral and the important associated integral, the convolution.

It would probably be rather too indigestible for most practising electron microscopists and diffractionists if we attempted a complete treatment of diffraction along these lines. Instead we relate our treatment where appropriate, to the more familiar ground of the differential equation methods. However we would like to suggest that these may be considered as parallels drawn at places to an underlying unified line of development and introduced in order to satisfy the needs of convenience or convention.

For ease of understanding, we have avoided over-rigorous arguments or mathematical complexity to the point where the purist may well criticise the logic of the development. In many cases we have duplicated mathematical statements with statements in words which are frequently less precise but may be of assistance for the less mathematically-minded reader.

To start with, we assume the reader is familiar with the use of a complex exponential to represent a wave function, the use of Fourier series to represent a periodic function and the elements of integration and differentiation. For those not having this background we recommend any one of a large number of text books of physical optics such as Ditchburn [1976] or Nussbaum and Phillips [1976]. Excellent treatments at a slightly more advanced level are given by Lipson and Lipson [1969] and Goodman [1968].

1.2. Wave equations and waves

1.2.1. Wave functions

As a means for obtaining the intensities of diffracted beams of radiation, we make use of the convenient conventional idea of a wave function. For none of the electro-magnetic radiations and particle beams we will be considering is it possible to observe any oscillatory wave motion. The wave function, a complex function of spatial coordinates which we write $\psi(\boldsymbol{r})$, is a convenient mathematical device for obtaining the observable quantity, the intensity or energy transfer in an experiment, given by $|\psi(\boldsymbol{r})|^2$. By analogy with water waves or waves in a string, we can visualize a wave function and appreciate the concepts of the wavelength λ, the wave vector $\boldsymbol{k}$ which indicates the direction of propagation and is of magnitude $2\pi/\lambda$, the frequency ν or angular frequency in radians per second, ω, the phase velocity of the wave v and the group velocity.

If we consider them as particles, the X-ray photons, electrons and neutrons have very different properties as shown by their collisions with other particles or, more practically, their generation and detection. However if we consider only their propagation through space and their scattering by matter or fields with no appreciable loss of energy all these radiations may be considered as waves, described by wave functions which are solutions of the same type of differential equation, the wave equation. Thus we can deal with a relatively simple semi-classical wave mechanics, rather than the full quantum mechanics needed for interactions of quanta involving changes of energy. The practical differences in experimental techniques and interpretation of measured intensities for the different radiations derive from the differing values of the parameters in the wave equation.

1.2.2. Electromagnetic waves

For electromagnetic waves, the wave equation is usually written for the electric of magnetic field vector $\boldsymbol{E}$ or $\boldsymbol{H}$; for example,

$$\frac{\partial^2 \boldsymbol{E}}{\partial x^2} + \frac{\partial^2 \boldsymbol{E}}{\partial y^2} + \frac{\partial^2 \boldsymbol{E}}{\partial z^2} \equiv \nabla^2 \boldsymbol{E} = \epsilon\mu \frac{\partial^2 \boldsymbol{E}}{\partial t^2}, \tag{1}$$

where ϵ is the permittivity or dielectric constant and μ is the magnetic permeability of the medium. The velocity of the waves in free space is $c = (\epsilon_0\mu_0)^{1/2}$.

For most purposes the vector nature of the wave amplitude will not be important. For simple scattering experiments with unpolarized incident radiation, for example, the only consequence of the vector properties is multiplication of the scattered intensity by a polarization factor, dependent only on the angle of scattering. Hence we may usually simplify our considerations by assuming that we are dealing with a scalar amplitude function $\psi(\boldsymbol{r})$ which is a solution to the wave equation,

$$\nabla^2 \psi = \epsilon\mu \frac{\partial^2 \psi}{\partial t^2}. \tag{2}$$

Except in special situations we may assume $\mu = 1$. The permittivity ϵ, the refractive index $n = (\epsilon/\epsilon_0)^{1/2}$ or the velocity $v = c/n$, is a function of position in space corresponding to the variation of electron density.

For purely elastic scattering, we need consider only a single frequency, ω. As we will see later, if more than one frequency is present, the intensity may be calculated for each wavelength separately and then these intensities are added. Hence we can assume that the wave-function is of the form

$$\psi = \psi_s \exp\{i\omega t\} .$$

Then, since $\partial^2\psi/\partial t^2 = -\omega^2\psi$, the wave equation (2) becomes

$$\nabla^2\psi = -\epsilon\mu\omega^2\psi$$

or

$$\nabla^2\psi + k^2\psi = 0 , \tag{3}$$

where k is $(\epsilon\mu)^{1/2}\omega$ or ω/v, the wave number.

For X-rays the dielectric constant and refractive index are very close to unity.

1.2.3. Particle waves

For electrons or other particles, the corresponding wave equation is the Schroedinger equation which may be written

$$\frac{\hbar^2}{2m}\nabla^2\psi + e\varphi(\boldsymbol{r})\psi = -i\hbar\frac{\partial\psi}{\partial t} , \tag{4}$$

where ψ is a wave function such that $|\psi|^2$ represents the probability of an electron being present in unit volume and $\varphi(\boldsymbol{r})$ is the electrostatic potential function which can be assumed to vary sufficiently slowly for our purposes even through the variation is caused by the distribution of positive and negative charges in the atoms and ions in matter. If, as before, we consider only a single frequency ω, and put $h\omega = eE$, where eE is the kinetic energy for $\varphi = 0$ i.e. in field-free space, then the time independent wave equation is

$$\nabla^2\psi + \frac{2me}{\hbar^2}\{E + \varphi(\boldsymbol{r})\}\psi = 0 , \tag{5}$$

where the bracket multiplied by e gives the sum of kinetic and potential energies of the electron.

This is of exactly the same form as (3), with

$$k^2 = (2me/\hbar^2)\{E + \varphi(\boldsymbol{r})\} .$$

Correspondingly the refractive index for electrons having kinetic (or incident) energy eE in a potential field $\varphi(\boldsymbol{r})$ is

$$n = \frac{k/\omega}{(k/\omega)_0} = \frac{(E + \varphi(\boldsymbol{r}))^{1/2}}{E^{1/2}} \approx 1 + \frac{\varphi}{2E} . \tag{6}$$

The convention for the sign of φ used here is such that $\varphi/2E$ is positive for an electron in a positive field which accelerates it. Hence the refractive index of matter for electrons is slightly greater than unity.

The simple, non-trivial solution of the wave equation (3) or (5), which can be verified by substitution, is

$$\psi = \psi_0 \exp\{-\mathrm{i}\boldsymbol{k}\cdot\boldsymbol{r}\} , \tag{7}$$

representing a plane wave proceeding in the direction specified by the vector $\boldsymbol{k}$ which has a magnitude $|\boldsymbol{k}| \equiv k = 2\pi/\lambda$. Adding the time dependence, this becomes

$$\psi = \psi_0 \exp\{\mathrm{i}(\omega t - \boldsymbol{k}\cdot\boldsymbol{r})\} . \tag{8}$$

The choice of the positive sign of the exponent is arbitrary. A negative sign before the i is equally valid. The choice as to whether the ωt or the $\boldsymbol{k}\cdot\boldsymbol{r}$ is given the minus sign is a matter of convention. It is important only to be consistent in the choice. The form of (8) implies that for t constant the phase decreases for increasing positive distance in the direction of $\boldsymbol{k}$ and for a fixed position the phase increases with time.

A further solution of importance, corresponding to a spherical wave radiating from a point source, taken to be the origin, is

$$\psi = \psi_0 \frac{\exp\{\mathrm{i}(\omega t - kr)\}}{r} , \tag{9}$$

where $r = |\boldsymbol{r}|$. Since the direction of propagation is always radial, $\boldsymbol{k}\cdot\boldsymbol{r} = kr$. This form is consistent with the inverse square law, since the energy per unit area, proportional to $\psi\psi^*$, will be proportional to r^{-2}.

1.3. Superposition and coherence

1.3.1. Superposition

We have seen that the wave functions for both electromagnetic waves and particles are solutions of linear differential equations such as (3). A fundamental property of the solutions of such equations is that the sum of any two solutions is also a solution. In terms of waves, we can state that if any number of waves, represented by the wave-functions ψ_n, coexist in space, the resultant disturbance will be represented by the wave-function $\psi = \Sigma_n \psi_n$. This property may be referred to as the Principle of Superposition. This is fundamental to all considerations of diffraction.

In general, waves may be considered to originate from sources having approximately the dimensions of atoms. X-rays arise when electrons jump from one energy level to another in an atom, or when an incident electron collides with an atom. Electrons may be considered to be emitted from the region of high electron density surrounding an atom. However the "uncertainty principle" implies that the origin of a photon or electron can not be specified with a precision of better than about half a wavelength, so that "point sources" must be of at least this size.

For all our present purposes, half a wavelength is sufficiently small to be neglected in comparison with all other dimensions considered so that we may think of the radiation as coming from one or more independently-emitting point sources. Hence, first of all, we must consider how we may treat the usual case of radiation emitted from a large number of such independent sources.

1.3.2. Independent point sources

We start by considering two sources at point Q_1 and Q_2. At some point of observation, P, the wave function will be

$$\psi_P = \psi_{1P} + \psi_{2P} = A_1 \exp\{i(\omega_1 t - \boldsymbol{k}_1 \cdot \boldsymbol{r}_1)\}$$
$$+ A_2 \exp\{i[(\omega_2 t - \boldsymbol{k}_2 \cdot \boldsymbol{r}_2) + \alpha(t)]\},$$

where the phase factor $\alpha(t)$ has been added to suggest that the phase of ψ_2 relative to ψ_1 will vary in a random manner with time as separate, independent quanta of radiation are generated at the two points. The amplitudes A_1, A_2 are assumed real. The intensity at P at any one time will be

$$I(t) = |\psi_{1P} + \psi_{2P}|^2$$
$$= A_1^2 + A_2^2 + 2A_1A_2 \cos\{(\omega_1 - \omega_2)t - (\boldsymbol{k}_1 \cdot \boldsymbol{r}_1 - \boldsymbol{k}_2 \cdot \boldsymbol{r}_2) - \alpha(t)\}. \tag{10}$$

The observed intensity will be the time average of $I(t)$ since the frequencies involved and the rate of fluctuations of the phase factor $\alpha(t)$ are both unobservably high; so that

$$I_{obs} = \langle I(t)\rangle_t = \langle A_1^2\rangle_t + \langle A_2^2\rangle_t$$

$$+ 2\langle A_1 A_2 \cos\{(\omega_1 - \omega_2)t - (\boldsymbol{k}_1\cdot\boldsymbol{r}_1 - \boldsymbol{k}_2\cdot\boldsymbol{r}_2) - \alpha(t)\}\rangle_t \,. \tag{11}$$

For different frequencies, the last term will fluctuate with a frequency $\omega_1 - \omega_2$ and so will average out to zero even if $A_1 = A_2, r_1 = r_2$ and $\alpha(t) = 0$, i.e. if the two waves come from one source. If the frequencies are the same for two separate sources, the phase term $\alpha(t)$ will vary in a random manner so that the cosine term will have positive and negative values with equal probability and so, again, will average to zero. Hence

$$I_{obs} = \langle A_1^2\rangle_t + \langle A_2^2\rangle_t$$

$$= I_1 + I_2 \,, \tag{12}$$

where I_1 and I_2 are the intensities at the point of observation due to the two sources separately.

From this rather rough, non-rigorous argument we may conclude that for any single source the observed intensity will be the sum of the intensities for different frequencies: for two or more independent sources having the same or different frequencies, the observed intensity will be the sum of the intensities given by the individual sources separately.

Hence we are justified in following the usual procedure of considering diffraction problems for the ideal case of a monochromatic point source either at a finite distance or at infinity (parallel, coherent illumination) and then generalizing, if necessary, to take account of the finite source size or frequency range. A source of finite size may be considered as made up of a collection of very small, independently-emitting points. For each one of these points the intensity at a point of observation is calculated taking account of the relative intensity of the source at that point. Then the intensities are added for all source points. If there is a finite range of frequencies, the intensity is calculated for each frequency and then the summation is made for all frequencies.

The actual intensity measurement recorded depends on the characteristics of the detector, including the variation of its response with the frequency of the incident radiation and also with the position and direction of the radiation.

Most, if not all, detectors may be considered incoherent in that they are not sensitive to phase relationships of the incident radiation. They record only incident energies. We are justified in considering the intensity at each point of the detector and for each wavelength separately and then adding intensities for all points of the detector and all wavelengths, taking the response functions of the detector into account, to find the actual measured intensity.

1.4. Huygens' principle

1.4.1. Kirchhoff's formulation

The original concept of Huygens pictures the propagation of waves through space as involving the generation of secondary waves at each point of a wave-front so that the envelope of the secondary waves becomes a new wave-front. This is a simple intuitive picture which permits an "understanding" or interpretation of the formula of Kirchhoff which may be derived directly from the wave equation by application of Green's theorem. This derivation is a standard one, reproduced in many physics texts and will not be repeated here.

The Kirchhoff formula may be written thus; the disturbance at a point P due to any wave-field u, representing a solution of the wave equation, is given by integrating over any closed surface containing P as

$$u_{\mathrm{P}} = \frac{1}{4\pi} \oint_S \left(\frac{\exp\{-ikr\}}{r} \cdot \operatorname{grad} u - u \cdot \operatorname{grad} \left[\frac{\exp\{-ikr\}}{r} \right] \right) \mathrm{d}S . \tag{13}$$

If we apply this to the wave field u due to a point source of unit strength at Q, we obtain

$$u_{\mathrm{P}} = \frac{1}{4\pi} \int \frac{\exp\{-ikr\}}{r} \cdot \frac{\exp\{-ikr_q\}}{r_q}$$

$$\times \left\{ \left(\frac{1}{r} + ik \right) \cos \widehat{\boldsymbol{nr}} - \left(\frac{1}{r_q} + ik \right) \cos \widehat{\boldsymbol{nr}}_q \right\} \mathrm{d}S ,$$

where $\widehat{\boldsymbol{nr}}$ and $\widehat{\boldsymbol{nr}}_q$ are the angles between the surface normal $\boldsymbol{n}$ and the vectors $\boldsymbol{r}$ and $\boldsymbol{r}_q$ to the points P and Q, as in Fig. 1.1 the usual convention in this case being that all distances are measured from the surface.

Making the approximation that all the dimensions involved are much greater than the wavelength, r^{-1} and r_q^{-1} can be neglected in comparison with k and we obtain

$$u_{\mathrm{P}} = \frac{\mathrm{i}}{2\lambda} \oint \frac{\exp\{-ikr_q\}}{r_q} \cdot \frac{\exp\{-ikr\}}{r} \{\cos \widehat{\boldsymbol{nr}} - \cos \widehat{\boldsymbol{nr}}_q\} \,\mathrm{d}S . \tag{14}$$

This may be interpreted in terms of the Huygens concept as implying that

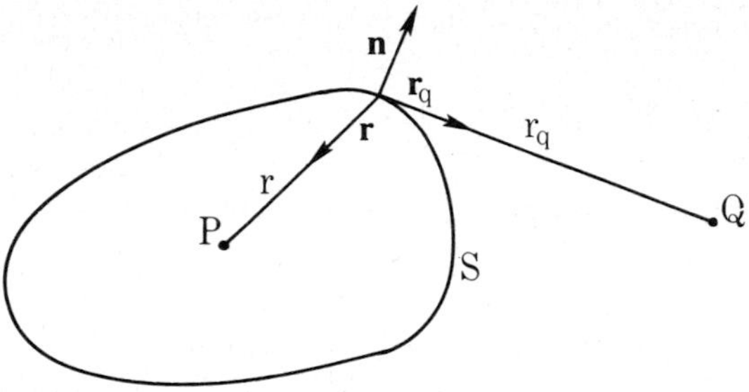

Fig. 1.1. Diagram illustrating Kirchhoff's derivation of the amplitude at P due to a source point Q.

the secondary spherical waves emitted from each surface element, $r^{-1} \exp\{-ikr\} \mathrm{d}S$, have relative amplitude proportional to the amplitude of the incident wave from Q, $r_q^{-1} \exp\{-ikr_q\}$.

There is a change of phase of $\pi/2$, represented by multiplying by i, a scale factor of λ^{-1}, and an obliquity factor, $\frac{1}{2}\{\cos \widehat{\boldsymbol{nr}} - \cos \widehat{\boldsymbol{nr}_q}\}$, which ensures that the waves add up with maximum amplitude in the forward direction and are not propagated backwards. Thus the expression (14) may be considered as a mathematical way of writing a slightly modified, more accurate, version of Huygens' Principle.

1.4.2. Application of the Kirchhoff formula

The main area of usefulness of the Kirchhoff formula is not for propagation through free space, as assumed for (14), but rather for cases in which the wave function $\psi(x, y)$ represents the wave modified by the presence of an object and the surface of integration is conveniently the exit surface of the object. We neglect the many complications which may arise, particularly for wavelengths comparable with the dimensions of the structure of the object, with confidence that these complications need not concern us.

For the simple, idealized case of a planar, two dimensional object placed between P and Q, we may define a "transmission function" $q(X, Y)$ which is multiplied by the incident wave function to represent the effect of the object on the amplitude and phase of the incident wave. Then, for a point source of radiation, the wave incident on the object is $r_q^{-1} \exp\{-ikr_q\}$, the wave emerging from the object is $q(X, Y) r_q^{-1} \exp\{-ikr_q\}$ and the wave-function $\psi(x, y)$ at a point on a plane of observation is given by

$$\psi(x, y) = \frac{\mathrm{i}}{2\lambda} \iint \frac{\exp\{-\mathrm{i}kr_q\}}{r_q}$$

$$\times\, q(X, Y) \frac{\exp\{-\mathrm{i}kr\}}{r} \{\cos \widehat{\boldsymbol{Zr}} + \cos \widehat{\boldsymbol{Zr}_q}\} \,\mathrm{d}X\, \mathrm{d}Y\,, \tag{15}$$

where X, Y and Z are coordinates referred to orthogonal axes. The surface S is considered to be the X, Y plane plus a closing surface at infinity. We follow the convention by which the Z axis is the direction of propagation.

Equations such as (15) may be applied directly to the elastic interaction of fast electrons (energies greater than about 20 keV) with matter since the wavelengths are less than 10^{-1} Å and so considerably smaller than atomic dimensions. They may also be applied to the scattering of X-rays or thermal neutrons (wavelength of the order of 1 Å) by accumulations of matter which are considerably larger than the wavelength i.e. for small-angle scattering experiments. However these equations can not be used for the consideration of the scattering of X-rays by electrons or of neutrons by nuclei. For such purposes we must rely on scattering theory, which follows a somewhat different line in its development from the wave equation, or from the fundamental postulates of quantum mechanics.

1.5. Scattering theory approach

1.5.1. Integral form of wave equation

Again, for simplicity, we consider only a scalar wave, neglecting the complications of treating vector quantities, and so deal with scattering theory as developed for the scattering of particles by a potential field. (See, for example, Wu and Ohumara [1962].)

The wave equation (5) may be written

$$[\nabla^2 + k_0^2 + \mu\varphi(\boldsymbol{r})]\, \psi = 0\,, \tag{16}$$

where k_0 represents the wave number for the incident wave in free space and μ is a parameter which specifies the strength of the interaction with the potential field.

As an alternative, most useful for scattering theory, we may write the equivalent integral equation by making use of the Green's function $G(\boldsymbol{r}, \boldsymbol{r}')$. For scattering radiation from a potential field, $G(\boldsymbol{r}, \boldsymbol{r}')$ represents the amplitude at a point of observation $\boldsymbol{r}$, due to a point of unit scattering strength at $\boldsymbol{r}'$ in the field, (see Fig. 1.2):

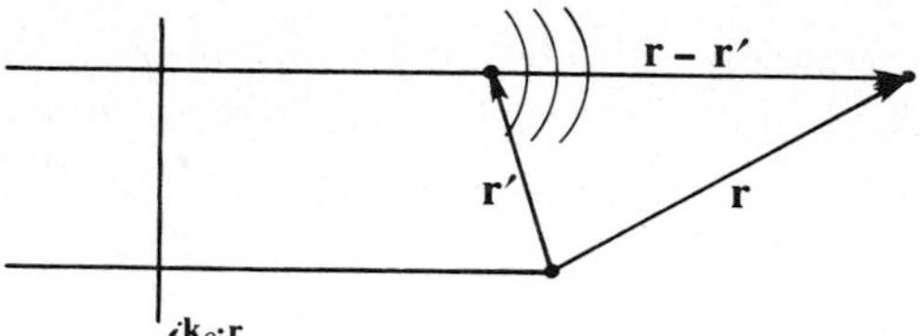

Fig. 1.2. Definition of distances for scattering problems.

$$\psi(\boldsymbol{r}) = \psi^{(0)}(\boldsymbol{r}) + \mu \int G(\boldsymbol{r}, \boldsymbol{r}')\, \varphi(\boldsymbol{r}')\, \psi(\boldsymbol{r}')\, \mathrm{d}\boldsymbol{r}' \tag{17}$$

where $\psi^{(0)}(\boldsymbol{r})$ represents the wave incident on the scattering field and the integral represents the scattered radiation. The appropriate form of the Green's function is

$$G(\boldsymbol{r}, \boldsymbol{r}') = \frac{\exp\{ik|\boldsymbol{r} - \boldsymbol{r}'|\}}{4\pi|\boldsymbol{r} - \boldsymbol{r}'|}\ , \tag{18}$$

which from (9) is the amplitude at $\boldsymbol{r}$ of a spherical wave emitted from the point $\boldsymbol{r}'$.

Thus the expression (17) may be compared with the expression (15) derived from the Kirchhoff integral. We may interpret it as indicating that each point of the scattering field gives off a spherical wave (18), and the strength of this wave depends on the value of the scattering potential $\varphi(\boldsymbol{r}')$ and the wave function $\psi(\boldsymbol{r}')$. We would have the exact three-dimensional equivalent of (15) if we could say that the amplitude of the scattered wave was proportional to the incident wave amplitude $\psi^{(0)}(\boldsymbol{r})$, but this is not possible in general because the scattered radiation itself forms part of the wave function $\psi(\boldsymbol{r})$. Hence we have an integral equation which is much more difficult to solve.

1.5.2. Born series

If the amplitude of the scattered wave is much less than the incident wave amplitude, we may assume as a first approximation that $\psi(\boldsymbol{r})$ in the integral may be replaced by $\psi^{(0)}(\boldsymbol{r})$, the incident wave amplitude. This is the "First Born Approximation". Higher order Born approximations are found by iteration. Thus the second order approximation is given by replacing $\psi(\boldsymbol{r})$ in the integral by $\psi^{(0)}(\boldsymbol{r}) + \psi^{(1)}(\boldsymbol{r})$ and so on.

For a plane incident wave $\psi^{(0)}(\boldsymbol{r}) = \exp\{-i\boldsymbol{k}_0 \cdot \boldsymbol{r}\}$, the first Born approximation gives

$$\psi^{(0)}(\boldsymbol{r}) + \psi^{(1)}(\boldsymbol{r}) = \exp\{-i\boldsymbol{k}_0 \cdot \boldsymbol{r}\}$$

$$+ \frac{\mu}{4\pi} \int \frac{\exp\{-ik|\boldsymbol{r} - \boldsymbol{r}'|\}}{|\boldsymbol{r} - \boldsymbol{r}'|}\, \varphi(\boldsymbol{r}') \exp\{-i\boldsymbol{k}_0 \cdot \boldsymbol{r}'\}\, \mathrm{d}\boldsymbol{r}'\ . \tag{19}$$

Under the usual conditions of a scattering experiment we may assume that the point of observation is at $\boldsymbol{r} = \boldsymbol{R}$ where $\boldsymbol{R}$ is very large compared with the dimensions of the scattering field. Then, writing $\boldsymbol{q} = \boldsymbol{k} - \boldsymbol{k}_0$, we obtain the asymptotic form,

$$\psi^{(0)}(\boldsymbol{r}) + \psi^{(1)}(\boldsymbol{r}) = \exp\{-\mathrm{i}\boldsymbol{k}_0 \cdot \boldsymbol{R}\}$$

$$+ \frac{\mu}{4\pi} \frac{\exp\{-\mathrm{i}k_0 R\}}{R} \int \varphi(\boldsymbol{r}') \exp\{-\mathrm{i}\boldsymbol{q} \cdot \boldsymbol{r}'\} \, \mathrm{d}\boldsymbol{r}' \, . \tag{20}$$

The characteristic scattering from the potential field may then be sorted out from this by defining a scattering amplitude $f(\boldsymbol{q})$ such that the right-hand side of (20) is

$$\exp\{-\mathrm{i}\boldsymbol{k}_0 \cdot \boldsymbol{R}\} + R^{-1} \exp\{-\mathrm{i}k_0 R\} f(\boldsymbol{q}) \, .$$

Then

$$f(\boldsymbol{q}) = \frac{\mu}{4\pi} \int \varphi(\boldsymbol{r}) \exp\{-\mathrm{i}\boldsymbol{q} \cdot \boldsymbol{r}\} \, \mathrm{d}\boldsymbol{r} \, . \tag{21}$$

This is the first Born approximation for the scattering amplitude, given by assuming that the incident wave amplitude is equal to the total wave amplitude in the scattering field: i.e. that the scattered amplitude is negligibly small. The scattered wave is made up of contributions scattered directly from the incident wave. Hence this is a single-scattering approximation.

In general this first Born approximation is very good and useful for weakly scattering fields or objects. For stronger scattering, further terms may be calculated in the Born series

$$\psi(\boldsymbol{r}) = \psi^{(0)}(\boldsymbol{r}) + \psi^{(1)}(\boldsymbol{r}) + \psi^{(2)}(\boldsymbol{r}) + \dots$$

by use of the recurrence relationship

$$\psi^{(n)}(\boldsymbol{r}) = \frac{\mu}{4\pi} \int \frac{\exp\{-\mathrm{i}k|\boldsymbol{r} - \boldsymbol{r}'|\}}{|\boldsymbol{r} - \boldsymbol{r}'|} \varphi(\boldsymbol{r}') \, \psi^{(n-1)}(\boldsymbol{r}') \, \mathrm{d}\boldsymbol{r}' \, . \tag{22}$$

However the convergence of this Born series is usually poor if the first order approximation fails. The addition of the second order term improves the approximation over a rather limited range of scattering strengths and is sometimes useful in suggesting the nature of the modifications needed when the first order approximation fails. But the higher order terms become rapidly more complicated and difficult to calculate and it is not often useful to evaluate them.

1.6. Reciprocity

One point of particular interest for the practical application of these mathematical treatments is illustrated by the formulas we have given. In the expressions derived from the Kirchhoff formula relating to radiation from a point Q and a point of observation P it is seen that the expressions are symmetrical with respect to P and Q. If the source were at P and the point of observation were at Q, so that the vectors $\boldsymbol{r}$ and $\boldsymbol{r}_q$ were interchanged, the same amplitude would result.

The expression (20) for single scattering, and the more general series (17) relate an incident plane wave defined by $\boldsymbol{k}_0$ and an out-going plane wave $\boldsymbol{k}$, corresponding to source and observation points at infinity. Again, if the source and observation points were interchanged there would be no change in the resulting amplitude. These observations are consistent with the general Reciprocity theorem of wave optics introduced by Helmholtz [1886]. This may be restated as follows:

The amplitude of the disturbance at a point P due to radiation from a point Q which has traversed any system involving elastic scattering processes only, is the same as the amplitude of the disturbance which would be observed at Q if the point source were placed at P.

The first application of this theorem to diffraction phenomena was made by von Laue [1935] who employed it to simplify the theoretical treatment of Kossel lines due to diffraction of radiation originating from point sources within a crystal. More recently Pogany and Turner [1968] showed its general applicability and usefulness in electron diffraction and electron microscopy under n-beam dynamical diffraction conditions. They further showed that the same relationship applies for intensities (but not amplitudes) if inelastic scattering is involved, provided that the energy loss in the inelastic scattering process is negligible.

The applications of reciprocity have recently multiplied in number and importance. We will meet several in later chapters.

1.7. Fresnel diffraction

1.7.1. Small angle approximation

From the general Kirchhoff formula it is possible to derive relatively simple forms appropriate to the conditions which typify particular classes of diffraction effects. "Fresnel diffraction" refers usually, although not exclusively, to the phenomena observed close to a two-dimensional object illuminated by plane parallel incident light. If the object plane is perpendicular to the direction of incidence we may replace the incident radiation in (15) by $\psi_0 = 1$, which represents a plane wave of unit amplitude having zero phase at $Z = 0$. Then the amplitude on any plane of observation a distance R beyond the object is

$$\psi(x,y) = \frac{\mathrm{i}}{2\lambda} \iint q(X,Y) \frac{\exp\{-\mathrm{i}kr\}}{r} (1 + \cos \widehat{\boldsymbol{Zr}})\, \mathrm{d}X\, \mathrm{d}Y \tag{23}$$

where $r^2 = (x - X)^2 + (y - Y)^2 + R^2$.

If, as was assumed earlier, the wavelength is small compared with the dimensions of the object the angles of deflection of the radiation will be small and we may make a small angle approximation, putting $\cos \widehat{\boldsymbol{Zr}} = 1$, $r = R$ in the denominator and, in the exponent,

$$\begin{aligned} r &= \{R^2 + (x - X)^2 + (y - Y)^2\}^{1/2} \\ &\approx R + \frac{(x - X)^2 + (y - Y)^2}{2R}, \end{aligned} \tag{24}$$

so that

$$\begin{aligned} \psi(x,y) &= \frac{\mathrm{i} \exp\{-\mathrm{i}\boldsymbol{k}\cdot\boldsymbol{R}\}}{R\lambda} \\ &\times \iint q(X,Y) \exp\left\{\frac{\mathrm{i}k[(x - X)^2 + (y - Y)^2]}{2R}\right\} \mathrm{d}X\, \mathrm{d}Y\,. \end{aligned} \tag{25}$$

For the special case that the object has a transmission function $q(X)$ which varies in one dimension only, as in the idealized cases of straight edges, slits and so on, the integral over Y can be carried out to give

$$\int_{-\infty}^{\infty} \exp\left\{\frac{-\mathrm{i}k(y - Y)^2}{2R}\right\} \mathrm{d}Y = \left(\frac{R\lambda}{\mathrm{i}}\right)^{1/2}$$

so that

$$\psi(x) = \left(\frac{\mathrm{i}}{R\lambda}\right)^{1/2} \exp\{-\mathrm{i}\boldsymbol{k}\cdot\boldsymbol{R}\} \int_{-\infty}^{\infty} q(X) \exp\left\{\frac{-\mathrm{i}k(x - X)^2}{2R}\right\} \mathrm{d}X\,. \tag{26}$$

1.7.2. Fresnel integrals

For the special cases of objects which are either completely opaque or transparent so that $q(X) = 1$ in some parts and $q(X) = 0$ elsewhere, this integral reduces to a sum of integrals of the form

$$\int_a^b \exp\left\{\frac{-ik(x-X)^2}{2R}\right\} dX .$$

These integrals can be expressed in terms of the Fresnel integrals

$$C(x) = \int_0^x \cos(\tfrac{1}{2}\pi u^2)\, du ,$$

$$S(x) = \int_0^x \sin(\tfrac{1}{2}\pi u^2)\, du .$$

For example let us consider a single slit of width a for which the transmission function in one dimension is

$$q(x) = \begin{cases} 1 & \text{if } |x| \leqslant a/2 \\ 0 & \text{if } |x| > a/2 . \end{cases}$$

Then the integral of (26) is

$$\int_{-a/2}^{a/2} \exp\left\{\frac{-ik(x-X)^2}{2R}\right\} dX .$$

Putting $u^2 = 2(x-X)^2/R\lambda$, this becomes

$$\int_A^B \cos(\tfrac{1}{2}\pi u^2)\, du - \mathrm{i} \int_A^B \sin(\tfrac{1}{2}\pi u^2)\, du$$

where

$$A^2 = 2(x+\tfrac{1}{2}a)^2/R\lambda \quad \text{and} \quad B^2 = 2(x-\tfrac{1}{2}a)^2/R\lambda ;$$

and this is equal to

$$\{C(B) - C(A)\} - \mathrm{i}\,\{S(B) - S(A)\} .$$

The Fresnel integrals may be evaluated numerically by reference to tabulations of these functions. Alternatively the amplitudes may be found graphically by use of the very ingenious device of the Cornu Spiral.

These methods and results for the most usual applications are described in detail in most text books of optics and so will not be reproduced here. The typical Fresnel fringes formed by diffraction at a straight edge are well known in both visible-light optics and electron optics and are used in electron microscopy as an aid to focussing.

1.7.3. Periodic objects – "Fourier images"

There are very few functions $q(X)$ for which the integral (26) can be evaluated analytically, but the few favorable cases include the important one of periodic functions which we consider in some detail here because of its interesting implications for the electronoptical imaging of crystals.

Consider a planar object with transmission function $q(X) = \cos(2\pi X/a)$. This would be difficult to achieve in practice, but not impossible since the negative sign required could be obtained by use of a "half-wave plate" which changes the phase of the incident radiation by π. The integral (26) becomes

$$\psi(x) = \left(\frac{i}{R\lambda}\right)^{1/2} \exp\{-ikR\} \int_{-\infty}^{\infty} \cos\left(\frac{2\pi X}{a}\right) \exp\left\{\frac{-ik(x-X)^2}{2R}\right\} dX\,, \tag{27}$$

or, putting $X = x - W$,

$$\psi(x) = \left(\frac{i}{R\lambda}\right)^{1/2} \exp\{-ikR\} \cos\left(\frac{2\pi x}{a}\right)$$

$$\times \int_{-\infty}^{\infty} \cos\left(\frac{2\pi W}{a}\right) \exp\left\{\frac{-ikW^2}{2R}\right\} dW\,.$$

The corresponding sine term is an integral over an odd function and so is zero. Making use of the standard integral form

$$\int_{-\infty}^{\infty} \exp\{-a^2x^2\} \cos bx \, dx = \frac{\pi^{1/2}}{a} \exp\{-b^2/4a^2\}\,, \tag{28}$$

we obtain

$$\psi(x) = \exp\{-ikR\} \exp\{i\pi R\lambda/a^2\} \cos(2\pi x/a)\,, \tag{29}$$

so that the intensity distribution on the plane of observation is

$$I(x) = \cos^2(2\pi x/a)\,. \tag{30}$$

This is independent of R. Hence on any plane of observation at any distance from the object, the intensity distribution will be exactly the same as at the exit face of the object. If such an object, illuminated by plane parallel light, were imaged, the image could never be made to go out of focus.

For a general periodic object (restricted for simplicity to functions symmetric about the origin) we write

$$q(X) = \sum_h F_h \cos(2\pi hx/a) .$$

Then we may use the result (29) for each cosine component giving

$$\psi(x) = \exp\{-ikR\} \sum_h F_h \exp\{i\pi R\lambda h^2/a^2\} \cos(2\pi hx/a) . \tag{31}$$

This is again a periodic distribution, having the same periodicity as the object but with relative phases of the Fourier coefficients which vary with the distance R.

For special values of R given by $R = 2na^2/\lambda$, the exponent in (31) is $2N\pi$ where $N = nh^2$ is an integer. Then

$$\psi(x) = \exp\{-ikR\} \sum_h F_h \cos(2\pi hx/a) , \tag{32}$$

and the intensity distribution is exactly the same as at the exit surface of the object.

For the values $R = na^2/\lambda$ where n is odd, the exponent will be $(2N + 1)\pi$, so that, since h is odd if nh^2 is odd,

$$\psi(x) = \exp\{-ikR\} \sum_h F_h \exp\{-i\pi h\} \cos(2\pi hx/a)$$

$$= \exp\{-ikR\} \sum_h F_h \cos \frac{2\pi h}{a}(x \pm \tfrac{1}{2}a) . \tag{33}$$

Thus again the intensity distribution is exactly the same as for $R = 0$ except that it is translated by half the periodicity.

Thus for parallel incident illumination the intensity distribution is exactly repeated at intervals of $R = na^2/\lambda$. This phenomenon was first reported by Talbot [1836], was partially explained by Rayleigh [1881], explored further by Weisel [1910] and Wolfke [1913] and "rediscovered" and examined in detail for possible applications to electron diffraction and electron microscopy by Cowley and Moodie [1957a,b,c; 1960] who named these self-images of a periodic object "Fourier Images". A number of different aspects of the phenomenon will be explored in various contexts in later pages. Here we mention only the case that the incident radiation comes from a point source at a finite distance, say R_1, from the periodic object. Then (26) is replaced by

$$\psi(x) = C \int_{-\infty}^{\infty} \exp\left\{\frac{-ikX^2}{2R_1}\right\} q(X) \exp\left\{\frac{-ik(x-X)^2}{2R}\right\} \mathrm{d}X\,, \tag{34}$$

where the first exponential function represents a spherical wave from the point source incident of the object having transmission function $q(X)$.

It is readily shown that the Fourier images will by magnified by a factor $(R+R_1)/R_1$ and will occur at positions given by

$$\frac{1}{R} + \frac{1}{R_1} = \frac{\lambda}{na^2}. \tag{35}$$

The real Fourier images occur at successive larger intervals in R up to some maximum positive value of n, beyond which λ/na^2 is less than $1/R_1$. Then there are virtual Fourier images on the source side of the object corresponding to negative values of n and to values from plus infinity down to the n value for which $R_1 - na^2/\lambda$, as suggested in Fig. 1.3.

1.8. Fraunhofer diffraction

The approximation to the general Kirchhoff formula which typifies the Fraunhofer diffraction condition is that the overall dimensions of the object should be very much smaller than the distances to source or point of observation; or, in the more usual parlance, the source and point of observation are effectively at infinity. Thus we assume an incident plane wave of unit amplitude and write

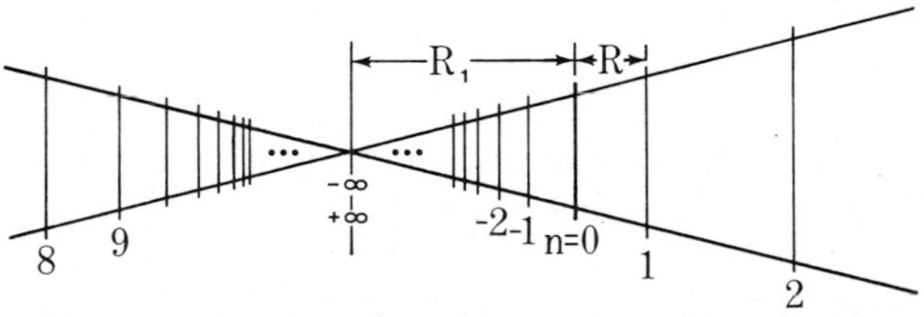

Fig. 1.3. Positions of the Fourier images of a periodic object, n = 0, illuminated by a point source.

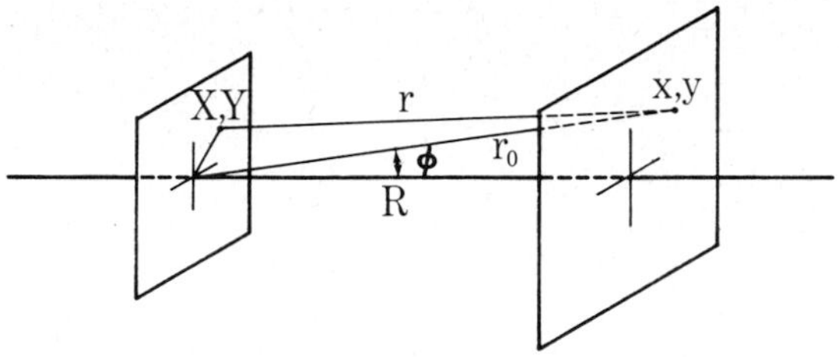

Fig. 1.4. Coordinate systems for the description of Fraunhofer diffraction.

$$\psi(x, y) = \frac{\mathrm{i}}{\lambda}\left(\frac{1+\cos\varphi}{2}\right)\iint q(X, Y)\,\frac{\exp\{-\mathrm{i}kr\}}{r}\,\mathrm{d}X\,\mathrm{d}Y\,. \tag{36}$$

The obliquity factor is taken outside the integral since it will not vary appreciably for the small range of X and Y considered. It may be included with the other constants and terms of modulus unity which are combined into a factor C and usually ignored when only relative, and not absolute intensities, are of interest. The distance r is compared to the distance r_0 from the origin of coordinates in the object to the point of observation, as suggested in Fig. 1.4. Then

$$r = \{R^2 + (x-X)^2 + (y-Y)^2\}^{1/2}$$

$$\approx \{(R^2 + x^2 + y^2) - 2(xX + yY)\}^{1/2}$$

$$\approx r_0 - \frac{x}{r_0}X - \frac{y}{r_0}Y\,.$$

Putting

$$x/r_0 = l = \sin\varphi_x\,,$$

$$y/r_0 = m = \sin\varphi_y\,,$$

where φ_x and φ_y are the components of the scattering angle, we have the amplitude as a function of the angular variables;

$$\psi(l, m) = C\iint_{-\infty}^{\infty} q(X, Y)\exp\{\mathrm{i}k(lX + mY)\}\,\mathrm{d}X\,\mathrm{d}Y \tag{37}$$

and

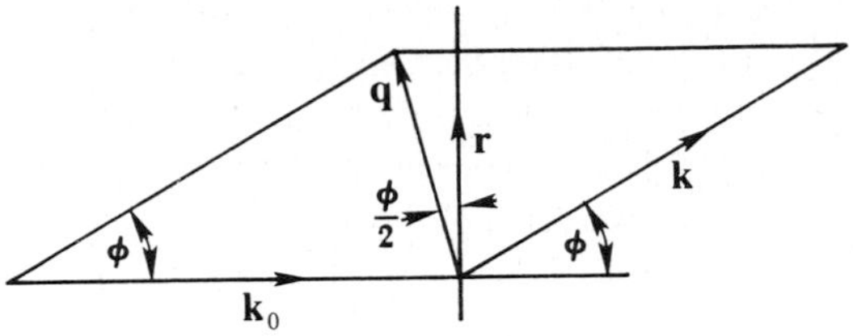

Fig. 1.5. Definition of scattering vectors.

$$C = (1 + \cos\varphi)\, \mathrm{i}\, \frac{\exp\{-\mathrm{i}kr_0\}}{2r_0\lambda} .$$

From this formula, all the well-known results on the Fraunhofer diffraction from one and two dimensional objects, such as slits, apertures, gratings etc., may be derived. However we note at this stage that the integral in (37) has the form of a Fourier transform integral. In the next chapter we outline the properties of the Fourier transform and work out the form of the diffraction pattern for a number of simple object as examples in the use of this type of transform which represents the basic tool for most kinematical diffraction theory and a great deal of dynamical theory.

At this stage we note the equivalence of (37) and the expression for the diffracted amplitude in the First Born approximation of scattering theory (20). If we consider the transmission function $q(X, Y)$ to be derived from a planar distribution of scattering potential $V(\boldsymbol{r}')$, the integral of (20) is confined to values of $\boldsymbol{r}'$ lying in the X, Y plane.

As seen from Fig. 1.5, the magnitude of $\boldsymbol{q}$ is

$$|\boldsymbol{q}| = 2k \sin(\varphi/2) ,$$

so that, for the X direction, for example,

$$\boldsymbol{q}\cdot\boldsymbol{r}' = 2kr' \sin(\varphi_x/2)\cos(\varphi_x/2) = kr \sin\varphi_x$$

$$= klX .$$

Hence the integrals of (20) and (37) are equivalent. The integral of (20) has the advantage that it may be immediately applied to three-dimensional distributions provided that the necessary conditions of weak scattering apply. The form (37) necessarily deals with objects which can be considered two-dimensional and is essentially a single-scattering form, since no distance can exist between double scattering events, but there is no restriction on the strength of the scattering. The presence of the obliquity factor in (37) is a consequence of the assumption of a planar distribution of the scattering function. It does not exist in (20) because no such restriction is placed on $V(\boldsymbol{r}')$.

The manner in which (37) may be used to develop a treatment for the scattering from extended three-dimensional distributions will be discussed in a later chapter. In this way we provide an additional approach to the problem of the many-beam dynamical theory for the diffraction from strongly scattering crystals.

Problems

1. Starting from the expression (34) show that for a point source of radiation Fourier images of a periodic object will occur at positions given by equation (35). Find the magnification of the images. {Note: make use of the standard integral, (28).}

2. Make use of a Cornu spiral to plot, to scale, the intensity distribution in a plane of observation 5000 Å behind an opaque straight edge illuminated by a parallel beam of electrons of wavelength 0.04 Å (and energy 87 keV). [Note: good Cornu spirals are given in many text books including Stone, *Radiation and Optics* and Joos, *Theoretical Physics* and in Jahnke and Emde, *Tables of Functions.*]

Give qualitative arguments to suggest how this intensity distribution would be affected if

(a) the plane of this Fresnel diffraction pattern is imaged by an electron microscope having a resolution of 10 Å, or,

(b) the incident electron beam comes from a finite source which subtends an angle of 10^{-3} radians at the straight edge.

CHAPTER 2

Fourier transforms and convolutions

2.1. Preliminaries

2.1.1. Delta-functions and discontinuities

In this chapter we provide the mathematical background for much of what follows. Most considerations of kinematical diffraction involve the use of the Fourier transform in one form or another. One of the most important properties of the Fourier transform involves the concept of a Convolution, or Convolution integral, and for both this and the Fourier transform it is convenient to use delta-functions. Hence we define and discuss this latter function first so that we may make use of it later.

In this, as in all subsequent mathematical exposition, we do not strive for mathematical rigor. We will assume that all functions considered are sufficiently well-behaved in a mathematical sense to allow us to use them in a way which makes physical sense. When we make use of delta functions and other discontinuous functions we do so as a matter of convenience. When we wish to represent real situations, in which there can be no discontinuities, we use the discontinuous mathematical functions as a short-hand notation for the continuous functions of physical reality to which they approximate.

For example, a Dirac delta function at $x = a$ is defined by

$$\delta(x-a) = \begin{cases} 0 & \text{for} \quad x \neq a \\ \infty & \text{for} \quad x = a \end{cases}$$

and

$$\int_{-\infty}^{\infty} \delta(x-a)\mathrm{d}x = 1 \ . \tag{1}$$

The delta function at $x = 0$, $\delta(x)$, can be considered as the limit of a set of real continuous functions, such as gaussians:

$$\delta(x) = \lim_{a\to\infty}\left[\frac{a}{\pi^{1/2}}\exp\{-a^2x^2\}\right]. \tag{2}$$

As a tends to infinity the gaussian function has a maximum value tending to infinity and a half-width $(1/a)$ tending to zero but the integral over the function is always unity. Then we can use a delta function as a convenient notation for any function of integral unity in the form of a sharp peak having a width so small that it is not experimentally significant.

Similarly a weighted delta function $c\cdot\delta(x)$ is used to indicate a sharp peak for which the integral is c. The definition of the function as the limit of a series of functions can be convenient to clarify or prove various relationships. For example,

$$\delta(bx) = \lim_{a\to\infty}\left[\frac{a}{\pi^{1/2}}\exp\{-a^2b^2x^2\}\right] = b^{-1}\cdot\delta(x)\,.$$

We may define a delta function in two dimensions, $\delta(x,y)$ which is zero except at $x = y = 0$ and for which

$$\iint\delta(x,y)\mathrm{d}x\,\mathrm{d}y = 1\,.$$

Similarly for any number of dimensions we may define $\delta(\boldsymbol{r})$ or $\delta(\boldsymbol{r}-\boldsymbol{a})$ where $\boldsymbol{r}$ and $\boldsymbol{a}$ are vectors in the n-dimensional space.

We note that in two dimensions $\delta(x)$ represents a line and in three dimensions $\delta(x)$ represents a plane.

For a detailed discussion of delta functions see, for example, Lighthill [1960] or Arsac [1966].

We note in passing the important definition of a delta function,

$$\delta(x) = \int_{-\infty}^{\infty}\exp\{2\pi\mathrm{i}xy\}\mathrm{d}y\,, \tag{3}$$

which will occur later in connection with Fourier transforms (equations (33) to (36)).

2.1.2. Convolutions

In one dimension, the Convolution integral (or, convolution product, or convolution, or faltung, or folding) of two functions $f(x)$ and $g(x)$ is defined as

$$C(x) = f(x) * g(x) \equiv \int_{-\infty}^{\infty} f(X)g(x-X)\,\mathrm{d}X\,. \tag{4}$$

By simple change of variable, we find that

$$f(x) * g(x) = \int_{-\infty}^{\infty} g(X)f(x-X)\,\mathrm{d}X = g(x) * f(x)\,. \tag{5}$$

For two or more dimensions we may use the vector form,

$$f(\boldsymbol{r}) * g(\boldsymbol{r}) = \int f(\boldsymbol{R})g(\boldsymbol{r}-\boldsymbol{R})\,\mathrm{d}\boldsymbol{R}\,. \tag{6}$$

The identity operation is the convolution with the Dirac delta function:

$$\begin{aligned} &f(x) * \delta(x) = f(x)\,,\\ &f(x) * \delta(x-a) = f(x-a)\,. \end{aligned} \tag{7}$$

2.1.3. Examples of convolutions

The convolution integral (4) or (6) appears with great frequency in many areas of scientific work, being fundamental to the interpretation of most experimental measurements and an essential component of many sophisticated theoretical developments such as the Green's function methods of theoretical physics. To get a clearer understanding of what is involved we analyse the integral (4) in detail. It may be written thus: the function $f(X)$ is multiplied by the function $g(X)$ which has been shifted to an origin at $X = x$ and inverted to give $g(x-X)$. The value of the product of $f(X)$ and $g(x-X)$ is integrated over X and the result plotted as a function of x to give $C(x)$.

This is exactly the process involved, for example, in measuring the intensity of a spectral line by scanning it with a detector having a finite slit as input aperture, as suggested in Fig. 2.1. The coordinate X may represent the angle of scattering of light by a prism or diffraction grating and the intensity distribution $I(X)$ shows the spectral lines of interest.

The intensity distribution in the spectrum is measured by recording the intensity passing through a slit whose transmission function is

$$g(X) = \begin{cases} 0 & \text{for} \quad |X| \geqslant a/2\,,\\ 1 & \text{for} \quad |X| < a/2\,, \end{cases}$$

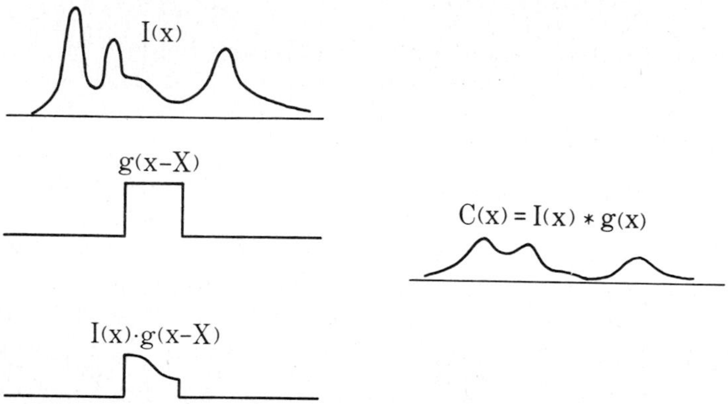

Fig. 2.1. A convolution operation. An intensity function $I(X)$ representing a spectral line, is multiplied by the transmission function $g(x-X)$ of a slit, centered at $X = x$; the product of these two functions is integrated to give the measured intensity $I_{obs}(x)$.

i.e. the slit transmits all radiation within a range of X of width a and excludes all other radiation. When the slit is placed at $X = x$, the transmitted intensity as a function of X is $I(X)g(x-X)$. What is recorded is the total transmitted intensity. If this is plotted as a function of x, the position of the slit, we get

$$I_{obs} = I(x) * g(x) = \int_{-\infty}^{\infty} I(X)g(x-X)\mathrm{d}X .$$

A single very sharp spectral line will give an observed intensity of unity for a range of x values equal to the slit width a;

$$I_{obs} = \delta(x) * g(x) = \int_{-\infty}^{\infty} \delta(X)g(x-X)\mathrm{d}X = g(x) .$$

For a general intensity distribution, $I(X)$, each sharp spectral line or each part of a broader spectral line will be "spread out" by the "spread-function" $g(X)$, so that the recorded intensity I_{obs} will be less sharply peaked or less well resolved than the original spectrum.

Likewise, the blurring of the image due to the imperfections of a camera lens may be described in terms of convolution of the ideally perfect image intensity with some function $g(x,y)$. For a point source of light, the ideal image would be a delta function. The spreading gives

$$I_{obs} = g(x,y) * \delta(x,y) = g(x,y) \,. \tag{8}$$

For a general object, consisting of a large number of independently emitting points sources, having an ideal image, $I_0(x,y)$,

$$I_{obs} = g(x,y) * I_0(x,y) \,. \tag{9}$$

Thus each point of the original intensity distribution is spread into a disc of intensity and the overlapping of these discs gives a blurring and loss of resolution of the image. This illustrates the concept that the "spread function" represents the response of the system to a delta function input, in this case a point source. This is the basis for the Green's function method useful in scattering theory and many other fields of physics and also for the analysis of the properties of an electronic circuit by measuring its response to a sharp voltage or current pulse.

An excellent example of a convolution is provided by Huygens' Principle as expressed by the Kirchhoff formulas. Each point on a wave front is considered to give rise to a spherical secondary wave having an initial amplitude proportional to the amplitude of the incident wave. Then the amplitudes of the secondary waves are added to give the amplitude on a plane of observation. Thus the amplitude function, $q(x,y)$, on the original wave front is spread out by a function which represents the spherical secondary wave emitted by a single point on the wave front

We have written this explicitly as a convolution integral in the equation (1.25) for Fresnel diffraction in the small angle approximation. This may then be re-written

$$\psi(x,y) = q(x,y) * \left[\frac{\mathrm{i}\exp\{-\mathrm{i}kR\}}{R\lambda} \exp\left\{ \frac{-\mathrm{i}k(x^2+y^2)}{2R} \right\} \right], \tag{10}$$

and the function in square brackets may be referred to as the "propagation function", or the wave function obtained for a point source, $q(x,y) = \delta(x,y)$.

Likewise the First Born approximation result given in eq. (1.19) may be re-written in terms of a convolution. The integral giving the singly scattered amplitude becomes,

$$V(\boldsymbol{r}) \exp\{-\mathrm{i}\boldsymbol{k}_0 \cdot \boldsymbol{r}\} * \frac{\exp\{-\mathrm{i}kr\}}{r} \,. \tag{11}$$

The first function here is the incident wave modified by the potential field $V(\boldsymbol{r})$. This is convoluted with the amplitude due to a point source, namely the amplitude of a spherical wave from the origin. Thus the equation (1.19) or (11) simply states that the observed amplitude is the sum of the amplitudes of spherical waves from all points of the scatterer, and the amplitude of scattering from each point is proportional to the product of the incident wave amplitude and the value of the potential function, $V(\boldsymbol{r})$ at the point.

2.2. Fourier transforms: general

2.2.1. Definitions

The Fourier transform of a one-dimensional function $f(x)$ is defined as

$$\mathcal{F}\,[f(x)] \equiv F(u) = \int_{-\infty}^{\infty} f(x) \exp\{2\pi \mathrm{i} ux\}\,\mathrm{d}x\,. \tag{12}$$

The inverse transform, $\mathcal{F}^{-1}$, is defined so that

$$f(x) = \mathcal{F}^{-1}\,[\mathcal{F}\{f(x)\}]$$

$$= \int_{-\infty}^{\infty} F(u) \exp\{-2\pi \mathrm{i} ux\}\,\mathrm{d}u\,. \tag{13}$$

Here we follow the convention of including 2π in the exponent. This is the convention commonly used in considerations of diffraction and is convenient in that it avoids the necessity of adding a constant multiplier in either (12) or (13). In other conventions, often used in solid-state physics, the 2π is omitted from the exponent. It must then be included as a constant; the integral in either (12) or (13) is multiplied by $(2\pi)^{-1}$ or both integrals are multiplied by $(2\pi)^{-1/2}$.

For more than one dimension we may use the vector form of (12)

$$F(\boldsymbol{u}) = \int f(\boldsymbol{r}) \exp\{2\pi \mathrm{i} \boldsymbol{u}\cdot\boldsymbol{r}\}\,\mathrm{d}\boldsymbol{r}\,. \tag{14}$$

The vector $\boldsymbol{u}$ may be considered as a vector in "Fourier transform space". For the three-dimensional case, for example, the vector $\boldsymbol{r}$ may be considered to have coordinates x, y, z and $\boldsymbol{u}$ may be considered to have coordinates u, v, w.

Then the scalar product is $\boldsymbol{u}\cdot\boldsymbol{r} = ux + vy + wz$ and

$$F(u, v, w) = \iiint_{-\infty}^{\infty} f(x, y, z) \exp\{2\pi \mathrm{i}(ux+vy+wz)\}\,\mathrm{d}x\,\mathrm{d}y\,\mathrm{d}z\,, \tag{15a}$$

and

$$f(x,y,z) = \iiint_{-\infty}^{\infty} F(u,v,w) \exp\{-2\pi i(ux+vy+wz)\} \,du\,dv\,dw \tag{15b}$$

We have seen in Chapter 1 that the amplitude of scattering from an object in the Fraunhofer diffraction approximation, whether derived from the Kirchhoff formulation or from scattering theory, is described by a Fourier transform integral. In (1.37) for example, we put $u = l/\lambda$, $v = m/\lambda$ to get the two-dimensional form of (15b). We may thus describe the diffraction amplitude in terms of a distribution in Fourier transform space, often referred to, as we shall see, as "Reciprocal space". This will be our most common means for deriving or understanding diffraction effects, so we now proceed to describe and illustrate the most important properties and behavior of the Fourier transform.

2.2.2. *Properties of Fourier transforms*

Instead of using the complex exponential, we may write (12) as

$$F(u) = \int_{-\infty}^{\infty} f(x) \cos(2\pi ux)\,dx + i \int_{-\infty}^{\infty} f(x) \sin(2\pi ux)\,dx\,. \tag{16}$$

If the function $f(x)$ is real and an even function, so that $f(-x) = f(x)$, the sine integral is zero so that

$$F(u) = \int_{-\infty}^{\infty} f(x) \cos(2\pi ux)\,dx = 2\int_{0}^{\infty} f(x) \cos(2\pi ux)\,dx\,, \tag{17}$$

and $F(u)$ is a real function.

If $f(x)$ is a real odd function so that $f(-x) = -f(x)$ then the cosine integral is zero and

$$F(u) = i\int_{-\infty}^{\infty} f(x) \sin(2\pi ux)\,dx = 2i\int_{0}^{\infty} f(x) \sin(2\pi ux)\,dx\,, \tag{18}$$

and the function $F(u)$ is pure imaginary.

Since any real function can be written as a sum of an even and an odd function,

$$f(x) = \tfrac{1}{2}\{f(x) + f(-x)\} + \tfrac{1}{2}\{f(x) - f(-x)\} = f_e(x) + f_0(x)\,, \tag{19}$$

we may write

$$F(u) = A(u) + \mathrm{i}B(u)$$

where $A(u)$ and $B(u)$ are real functions given by

$$A(u) = 2\int_0^\infty f_\mathrm{e}(x)\cos(2\pi ux)\,\mathrm{d}x\ ,$$

and

$$B(u) = 2\int_0^\infty f_0(x)\sin(2\pi ux)\,\mathrm{d}x\ . \tag{20}$$

It is these cosine and sine integrals that are tabulated, for the most part, in the lists of Fourier integrals given, for example, in the compilations of Erdeyli [1954] and Sneddon [1951]. Photographic representations of Fourier transforms, obtained by optical diffraction, are given by Harburn et al. [1975].

A number of general relationships may be written for any function $f(x)$, real or complex, thus;

Real space	*Fourier transform space*	
$f(x)$	$F(u)$	(21)
$f(-x)$	$F(-u)$	(22)
$f^*(x)$	$F^*(-u)$	(23)
$f(ax)$	$\frac{1}{a}F(u/a)$	(24)
$f(x) + g(x)$	$F(u) + G(u)$	(25)
$f(x-a)$	$\exp\{2\pi\mathrm{i}au\}\,F(u)$	(26)
$\frac{\mathrm{d}}{\mathrm{d}x}f(x)$	$(-2\pi\mathrm{i}u)\,F(u)$	(27)
$\frac{\mathrm{d}^n}{\mathrm{d}x^n}f(x)$	$(-2\pi\mathrm{i}u)^n F(u)$.	(28)

These relationships may be readily proved by writing out the relevant integrals. For (24):

$$\int_{-\infty}^{\infty} f(ax)\exp\{2\pi i u x\}\,\mathrm{d}x$$

$$=\frac{1}{a}\int_{-\infty}^{\infty} f(X)\exp\ 2\pi i\frac{uX}{a}\ \mathrm{d}X$$

$$=\frac{1}{a}F(u/a)\ .$$

For (26):

$$\int_{-\infty}^{\infty} f(x-a)\exp\{2\pi i u x\}\,\mathrm{d}x$$

$$=\int_{-\infty}^{\infty} f(x)\exp\{2\pi i(uX+ua)\}\,\mathrm{d}X$$

$$=F(u)\exp\{2\pi i u a\}\ .$$

For (27):

$$\int\frac{\mathrm{d}}{\mathrm{d}x}f(x)\exp\{2\pi i u x\}\,\mathrm{d}x$$

$$=\iint\frac{\mathrm{d}}{\mathrm{d}x}\left[F(v)\exp\{-2\pi i v x\}\,\mathrm{d}v\right]\exp\{2\pi i u x\}\,\mathrm{d}x$$

$$=\int(-2\pi i v)F(v)\int\exp\{2\pi i(u-v)x\}\,\mathrm{d}v\,\mathrm{d}x$$

$$=(-2\pi i u)F(u)\quad\text{since}\quad\int\exp\{2\pi i(u-v)x\}\,\mathrm{d}x=\delta(u-v)\ .$$

The relation (28) follows by repetition of the derivation of (27).

2.2.3. Multiplication and convolution

We add the two important relationships, the Multiplication theorem

$$\mathcal{F}[f(x)\cdot g(x)]=F(u)*G(u)\ , \tag{29}$$

i.e. the Fourier transform of a product of two functions is the convolution of their Fourier transforms, and the Convolution theorem,

$$\mathcal{F}[f(x)*g(x)]=F(u)\cdot G(u)\ , \tag{30}$$

i.e. the Fourier transform of the convolution of two functions is the product of their Fourier transforms.

Here we have followed the convention that functions in real space are represented by small letters and the Fourier transforms are represented by the corresponding capital letters.

These theorems are again easily proved, using manipulations which are in general non-rigorous but are valid for the types of functions which we will be employing. For example, to derive (30), putting $x - X = y$ in the left side of the integral form gives,

$$\iint f(X) g(x-X) \mathrm{d}X \cdot \exp\{2\pi i u x\} \mathrm{d}x$$

$$= \int f(X) g(y) \exp\{2\pi i u (X+y)\} \mathrm{d}X \mathrm{d}y$$

$$= \int f(X) \exp\{2\pi i u X\} \mathrm{d}X \int g(y) \exp\{2\pi i u y\} \mathrm{d}y$$

$$= F(u) \cdot G(u) .$$

2.2.4. *Space and time*

In addition to providing the relationship between spatial distributions $f(\boldsymbol{r})$ and the diffraction amplitudes, $F(\boldsymbol{u})$ the Fourier transform also relates the variation of a function in time, $f(t)$, and the corresponding frequency distribution. Thus we may write

$$F(\nu) = \int_{-\infty}^{\infty} f(t) \exp\{2\pi i \nu t\} \mathrm{d}t , \tag{31}$$

and

$$f(t) = \int_{-\infty}^{\infty} F(\nu) \exp\{-2\pi i \nu t\} \mathrm{d}\nu , \tag{32}$$

where we use the frequency ν, rather than the angular frequency ω. To make the analogy complete it is necessary to introduce the artifice of negative frequencies. While this is inconsistent with common parlance, it is readily seen that a negative frequency can correspond to a negative progression of the phase of a wave in time, i.e. to a wave going "backwards".

For a function of both space and time, $f(\boldsymbol{r}, t)$, a Fourier transform may be made with respect to any one or all of the coordinates or with respect to time, or with respect to both spatial coordinates and time. Hence where any ambiguity is possible it is necessary to specify the variable or variables involved in the Fourier transform, possibly by use of a subscript on $\mathcal{F}$. For example,

$$\mathcal{F}_{x,t}[f(x,y,z,t)] = F(u,y,z,\nu)$$

$$= \iint_{-\infty}^{\infty} f(x,y,z,t) \exp\{2\pi i(ux+\nu t)\} \, dx \, dt \, .$$

2.3. Fourier transforms and diffraction: examples

We now provide a series of examples to serve the two functions of familiarizing the reader with common Fourier transforms and demonstrating the application of Fourier transforms to kinematical diffraction. We refer for the most part to diffraction from simple one- or two-dimensional objects.

2.3.1. Point source or point aperture

The amplitude distribution of a very small source or the transmission through a very small aperture (or slit) in one dimension may be described as $\delta(x)$, or by $\delta(x-a)$ when it is not at the origin. The Fourier transform used to derive the Fraunhofer diffraction pattern is

$$\mathcal{F}\,\delta(x) = 1 \, ,$$
$$\mathcal{F}\,\delta(x-a) = \exp\{2\pi iua\} \, . \tag{33}$$

To show this we write the integral

$$\int_{-\infty}^{\infty} \delta(x-a) \exp\{2\pi iux\} \, dx \, .$$

The integrand is zero except for $x = a$. Hence the integral may be written

$$\exp\{2\pi iua\} \int_{-\infty}^{\infty} \delta(x-a) \, dx = \exp\{2\pi iua\} \, .$$

The amplitude of a diffraction pattern will be proportional to $F(u) = \mathcal{F}\,\delta(x-a)$ where $u = l/\lambda$. The intensity observed will be proportional to $|F(u)|^2 = 1$.

Thus, as is well known, the Fraunhofer diffraction pattern from a point source has uniform intensity, apart from the factor $1/R^2$ and the obliquity factor (if applicable) which have been omitted.

2.3.2. A plane wave: the inverse of 2.3.1

Fourier transform of a plane wave (equation 1.8) with respect to t only gives

$$\mathcal{F}_t \exp\{2\pi i(\nu_1 t - x/\lambda_1)\} = \delta(\nu+\nu_1) \exp\{-2\pi i x/\lambda_1\}, \tag{34}$$

i.e. a delta function in frequency and a plane wave in real space. Fourier transform with respect to x gives

$$\mathcal{F}_x \exp\{2\pi i(\nu_1 t - x/\lambda_1)\} = \exp\{2\pi i\nu_1 t\}\, \delta(u - 1/\lambda_1), \tag{35}$$

i.e. a sinusoidal variation in time and a delta function in reciprocal space. Transforming with respect to both variables gives,

$$\mathcal{F}_{x,t} \exp\{2\pi i(\nu_1 t - x/\lambda_1)\} = \delta(\nu+\nu_1)\cdot\delta(u - 1/\lambda_1), \tag{36}$$

which is a delta function in both frequency and reciprocal space.

If we take the Fourier transform of all possible waves which can exist in a given medium with respect to both x and t we obtain a set of points defining the relationship between the frequency ν and λ^{-1} (or between the angular frequency ω and k) which is known as the dispersion relation for that type of wave and the particular medium.

2.3.3. Translation of an object

$$\mathcal{F} f(x-a) = \mathcal{F}[f(x) * \delta(x-a)] = F(u) \exp\{2\pi i u a\}, \tag{37}$$

where we have used (33) and the convolution theorem (30).

Thus translation of the object in real space has the effect of multiplying the amplitude in reciprocal space by a complex exponential. The intensity distribution of the Fraunhofer diffraction pattern is given by $|F(u)|^2$, which is independent of the translation.

2.3.4. Slit function

The transmission function of a slit of width a in an opaque screen is given by

$$f(x) \equiv q(x) = \begin{cases} 0 & \text{if} \quad |x| > a/2\,, \\ 1 & \text{if} \quad |x| \leqslant a/2\,. \end{cases}$$

The Fourier transform of this is

$$F(u) = \int_{-a/2}^{a/2} \exp\{2\pi i u x\}\,\mathrm{d}x = \frac{\sin(\pi a u)}{\pi u}\,. \tag{38}$$

Putting $u = l/\lambda$, we obtain the well-known form of the diffraction pattern

$$F(l) = a \sin\left(\frac{\pi l a}{\lambda}\right) \Big/ \left(\frac{\pi l a}{\lambda}\right), \tag{39}$$

with an intensity distribution

$$I(l) = a^2 \sin^2\left(\frac{\pi l a}{\lambda}\right) \Big/ \left(\frac{\pi l a}{\lambda}\right)^2, \tag{40}$$

which has a central maximum value a^2 for $l = 0$, and has decreasing subsidiary maxima with increasing $|l|$ with zero values for $l = n\lambda/a$ for $n \neq 0$ (see Fig. 2.2).

2.3.5. *Slit function-alternative*

To illustrate the application of (27), we note that for the slit function defined in 2.3.4,

$$g(x) = \frac{\mathrm{d}}{\mathrm{d}x} q(x) = \delta(x + \tfrac{1}{2}a) - \delta(x - \tfrac{1}{2}a)\,,$$

as suggested in Fig. 2.2. Then

$$G(u) = \exp\{-\pi i a u\} - \exp\{\pi i a u\} = -2i \sin(\pi a u)\,.$$

But, since from (27)

$$G(u) = (-2\pi i u)\,F(u)\,,$$

it follows that, as in equation (38),

$$F(u) = \frac{\sin(\pi a u)}{\pi u}\,.$$

2.3.6. *Straight edge*

The transmission function is

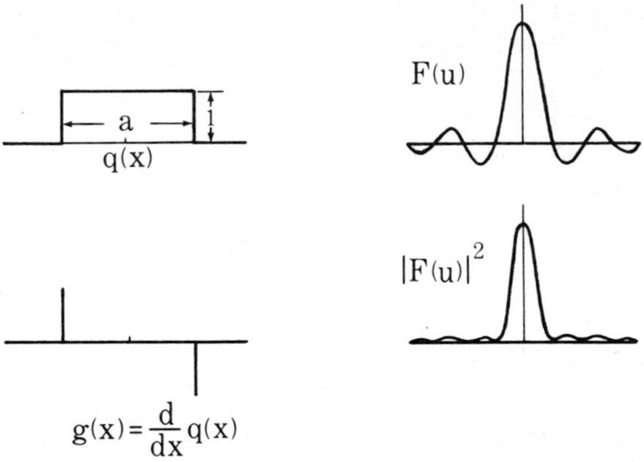

Fig. 2.2. Derivation of the Fourier transform of a slit function by consideration of its differential.

$$f(x) = \begin{cases} 0 & \text{if } x < 0\,, \\ 1 & \text{if } x \geqslant 0\,. \end{cases}$$

Using the same procedure as in 2.3.5 we put

$$g(x) = \frac{\mathrm{d}}{\mathrm{d}x} f(x) = \delta(x)\,.$$

Then

$$G(u) = 1$$

and

$$F(u) = (2\pi \mathrm{i} u)^{-1}\,.$$

But an indeterminate constant term has been omitted in the integration. We note that

$$\int_{-\infty}^{\infty} (f(x) - \tfrac{1}{2})\,\mathrm{d}x = 0\,.$$

This suggests that the missing constant term is 1/2. Inserting this gives the correct result,

$$F(u) = -\tfrac{1}{2}\delta(u) + (2\pi \mathrm{i}u)^{-1} . \tag{41}$$

2.3.7. Rectangular aperture

In the two dimensional form of 2.3.4 we define the transmission function of a rectangular aperture as

$$f(x,y) = \begin{cases} 1 & \text{if} \quad |x| < a/2 \text{ and } |y| < b/2 , \\ 0 & \text{elsewhere} . \end{cases}$$

Then

$$F(u,v) = \int_{-a/2}^{a/2} \exp\{2\pi \mathrm{i}ux\}\mathrm{d}x \int_{-b/2}^{b/2} \exp\{2\pi \mathrm{i}vy\}\mathrm{d}y$$

$$= ab \frac{\sin(\pi au)}{\pi au} \frac{\sin(\pi bv)}{\pi bv} , \tag{42}$$

so that, for diffraction from a rectangular aperture the intensity distribution is

$$I(u,v) = a^2 b^2 \frac{\sin^2(\pi au)}{(\pi au)^2} \frac{\sin^2(\pi bv)}{(\pi bv)^2} . \tag{43}$$

The maximum intensity at $u = v = 0$ is a^2b^2. Zeros occur at intervals of a^{-1} along the u direction, parallel to the x axis and b^{-1} along the v direction, as suggested in Fig. 2.3. Thus the intensity function falls off through decreasing oscillations in each direction. The dimensions of the distribution in reciprocal space are inversely proportional to the dimensions in real space. Integrating over the whole peak gives an integrated intensity equal to ab, the area of the aperture, as expected.

2.3.8. Circular aperture

The transmission function for a circular aperture is written

$$f(x,y) = \begin{cases} 1 & \text{if} \quad (x^2+y^2)^{1/2} < a/2 , \\ 0 & \text{elsewhere} . \end{cases}$$

The Fourier transform is best carried out by writing the Fourier integral in plane polar coordinates. The result is

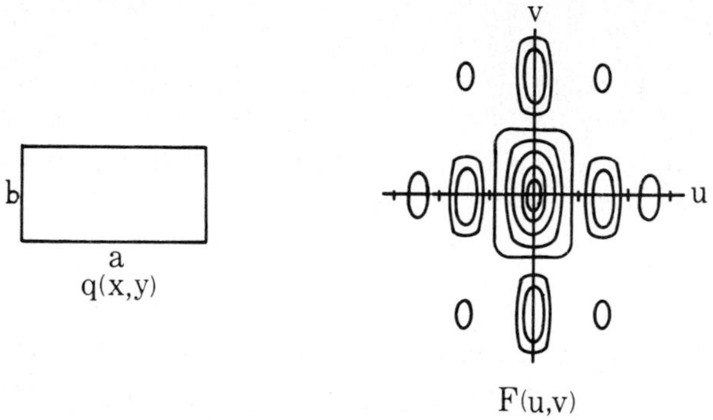

Fig. 2.3. Diagram suggesting the form of the Fourier transform of a rectangular aperture, equation (2.43).

$$F(u) = \left(\frac{\pi a^2}{2}\right)\frac{J_1(\pi a u)}{\pi a u}, \tag{44}$$

where u is a radial coordinate and $J_1(x)$ is the first order Bessel function. The function $J_1(x)/x$ is similar in form to $(\sin x)/x$ but has a somewhat broader central maximum with the first zero at $1.22\,a^{-1}$ instead of a^{-1}.

2.3.9. Two very narrow slits

For two slits a distance A apart we take the origin half-way between them and write

$$f(x) = s(x+A/2) + s(x-A/2)\,,$$

where $s(x)$ is the transmission function for one slit of width a as in 2.3.4. For very narrow slits we let a go to zero. But to keep the intensity finite we imagine the incident amplitude to be proportional to $1/a$. Then $s(x)$ becomes $\delta(x)$. We put

$$f(x) = \delta(x+A/2) + \delta(x-A/2)\,,$$

and

$$F(u) = \exp\{-\pi \mathrm{i} A u\} + \exp\{\pi \mathrm{i} A u\}$$
$$= 2\cos(\pi A u)\,, \tag{45}$$

and the diffraction pattern intensity is

$$I(l) = 4\cos^2(\pi A l/\lambda)\ . \tag{46}$$

Thus we get simple sinusoidal fringes of uniform amplitude. Comparison with the Fresnel diffraction pattern shows that for this particular object the diffraction pattern is independent of the approximation made to the general Kirchhoff integral (see Problem 2).

2.3.10. Two slits of appreciable width

For two slits of width a, a distance A apart, we may write

$$f(x) = s(x) * [\delta(x+A/2) + \delta(x-A/2)]\ ,$$

where $s(x)$ is the transmission function for a single slit defined as in 2.3.4 Using the convolution theorem and the results (39) and (45)

$$F(u) = 2a\,\frac{\sin(\pi a u)}{\pi a u}\cos(\pi A u)\ . \tag{47}$$

The intensity of the diffraction pattern is then as shown in Fig. 2.4; $\cos^2$ fringes of period $1/2A$ are modulated by a $(\sin^2 x)/x^2$ function which goes to zero for $u = a^{-1}$. This is a description of the result of the Young's fringe experiment, well known as of fundamental significance for the development of physical optics.

2.3.11. Finite wave train

The inverse of 2.3.9 gives

$$\mathcal{F}\{2\cos\pi A x\} = \delta(u+A/2) + \delta(u-A/2)\ .$$

If the wave train represented by the cosine function is cut to a finite length by multiplying it by a slit function of width B we have

$$f(x) = 2\cos(\pi A x)\, s_B(x)\ ,$$

so that, using the multiplication theorem,

$$F(u) = \{\delta(u+A/2) + \delta(u-A/2)\} * B\,\frac{\sin(\pi B u)}{\pi B u}\ . \tag{48}$$

Thus the amplitude distribution in reciprocal space will be the sum of two functions of the form $S(u) = \mathcal{F}s_B(x)$ centered on $u = \pm A/2$ (Fig. 2.5). If B is much greater than the periodicity $2/A$ of the wave train, the two peaks in $F(u)$

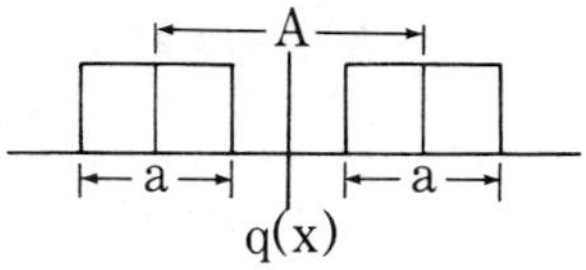

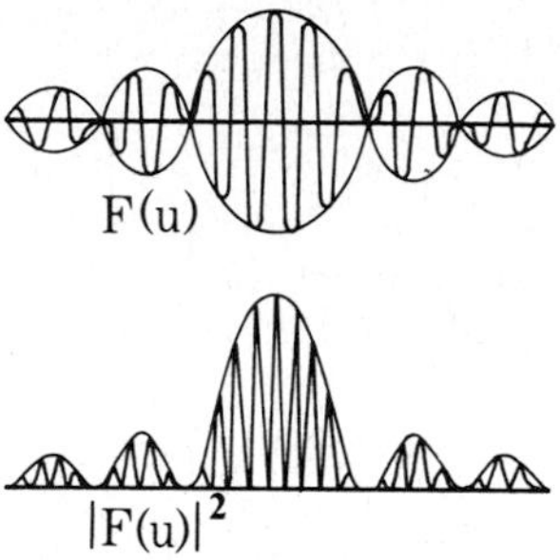

Fig. 2.4. Diagram suggesting the diffraction pattern from two parallel slits, equation (2.47).

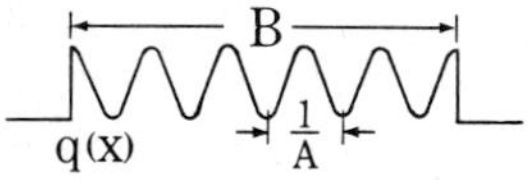

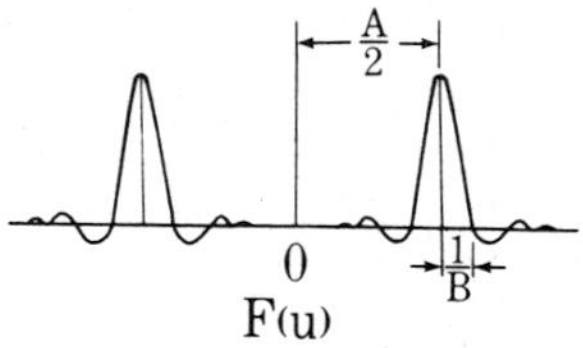

Fig. 2.5. A wave train cut off by multiplying by a slit function and its Fourier transform, equation (2.48).

will be much narrower than their separation and they will not overlap appreciably: Then the intensity distribution $|F(u)|^2$ will be very nearly given by

$$I(u) = B^2 \frac{\sin^2(\pi Bu)}{(\pi Bu)^2} * \{\delta(u+A/2) + \delta(u-A/2)\} .$$

This is a useful approximation for many purposes, but it can not be used if the length of the wave train B is only a few times the periodicity.

2.3.12. Periodic array of narrow slits

We assume a periodic array of slits having zero width and repeated at regular intervals, a indefinitely. Then

$$f(x) = \sum_{n=-\infty}^{\infty} \delta(x-na) .$$

The Fourier transform is, using (33),

$$F(u) = \sum_{-\infty}^{\infty} \exp\{2\pi i u n a\} .$$

This summation of a Fourier series with all coefficients unity gives a well-known result which we can derive by elementary methods as follows.

Since

$$\sum_{0}^{\infty} x^n = (1-x)^{-1} ,$$

we may write

$$F(u) = \sum_{0}^{\infty} [\exp\{2\pi i u a\}]^n + \sum_{0}^{\infty} [\exp\{-2\pi i a u\}]^n - 1$$

$$= [1 - \exp\{2\pi i u a\}]^{-1} + [1 - \exp\{-2\pi i u a\}]^{-1} - 1$$

$$= 0 ,$$

except that

$$F(u) = \infty \quad \text{for} \quad \exp\{2\pi i u a\} = 1 ,$$

i.e. if $2\pi u a = 2h\pi$ where h is an integer, or $u = h/a$. Then

$$F(u) = a^{-1} \sum_{h} \delta(u - h/a) , \tag{49}$$

where the factor a^{-1} gives the delta functions the correct weight.

Hence the Fourier transform is a set of equally spaced delta functions of period a^{-1} in reciprocal space.

2.3.13. *Arbitrary periodic function*

For an object with an arbitrary periodic transmission function we write

$$f(x) = \sum_{-\infty}^{\infty} F_h \exp\{-2\pi i h x/a\} . \tag{50}$$

Then

$$F(u) = \sum_{-\infty}^{\infty} F_h \int_{-\infty}^{\infty} \exp\left\{2\pi i\left(-\frac{hx}{a} + ux\right)\right\} dx \,.$$

From (7) this is

$$F(u) = \sum_{-\infty}^{\infty} F_h \,\delta(u - h/a) \,. \tag{51}$$

Hence the diffracted amplitude is represented by a set of delta functions equally spaced with separation a^{-1} in u, each delta function having the "weight" F_h equal to the corresponding Fourier coefficient of (50). This result and those which follow form the basis for much of our consideration of diffraction of X-rays and electrons by crystals.

2.3.14. Diffraction grating: thin slits

A set of N parallel, equally spaced thin slits constitutes a primitive form of diffraction grating. The transmission function may be written;

$$f(x) = \sum_{-(N-1)/2}^{(N-1)/2} \delta(x - na) \,. \tag{52}$$

Then

$$F(u) = \sum_{-(N-1)/2}^{(N-1)/2} \exp\{2\pi i u n a\}$$

$$= \exp\{-\pi i u (N-1) a\} \sum_{0}^{N-1} \exp\{2\pi i u n a\}$$

$$= \exp\{-\pi i u (N-1) a\} \,\frac{\exp\{2\pi i u N a\} - 1}{\exp\{2\pi i u a\} - 1} \,,$$

i.e.

$$F(u) = \frac{\sin(\pi N a u)}{\sin(\pi a u)} \,. \tag{53}$$

On the other hand, we may use the result of (49) and write the transmission function as

$$f(x) = s(x) \sum_{n=-\infty}^{\infty} \delta(x-na) , \tag{54}$$

where $s(x)$ is a slit function of width Na, which cuts off the transmission of all but N of the thin slits. Then, as in (48) we may write

$$F(u) = \sum_h \delta(u-h/a) * Na \frac{\sin(\pi Nau)}{\pi Nau} . \tag{55}$$

The functions (53) and (55) are similar (see Fig. 2.6). For each there are sharp peaks with side ripples, of the form $(\sin x)/x$, at intervals $u = a^{-1}$. The width of the peak, the distance from the maximum to the first zero, is $1/Na$.

It is not obvious from the form of the functions (53) and (55) that they are identical, but since the functions in real space, (52) and (54) are identical, the transforms must be also.

2.3.15. *Diffraction grating: general*

The individual lines of a diffraction grating, used in transmission, may be slits of finite width, rulings on glass, replicas in plastic of rulings on metal and so on. The transmission function of the whole grating will be given by the convolution of the transmission function of a single ruling $g(x)$ with a set of N delta functions, (52) or (54)

$$f(x) = \sum_{-(N-1)/2}^{(N-1)/2} \delta(x-na) * g(x) \tag{56}$$

so that

$$F(u) = G(u) \frac{\sin(\pi Nau)}{\sin(\pi au)} , \tag{57}$$

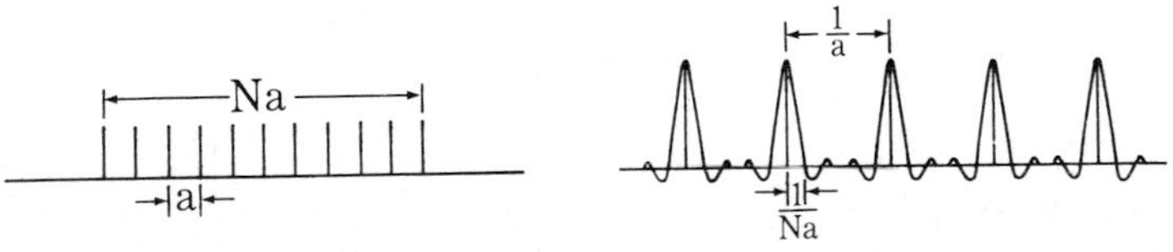

Fig. 2.6. The amplitude distribution for diffraction from a large number of parallel slits, as given by equation (2.53) or (2.55).

or

$$F(u) = G(u)\left[\sum_h \delta\left(u - \frac{h}{a}\right) * Na\,\frac{\sin(\pi Nau)}{\pi Nau}\right]. \tag{58}$$

Thus the height of each of the diffraction maxima is proportional to the value of the Fourier transform of $g(x)$ at that u value.

Alternatively we could say that the diffraction grating may be represented by cutting off a periodic transmission function by multiplying it with a slit function $s(x)$ of width Na. This is not the same as (56) since in this case we may be using $s(x)$ to cut off a continuous function. Correspondingly, the diffraction amplitude, written from (51) as

$$F(u) = \sum_h F_h\,\delta\left(u - \frac{h}{a}\right) * Na\,\frac{\sin(\pi Nau)}{\pi Nau}, \tag{59}$$

is not exactly the same as (57) or (58).

2.3.16. *Gaussian function*

A Gaussian function rarely occurs in practice as the transmission function of an object but is frequently used as an approximation, to "round-off" a discontinuity, or to achieve a suitable convergence of an analytic function, because of its desirable properties in relation to Fourier transform and convolution.

Putting

$$f(x) = \exp\{-a^2x^2\}, \tag{60}$$

we make use of the standard integral (1.28) to obtain

$$F(u) = \frac{\pi^{1/2}}{a}\exp\left\{\frac{-\pi^2u^2}{a^2}\right\}. \tag{61}$$

Hence, if $f(x)$ is a Gaussian of half-width a^{-1} then its Fourier transform is also a Gaussian having half-width a/π.

We note also that, if

$$f(x) = \exp(-a^2x^2) * \exp\{-b^2x^2\},$$

$$F(u) = \frac{\pi}{ab}\exp\left\{-\pi^2u^2\left(\frac{a^2+b^2}{a^2b^2}\right)\right\},$$

and, applying the inverse Fourier transform

$$f(x) = \left(\frac{\pi}{a^2+b^2}\right)^{1/2} \exp\left\{-\left(\frac{a^2b^2}{a^2+b^2}\right)x^2\right\}, \tag{62}$$

i.e. the convolution of two Gaussians is again a Gaussian.

2.3.17. Row of circular holes

The extension to two dimensions allows the above results to be applied to calculate the diffraction patterns from many simple two-dimensional distributions. However, in using delta functions and convolutions it is important to distinguish carefully the delta functions and convolutions in one and two dimensions.

We take as example the case of a linear row of equally spaced circular holes in an opaque screen. For this the transmission function may be written

$$f(x,y) = \sum_n \delta(x-na,y) \underset{x,y}{*} O(x,y) \tag{63}$$

where we have used $\delta(x-na,y)$ to indicate a delta function at $x = na, y = 0$, and $O(x,y)$ is the transmission function for a circular aperture as defined in 2.3.8, above. See Fig. 2.7.

The Fourier transformation of the set of delta functions with respect to x is

$$\sum_h \delta(u-h/a)$$

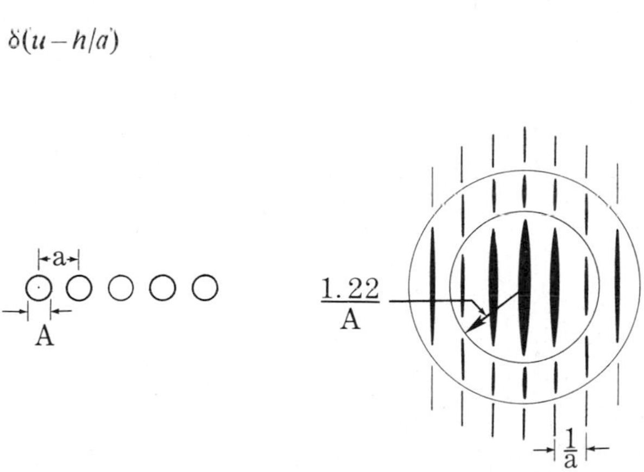

Fig. 2.7. The diffraction pattern from a row of circular holes equally spaced in an opaque screen. The widths of the lines are made to suggest their intensities.

and the Fourier transform with respect to y is unity. Hence the Fourier transform in x and y is a set of straight lines parallel to the v axis and equally spaced at intervals of a^{-1} in the u direction. This distribution will be modulated by multiplying by the Fourier transform of $O(x,y)$, namely

$$\left(\frac{\pi A^2}{2}\right)\frac{J_1(\pi AU)}{(\pi AU)},$$

where A is the diameter of the holes and $U = (u^2 + v^2)^{1/2}$.

This gives the set of lines of varying intensity as suggested by Fig. 2.7 where the width of the line has been used as an indication of relative intensity.

This example has some relevance in radio astronomy. A common form of radio interferometer used for detecting the radio waves from distant radio stars, consists of an equally spaced line of paraboloidal antennas or "dishes" each of which sums the incident amplitude over the circular aperture of the dish. It can be seen, by application of the reciprocity theorem that the amplitude obtained by summing the amplitudes from all dishes coherently will be exactly the same as the amplitude which would be observed at the distant radio star if the apertures of the dishes were to be illuminated from behind by a plane parallel incident wave. Thus, as a radio star moves across the sky, the amplitude measured by the interferometer will correspond to a linear trace across the diagram of the diffraction pattern in Fig. 2.7.

2.3.18. Complementary objects – Babinet's Principle

Consider two objects for which the transmission functions are complementary so that

$$q_1(x,y) + q_2(x,y) = 1\,. \tag{64}$$

For black and transparent objects, this means that the black areas of the first are the transparent parts of the second and vice versa.

Then, for the second object

$$q_2(x,y) = 1 - q_1(x,y)\,.$$

The diffraction pattern amplitude is

$$Q_2(u,v) = \delta(u,v) - Q_1(u,v)\,,$$

so that the intensity in the diffraction pattern is proportional to $|Q_2(u,v)|^2$ which is equal to $|Q_1(u,v)|^2$ except at the origin.

This exception, the exclusion of the origin point, was overlooked in the original formulation, known as Babinet's Principle or Babinet's Theorem.

2.3.19. Total intensities: Parseval's Theorem

In a diffraction experiment, conservation of energy implies that the total intensity in the diffraction pattern is equal to the total intensity at the exit surface of the object, i.e.

$$\iint_{-\infty}^{\infty} |q(x,y)|^2 \mathrm{d}x\,\mathrm{d}y = \iint_{-\infty}^{\infty} |Q(u,v)|^2 \mathrm{d}u\,\mathrm{d}v\,. \tag{65}$$

This is a special case of a general relationship of Fourier transform theory

$$\int |f(\boldsymbol{r})|^2 \mathrm{d}\boldsymbol{r} = \int |F(\boldsymbol{u})|^2 \mathrm{d}\boldsymbol{u}\,, \tag{66}$$

which in turn is a special case of the more general Parseval's Theorem relating to any two functions $f(\boldsymbol{r})$ and $g(\boldsymbol{r})$;

$$\int f(\boldsymbol{r})\, g^*(\boldsymbol{r})\mathrm{d}\boldsymbol{r} = \int F(\boldsymbol{u})\, G^*(\boldsymbol{u})\mathrm{d}\boldsymbol{u}\,. \tag{67}$$

From this theorem we see that

$$\int f(\boldsymbol{R}) f(\boldsymbol{r}+\boldsymbol{R})\mathrm{d}\boldsymbol{R} = \int |F(\boldsymbol{u})|^2 \exp\{2\pi \mathrm{i}\boldsymbol{u}\cdot\boldsymbol{r}\}\mathrm{d}\boldsymbol{u}\,, \tag{68}$$

which is a special case of the inverse of the multiplication theorem (29);

$$\mathcal{F}\,[F(u) F^*(u)] = f(r) * f(-r)\,. \tag{69}$$

This convolution of a function with the same function inverted is the auto-correlation function which has considerable application in many fields of science and, as we will see later, has special significance for diffraction under the title of a "generalized Patterson function".

Problems

1. Write down an expression for, and sketch the form of, the diffraction pattern given when an incident plane wave is diffracted by:

(1) a row of N circular holes, equally spaced in an opaque screen,

(2) a similar row of opaque circular discs,

(3) a similar row of holes in an opaque screen which are alternately circular and square,

(4) a row of circular holes in an opaque screen when each second one is covered by a half-wave plate (phase change π),

(5) two parallel rows of circular holes.

2. Find the Fresnel diffraction pattern given by a pair of parallel, very fine slits. Compare this with the Fraunhofer diffraction pattern.

3. A radio-interferometer consists of two perpendicular, intersecting lines of regularly spaced circular dishes (paraboloidal antennas) forming an equi-armed cross (known as a "Chris-cross"). If the signals from all dishes are added in phase, how does the response of the interferometer vary with the position of a radio source in the sky? By what manipulation of the electronic signals could the interferometer be "pointed" to receive maximum radiation from a particular direction?

4. Compare the response of the radio interferometer described in 3 with that of a complete two-dimensional square array of $N \times N$ paraboloidal antennas.

5. In a Young's fringe experiment, two fine parallel slits, separation a, are illuminated by monochromatic light from a distant, parallel slit of width d, which may be regarded as an incoherent source, so that intensities due to each point of the source are added on the plane of observation. Find the visibility (defined as $(I_{max}-I_{min})/(I_{max}+I_{min})$) of the interference fringes produced as a function of a and d. Taking the visibility of the fringes as a measure of the degree of coherence of the illumination at the two slits, verify the Zernike–van Cittart theorem which states that the degree of coherence is given by the Fourier transform of the intensity distribution of the source.

CHAPTER 3

Imaging and diffraction

3.1. Wave theory of imaging

3.1.1. Coherent wave optics

While, as we shall see, lenses may be used to obtain either diffraction patterns or images, the major development of the theory of lens action has been directed towards imaging. We give some brief account of this development here but prefer a somewhat different approach which allows the diffraction patterns and images to be treated with equal facility and stresses the relationship between them. This has considerable practical importance, especially in electron microscopy.

For many years, imaging was considered in terms of the approximations and from the point of view of geometric optics. The wave nature of light was taken into account only as a refinement when resolution was affected by diffraction effects due to finite aperture size. The formulation of imaging theory completely in terms of waves was introduced by Duffieux [1946] and subsequently developed by many authors including Hopkins [1953, 1950], Fellgett and Linfoot [1955] and Linfoot [1955]. An outline of the development has been given by Born and Wolf [1975]. This approach to imaging theory is based essentially on the Fourier transform. Here we describe first of all a more graphic version of the theory and then a more formal and elegant form.

We consider first the system suggested by Fig. 3.1. Incident radiation, passing through a small object or a small part of an object gives an amplitude distribution which, over a limited region in one dimension, is given by the function $f(x)$. Fraunhofer diffraction then gives rise to a distribution $F(u)$ on a spherical reference surface at the entrance to a lens system. The coordinate u measured along the spherical surface is φ/λ, where φ is the scattering angle, or s/R where s is the distance measured on the sphere.

The function of the lens is to transfer the distribution $F(u)$ into the distribution $F'(u')$ on a spherical reference surface centered on the part of the image, $\psi'(x)$, corresponding to $f(x)$. Following a common convention, quanti-

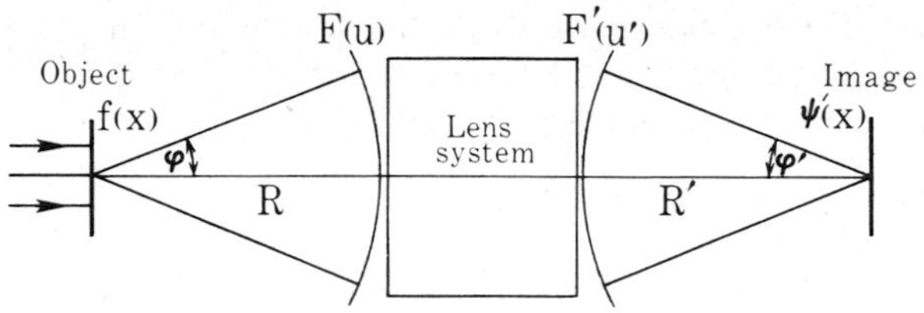

Fig. 3.1. Diagram suggesting the imaging process in a optical system. The complex amplitude distribution on a spherical reference surface in object space is converted into an amplitude distribution on a spherical reference surface in image space.

ties referring to the image space are primed. If the transfer of F to F' were perfect, with no perturbation of the function or aperture limitation, we would have $F'(s) = F(s)$. From the geometry of the situation,

$$s = uR = u'R'$$

so that, for this case,

$$F'(u) = F(u'R'/R) \, . \tag{1}$$

The distribution $F(u)$ is $\mathcal{F}f(x)$. By symmetry we put $F'(u') = \mathcal{F}\,\psi'(x)$. The transfer from F' to ψ' might be thought of as an inverse transform but since the direction of propagation from ψ' to F' is opposite to that in object space, the sign of the exponent is changed and a positive sign in the exponential of the transform of F' to ψ' makes it a direct Fourier transform.

Hence

$$\psi'(x) = \mathcal{F}\,F(R'u'/R)$$

$$= \int\!\!\int f(X) \exp\{2\pi \mathrm{i}(u'R'/R)X\}\,\mathrm{d}X \exp\{2\pi \mathrm{i}u'x\}\,\mathrm{d}u'$$

$$= \int f(X)\,\delta(X + Rx/R')\,\mathrm{d}X = f(-Rx/R') \, , \tag{2}$$

i.e., the image is a recreation of $f(x)$, inverted and magnified by a factor R'/R.

In practice, of course, the transfer from $F(u)$ to $F'(u')$ is not perfect. There is always a limitation due to the finite aperture size. In addition, the lens aberrations give rise to phase changes which vary as a function of u'. Classically, this phase change is described by a power series in u' or, in two dimensions, by a power series in ρ (proportional to $(u^2 + v^2)^{1/2}$) and the polar angle ϕ. In this way the tranditional aberration coefficients are generated, the most important for our purposes being the Third Order Spherical Aberration coefficient which derives from the coefficient of the ρ^4 term.

In the wave-theory treatment, the changes of amplitude and phase due to the limitations of the lens are represented by an Optical Transfer Function $T(u')$, characteristic of the lens. Then

$$F'(u') = F(R'u'/R)\, T(u') \,. \tag{3}$$

Correspondingly the modification of the image is represented by

$$\psi'(x) = f(-Rx/R') * t(x) \,. \tag{4}$$

For the simplest case, in two dimensions, of limitation by a circular aperture only, we have that, as in (2.44),

$$T(u', v') = \begin{cases} 1 & \text{if } (u'^2 + v'^2)^{1/2} < u_0/2 \\ 0 & \text{elsewhere} \end{cases}$$

so that

$$t(x, y) \equiv t(r) = J_1(\pi u_0 r)/u_0 r \,,$$

and the intensity distribution in the image is

$$I'(x, y) = \psi\psi^* = \left| f\left(-\frac{Rx}{R'}, -\frac{Ry}{R'}\right) * \frac{J_1(\pi u_0 r)}{\pi r} \right|^2 , \tag{5}$$

which represents a loss of resolution due to a blurring of the amplitude distribution.

When, in addition to the aperture limitation, the aberrations of the lens are appreciable, the optical transfer function includes the phase change term so that (3) may be written

$$F'(u') = F(R'u'/R)\, T_a(u') \exp\{i\Phi(u')\} \,,$$

where $T_a(u')$ is the aperture function and $\Phi(u)$ represents the change of phase, which may be expanded as a power series in ρ and ϕ to introduce the aberration coefficients. Then the observed intensity is given by

$$I'(x, y) = \left| f\left(-\frac{Rx}{R'}, -\frac{Ry}{R'}\right) * \frac{J_1(\pi u_0 r)}{\pi r} * \mathcal{F}[\exp\{i\Phi(u)\}] \right|^2 .$$

The effects of the added convolution here are not immediately apparent. They may be calculated in detail for particular cases but a simple intuitive understanding is difficult. We will discuss these effects later in terms of the Abbe treatment.

3.1.2. Incoherent wave imaging

The same sort of treatment applies if we consider incoherent imaging, i.e., imaging of an object which is self-luminous or which is illuminated by incoherent incident light or scatters the incident light incoherently. Then we consider the imaging of light from each point of the object separately and add the resulting intensities.

Referring to (4), the amplitude distribution for a point source on the axis of the lens will be $t(x)$ and the intensity distribution will be $|t(x)|^2$. Then the image intensity will

$$I'(x,y) = I\left(-\frac{Rx}{R'}, -\frac{Ry}{R'}\right) * |t(x)|^2 . \tag{6}$$

The spread function $|t(x)|^2$ can be referred to an Optical Contrast Transfer Function which characterizes the lens;

$$O(u') = T(u') * T^*(-u') . \tag{7}$$

For the simple case of a circular aperture, the image intensity is convoluted by the spread function

$$|t(x,y)|^2 = \frac{J_1^2(\pi u_0 r)}{(\pi r)^2} .$$

If we then apply the Rayleigh criterion for the resolution of two adjacent point sources, that the images can be resolved if the maximum of one point image falls at the first minimum of the other, i.e., at $r = 1.22/u_0$, we obtain the least resolvable distance for the lens as

$$\Delta x = \frac{1.22\lambda}{2\varphi_0} = \frac{1.22\lambda}{\alpha} \tag{8}$$

where α is the angular aperture subtended by the lens at the object. For convenience in considerations of microscopy we have transferred the dimensions back into object space.

The more formal description of the imaging process eliminates the need for the picture of a transfer from one reference surface to another as suggested in Fig. 3.1. The spread function, $t(x,y)$ for amplitudes for coherent imaging, or $|t(x,y)|^2$ for the incoherent case, is defined as the response to a point source in the object. The Fourier transform, $T(u,v)$ or $O(u,v) = T(u,v) * T^*(-u,-v)$, is the appropriate transfer function which is characteristic of the lens. The transfer function may be derived, independently of any aberration theory, by studying the image intensity distribution for particular test objects; for example, the contrast of the images of objects with transmission function $1 + \cos(2\pi a x)$ give the amplitudes and phases of the transfer for particular values of the "spatial frequency", a. From the transfer function the spread function is derived by Fourier transform.

3.2. Abbe theory

For our purposes, the formal use of Fourier transforms in imaging theory is interesting but limited, in the form described in the previous section, since we are interested in the concrete embodiment of the Fourier transform in the form of the diffraction pattern as well as in the image. We turn therefore to the imaging theory originating from the ideas of Abbe. Although this is essentially a wave theory, it is conveniently illustrated by reference to the geometric-optics diagram of Fig. 3.2.

Parallel light is incident on an object having transmission function $f(x, y)$ which is imaged by a lens having focal length, f. Light scattered by a point A of the object is brought to a focus at A′ in the image plane. From the geometry of the diagram the image is inverted and magnified by a factor R'/R so that the image is

$$\psi(x, y) = f\left(-\frac{Rx}{R'}, -\frac{Ry}{R'}\right) .$$

Also it may be noted that all light scattered through an angle φ is brought to a focus at one point in the back-focal plane. This is equivalent to interference at a point at infinity. Hence the amplitude distribution on the back-focal plane

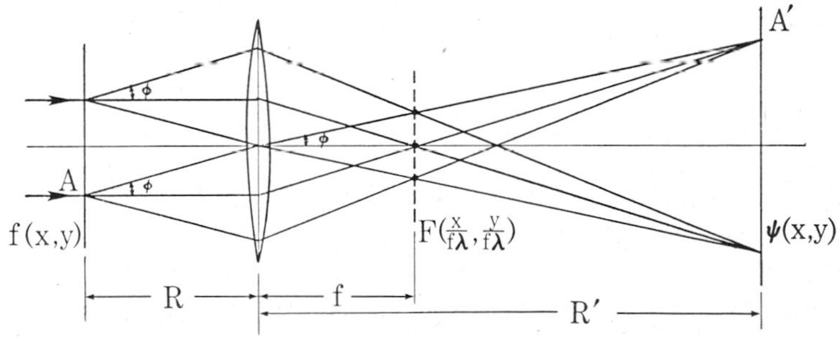

Fig. 3.2. Geometric optics ray diagram used to suggest the plausibility of the wave-optics description of the imaging process in the Abbe theory.

is that of the Fraunhofer diffraction pattern given by the Fourier transform function, $F(u, v)$. In this case $u = (\sin \varphi_x)/\lambda$ and if φ is not too large we may write $u = x/f\lambda$, $v = y/f\lambda$. Thus the imaging process may be described in terms of two Fourier transforms: The scattered radiation from the object interferes on the back-focal plane to give the Fraunhofer diffraction pattern described by a Fourier transform: then the radiation from the back-focal plane again forms an interference pattern on the image plane which is effectively at infinity so that the amplitude distribution in the image is given by Fourier transform of that in the back-focal plane.

Two immediate consequences are of interest. The first is that lenses may be used for light or electrons to give the Fraunhofer diffraction pattern of an object at a convenient location and on a convenient scale, depending on the focal length of the lens. Secondly, the limitations of the lens system in reproducing the object transmission function on the image plane may be described in terms of modifications of the amplitude and phase of the distribution on the back-focal plane.

Apart from their direct interest for consideration of optical and electron-optical imaging and diffraction, these points are of importance for X-ray diffraction because of the very graphic use made of the optical diffractometer by Taylor and Lipson [1964] to simulate the production of X-ray diffraction patterns and the reconstruction of crystal structures from diffraction amplitudes.

3.3. Small angle approximation

Although it is strictly limited in its validity, the small angle approximation which we introduced for the discussion of Fresnel Diffraction forms a very convenient basis for describing the essential behavior of imaging systems. It provides a model which reproduces all important features of the properties of optical systems with a relative mathematical simplicity and wide versatility.

All objects are considered to be composed of planar distributions having transmission functions $q_n(x, y)$. Propagation through a medium of constant refractive index is given by convolution with a propagation function which, in the small angle approximation, is $(\mathrm{i}/R\lambda) \exp\{-\mathrm{i}k(x^2 + y^2)/2R\}$. We introduce the concept of an ideal thin lens, which is a planar object having a transmission function $\exp\{\mathrm{i}k(x^2 + y^2)/2f\}$. It is readily confirmed that

$$\exp\{\mathrm{i}k(x^2 + y^2)/2f\} * \exp\{-\mathrm{i}k(x^2 + y^2)/2f\} = \delta(x, y) , \tag{9}$$

i.e., if a plane wave, amplitude unity, passes through the ideal thin lens, propagation of the wave through the focal length f gives a delta-function or a point cross-over. Similarly a point source placed a distance f before the ideal thin lens gives a plane wave:

$$\left[\delta(x,y) * \exp\left\{\frac{-ik(x^2+y^2)}{2f}\right\}\right] \exp\left\{\frac{ik(x^2+y^2)}{2f}\right\} = 1 . \tag{10}$$

In this approximation the amplitude given on a plane of observation when a plane wave passes through an object of transmission function $q(x, y)$ and then an ideal thin lens, as suggested in Fig. 3.3, is, in one dimension and omitting constant multipliers,

$$\psi(x) = \left[\left[\left[q(x) * \exp\left\{\frac{-ikx^2}{2R}\right\}\right] \exp\left\{\frac{ikx^2}{2f}\right\}\right] * \exp\left\{\frac{-ikx^2}{2R'}\right\}\right], \tag{11}$$

where the operations in the successive brackets represent propagation through a distance R, transmission through the lens, and propagation through a distance R'. By writing out the convolution integrals in detail it is readily shown that if $R' = f$, $\psi(x) = Q(x/f\lambda)$, and if $(1/R) + (1/R') = 1/f$, then $\psi(x) = q(-Rx/R')$. To prove the second case, (11) is written in detail, with dummy variables X and Y,

$$\psi(x) = \int\int q(Y) \exp\left\{\frac{-ik(X-Y)^2}{2R}\right\} \exp\left\{\frac{ikX^2}{2f}\right\} \exp\left\{\frac{ik(x-X)^2}{2R'}\right\} \mathrm{d}Y\,\mathrm{d}X .$$

Gathering the related exponents,

$$\psi(x) = \int\int q(Y) \exp\left\{\frac{-ikY^2}{2R}\right\} \exp\left\{\frac{-ik}{2}\left(\frac{1}{R}+\frac{1}{R'}-\frac{1}{f}\right)X^2\right\}$$

$$\times \exp\left\{\frac{-ikx^2}{2R'}\right\} \exp\left\{\frac{ik}{2}\left(\frac{Y}{R}+\frac{x}{R'}\right)X\right\} \mathrm{d}X\,\mathrm{d}Y .$$

Then if $(1/R) + (1/R') = 1/f$ the integral over X is seen to be $\delta(Y + Rx/R')$ and

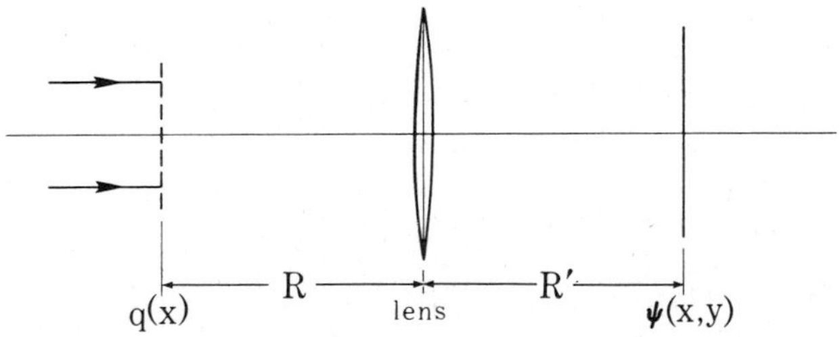

Fig. 3.3. Diagram defining the imaging system described by equation (3.11).

the integral over Y then gives $\psi(x) = Cq(-Rx/R')$, where C has modulus unity and may be absorbed with other similar terms omitted from the propagation functions.

The proof of the first case, for $R' = f$, is left as an exercise for the reader.

Thus the properties of producing diffraction patterns and images are reproduced. Obviously the action of any combination of sources, object and lenses may be reproduced by writing down the appropriate series of operations of convolution with a propagation function and multiplication with a transmission function. For example, if there is a point source at $x = X$ a distance R_0 before the object of Fig. 3.3, the amplitude distribution on the plane of observation is

$$\psi(x) = \left[\left[\left[\left[\delta(x-X) * \exp\left\{\frac{-ikx^2}{2R_0}\right\}\right] q_0(x)\right] * \exp\left\{\frac{-ikx^2}{2R}\right\}\right]\right.$$

$$\left.\times \exp\left\{\frac{ikx^2}{2f}\right\}\right] * \exp\left\{\frac{-ikx^2}{2R'}\right\}. \qquad (12)$$

By evaluation of the integrals it can be shown that the diffraction pattern is given for $1/(R_0 + R) + (1/R') = 1/f$, and the image is given, as before, $(1/R) + (1/R') = 1/f$. The effect on the diffraction pattern or on the image of incomplete coherence of the incident radiation, i.e., of illumination of the object by an incoherent source of appreciable extent, is given by summing the intensities for each point of the source separately. Thus the intensity for the source point $x = X$, given by $|\psi_X(x)|^2$, from (12), is calculated and multiplied by the intensity of the source point, $I_0(X)$. Then the observed intensity distribution is derived by integrating over X.

The effect of the limitation of the aperture of the lens on the image is investigated both experimentally and theoretically by placing the aperture in the back-focal plane to multiply the diffraction pattern by the transmission function of the aperture. The result is exactly the same as expressed in equations (5) and (6).

Excellent examples of the effects on the image of the limitation of the diffraction pattern by apertures of various sizes and shapes, produced by use of the optical diffractometer, are given by Taylor and Lipson [1964] (their Plates 43 to 46).

The effect of defocussing of the lens is readily deduced. If the plane of observation is a distance Δ' away from the plane of the in-focus image, which has amplitude distribution $\psi_0(x)$, then,

$$\psi(x) = \psi_0(x) * \exp\{-ikx^2/2\Delta'\}. \qquad (13)$$

Alternatively we may say that the amplitude on a plane a distance Δ away from the object is brought to a focus on the plane of observation so that the distribution which is imaged is

$$q(x) * \exp\{-ikx^2/2\Delta\} . \tag{14}$$

The amplitude distribution in the back-focal plane, given by Fourier transform, is then

$$Q(u) \exp\{\pi i \lambda \Delta u^2\} .$$

Thus the effect of defocus can be considered as equivalent to the addition of a second order phase term in the back-focal plane. Higher order terms in the exponent are introduced by aberrations of the lens. The third-order spherical aberration, for example, adds a term proportional to u^4.

3.4. Phase contrast

3.4.1. Phase and amplitude objects

A pure phase object is an idealized concept. It is essentially a two-dimensional object which changes the phase but not the amplitude of the incident wave. We may use it for convenience to describe a thin object having a varying refractive index when the angles of refraction of the incident waves are so small that the lateral spread of the waves within the object thickness can be ignored. In the one-dimensional form which we use for convenience, the transmission function is of the form.

$$q(x) = \exp\{i\varphi(x)\} \tag{15}$$

where the phase change $\varphi(x)$ depends on the thickness and refractive index of the material.

If a plane incident wave falls on this object, the transmitted intensity is $|q(x)|^2 = 1$. Similarly, an ideally perfect lens would give an exact re-creation of $q(x)$ so that the intensity distribution of the image would be

$$|\psi(x)|^2 = |q(-Rx/R')|^2 = 1 . \tag{16}$$

Thus there is no contrast in the image corresponding to the structure of the object. For most objects some absorption and multiple scattering takes place so that the object, if thin enough, must be regarded as a mixed phase and amplitude object with a transmission function which may be written

$$q(x) = \exp\{i\varphi(x) - \mu(x)\} , \tag{17}$$

$$|q(x)|^2 = \exp\{-2\mu(x)\}\,. \tag{18}$$

However for a large and important class of objects the pure phase object forms a reasonable approximation. Most thin biological samples are very nearly phase objects for visible light.

All electron microscope specimens of less than a certain thickness are essentially phase objects for electrons. The refractive index for electrons, given by equation (1.6), varies with the value of the electrostatic potential.

While for both light and electrons staining techniques have been developed to increase the effective absorption and so provide contrast in the in-focus image, these techniques are of limited use for some purposes and may introduce artifacts which complicate the image interpretation. We therefore consider in some detail the so-called "phase-contrast" techniques for obtaining contrast in the image of a pure phase object.

3.4.2. Out-of-focus contrast

This is the most common form of phase contrast, especially in electron microscopy. It is a well-known experimental observation that the contrast is a minimum near exact focus. Contrast appears off-focus and reverses when one goes through focus.

We consider the amplitude distribution on a plane a distance Δ from the object, written as

$$\psi(x) = q(x) * \exp\{-\mathrm{i}kx^2/2\Delta\}\,. \tag{19}$$

Instead of the form (15), it is more convenient to work with the Fourier transform function $\Phi(u)$, defined by

$$q(x) = \exp\{\mathrm{i}\varphi(x)\} \equiv \int \Phi(u) \exp\{-2\pi \mathrm{i}ux\}\,\mathrm{d}u\,. \tag{20}$$

The convolution in (19) is

$$\int\!\!\int \Phi(u) \exp\{-\mathrm{i}kX^2/2\Delta\} \exp\{-2\pi \mathrm{i}u(x - X)\}\,\mathrm{d}X\,\mathrm{d}u\,.$$

Making use of the standard integral (1.28) gives us

$$\int \Phi(u) \exp\{-2\pi \mathrm{i}ux\} \exp\{\mathrm{i}\pi\Delta\lambda u^2\}\,\mathrm{d}u\,. \tag{21}$$

Then if $\Delta\lambda$ is sufficiently small we may put

$$\exp\{\mathrm{i}\pi\Delta\lambda u^2\} \approx 1 + \mathrm{i}\pi\lambda\Delta u^2\,,$$

and we obtain

$$\psi(x) = \int \Phi(u) \exp\{-2\pi i u x\}\,(1 + i\pi\lambda\Delta u^2)\,du$$

$$= \exp\{i\varphi(x)\} + i\pi\Delta\lambda \int u^2 \Phi(u) \exp\{-2\pi i u x\}\,du\,. \qquad (22)$$

The relation (2.28) suggests that the integral in this expression may correspond to a second differential and, in fact

$$\frac{d^2}{dx^2}[\exp\{i\varphi(x)\}] = \int (-4\pi^2 u^2)\,\Phi(u) \exp\{-2\pi i u x\}\,du\,;$$

but straight-forward differentiation gives

$$\frac{d^2}{dx^2}[\exp\{i\varphi(x)\}] = -\exp\{i\varphi\}[\{\varphi'(x)\}^2 + i\varphi''(x)]\,,$$

where the primes represent differentiation with respect to x. Hence (22) is

$$\psi(x) = \exp\{i\varphi(x)\}\left[1 + \frac{\Delta\lambda}{4\pi}\varphi''(x) + \frac{i\Delta\lambda}{4\pi}\{\varphi'(x)\}^2\right]. \qquad (23)$$

Then the intensity of the out-of-focus object is, to first order in the small quantity, $\Delta\lambda$,

$$I(x) = 1 + \frac{\Delta\lambda}{2\pi}\varphi''(x)\,. \qquad (24)$$

Thus the contrast depends on the second differential of the phase function $\varphi(x)$, and is reversed when the sign of the defocus, Δ, is reversed.

As we noted previously, going out of focus is equivalent to changing the phase of the amplitude of the diffraction pattern in the back focal plane by a factor proportional to u^2. The effect of lens aberrations can be represented by changing the phase of the diffraction pattern by a function in the form of a power series in even orders of u and defocus is included as a first-order aberration, coming from the second-order term in u. The fourth-order terms include third-order spherical aberration.

When the lens is exactly focussed the phase changes due to these higher order terms in u will remain and will provide some amplitude contrast. To an even greater extent than with the second-order defocus terms, these phase shifts will be small for small u but will increase rapidly for large u so that in the image contrast most contributions will come from the outer part of the diffraction patterns. Hence contrast fluctuations may be expected where $\varphi(x)$ changes rapidly.

The treatment given above in equations (20) to (24) may be extended to include higher order terms in u, but in practice the result contains combinations of higher-order differentials of $\varphi(x)$ which are not so readily visualized and so are less useful.

For out-of-focus images, the phase changes due to spherical aberration are either added to or subtracted from the phase changes for defocus, depending on the direction of defocus, and will either assist or hinder the phase contrast imaging. This is an important consideration for electron microscopy of thin objects and will be discussed in more detail in Chapter 13.

3.4.3. *Aperture limitation*

For a phase object the formation if an image of zero contrast depends on interference of waves from the backfocal plane with exactly the right amplitude and relative phase. Any modification of the back-focal plane amplitude distribution will upset this balance and produce some contrast.

If, for example, a circular aperture is placed centrally in the back-focal plane, the image will be of the form

$$I(x, y) = \left| \exp\{\mathrm{i}\varphi(x, y)\} * \frac{J_1(\pi ar)}{\pi r} \right|^2 . \tag{25}$$

It is not immediately obvious that this represents amplitude contrast. However from qualitative reasoning we may predict that, firstly, since the outer parts of the diffraction pattern are affected, only the higher-order Fourier components of $q(x, y)$ will be modified; hence contrast will appear where there are sudden changes in $\varphi(x, y)$ as at the edges of particles. In fact the image contrast will show some similarity to the function $|\varphi'(x)|$. Secondly, since the image is convoluted by a spread function the resolution will not be better than the width of the spread function. Hence again, the gain in contrast is made at the expense of resolution.

Other forms of aperture limitation include the well-known case of "Schlieren" optics used for visualizing air flow in wind tunnels. A straight edge is introduced in the back-focal plane to cut off half the diffraction pattern. From the relation (2.41) it is seen that the effect will be represented in part by convolution of the image $\exp\{\mathrm{i}\varphi(x)\}$ with a function of the form $(2\pi \mathrm{i}x)^{-1}$, which gives contrast somewhat resembling $\varphi'(x)$.

3.4.4. *Zernike phase contrast*

The form of phase contrast introduced by Zernike is the most effective in that it produces maximum contrast with no loss of resolution and the contrast is proportional to the phase function $\varphi(x)$ itself, rather than to any differential of this function. It is most easily understood for the case of a small phase change, $|\varphi(x)| \ll 1$. For this condition we can assume that all higher powers of $\varphi(x)$ are negligible and write

$$q(x) = 1 + \mathrm{i}\varphi(x) . \tag{26}$$

Then in the back-focal plane the amplitude is

$$Q(u) = \delta(u) + \mathrm{i}\Phi(u) , \tag{27}$$

where the delta function represents the directly transmitted plane-parallel incident beam and $\Phi(u)$ is the radiation scattered into the rest of the diffraction pattern. No contrast results when these two portions are re-united in the image because they are $\pi/2$ out of phase, giving

$$\psi(x) = 1 + \mathrm{i}\varphi(x)$$

and $\psi\psi^* = 1$ to first order terms only. The central beam and the rest of the diffraction pattern can be brought back into phase if the phase of the central beam is changed by $\pi/2$ by inserting a quarter-wave plate. Then (27) becomes

$$Q'(u) = \mathrm{i}[\delta(u) + \Phi(u)]$$

and

$$\psi(x) = \mathrm{i}[1 + \varphi(x)] ,$$

so that, to the first order in $\varphi(x)$,

$$I(x) = \psi\psi^* = 1 + 2\varphi(x) . \tag{28}$$

This simple description does not often represent the practical situation. If the phase change is not small, as in many important cases in both optical and electron microscopy, the theory becomes much more complicated. Also it is impossible to change the phase of a delta-function central beam only. The phase plate has finite size and so changes the phase of part of the diffraction pattern also. Further, we have assumed a plane incident wave. This is a reasonable approximation for the usual conditions of electron microscopy. It is not so in optical microscopy since for a plane wave the intensity of the image is insufficient. Instead, a hollow cone of illumination is used with an annular phase-plate near the back-focal plane.

3.5. Multi-component systems

The treatment of optical systems in the small angle approximation reproduces most of the properties of real optical systems and is a very good approximation for electron optics of medium- and high-energy systems since the scattering of electrons of these energies by atoms is essentially a small-angle phenomenon. In Section 3.3 we showed how to write down the expressions for the diffraction patterns, images or the phase distributions on any plane in a simple ideal-thin-lens system, using this approximation. We now consider the extension of this treatment to multi-component systems. We consider, for brevity and convenience, only one-dimensional objects. The extension to two-dimensions is obvious.

Radiation from a source with amplitude distribution $q_0(x)$ passes through a series of planar objects having transmission functions $q_n(x)$. Propagation over the distance R_n, from the nth to the $(n+1)^{\text{th}}$ object is represented by convolution with the propagation function $p_n(x)$ (see Fig. 3.4). In the small-angle approximation,

$$p_n(x) = \left(\frac{\mathrm{i}}{R_n\lambda}\right)^{1/2} \exp\{-\mathrm{i}kR\} \exp\{-\mathrm{i}kx^2/2R\}\,, \tag{29}$$

and neglecting the exp $\{-\mathrm{i}kR\}$ term, the Fourier transform of $p_n(x)$ is

$$P_n(u) = \exp\{\pi \mathrm{i} R_n \lambda u^2\}\,. \tag{30}$$

Then the amplitude on the plane of observation, which may be regarded as the plane of the $(N+1)$th object is written

$$\psi_{N+1}(x) = q_N(x)\underset{N}{[}\ldots\underset{3}{[}q_2(x)\underset{2}{[}q_1(x)\underset{1}{[}q_0(x) * p_0(x)\underset{1}{]} * p_1(x)\underset{2}{]} * p_2(x)\underset{3}{]}\ldots\underset{N}{]} * p_N(x)\,, \tag{31}$$

where the brackets have been numbered for clarity. The contents of the N bracket represent the amplitude of the wave incident on the Nth object. This amplitude is multiplied by the transmission function $q_N(x)$ and the product is then convoluted with $p_N(x)$.

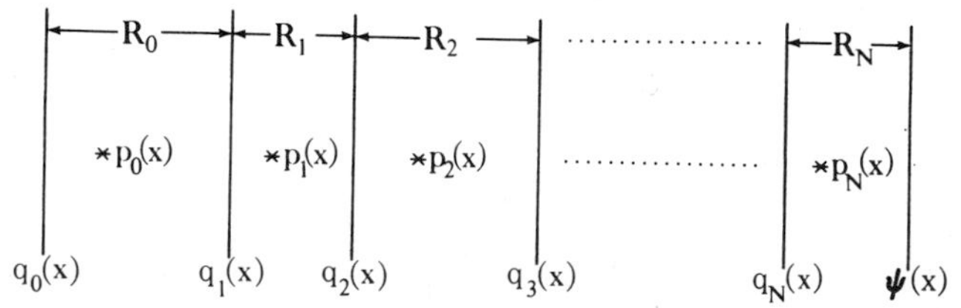

Fig. 3.4. Diagram illustrating the description of wave propagation through a multi-component system.

By making use of the convolution and multiplication theorems of Fourier transforms, the Fourier transform of (31) is written

$$\Psi_{N+1}(u) = \underset{N}{[}Q_N(u) * \ldots \underset{2}{[}Q_2(u) * \underset{1}{[}Q_1(u) * Q_0(u)\,P(u)\underset{1}{]}P_1(u)\underset{2}{]}P_2(u)\ldots\underset{N}{]}P_N(u)\,, \tag{32}$$

where again, the multiplication takes place before the convolution. This represents the Fraunhofer diffraction patterm given by the radiation emerging from the N objects.

In the case of the small angle approximation using (29) and (30), the expression for either the amplitude, (31), or the diffraction pattern, (32), may be written in terms of any combination of the functions in real space, $q_n(x)$, $p_n(x)$, or the functions of Fourier transform space, $Q_n(u)$, $P_n(u)$. This follows from the special properties of the complex exponentials with respect to Fourier transform and convolution.

Thus we have

$$q(x) * \exp\{-\mathrm{i}\alpha x^2\} = \int q(X) \exp\{-\mathrm{i}\alpha(x-X)^2\}\,\mathrm{d}X$$

$$= \exp\{-\mathrm{i}\alpha x^2\} \int q(X) \exp\{-\mathrm{i}\alpha X^2\} \exp\{2\mathrm{i}\alpha x X\}\,\mathrm{d}X$$

$$= \left(\frac{\pi}{\mathrm{i}\alpha}\right)^{1/2} \exp\{-\mathrm{i}\alpha x^2\}\left[Q\left(\frac{\alpha x}{\pi}\right) * \exp\{\mathrm{i}\alpha x^2\}\right]. \tag{33}$$

Similarly

$$Q(u) * \exp\{\mathrm{i}\alpha u^2\} = \left(\frac{\pi \mathrm{i}}{\alpha}\right)^{1/2} \exp\{\mathrm{i}\alpha u^2\}\left[q\left(\frac{\alpha u}{\pi}\right) * \exp\{-\mathrm{i}\alpha u^2\}\right]. \tag{34}$$

Thus any bracketed term of (31) or (32) may be converted to a form involving Q instead of q or q instead of Q. This is of particular value, for example, for perioduc objects for which it is often more convenient to deal with the Fourier transform, a set of weighted delta functions, rather than a continuous function.

Another useful relationship is

$$[q(x) \exp\{-\mathrm{i}\alpha x^2\}] * \exp\{-\mathrm{i}\beta x^2\}$$

$$= \left(\frac{\mathrm{i}\pi}{\alpha+\beta}\right)^{1/2} \exp\{-\mathrm{i}\beta x^2\}\left[Q\left(\frac{\beta x}{\pi}\right) * \exp\left\{\frac{\mathrm{i}\beta^2 x^2}{\alpha+\beta}\right\}\right]. \tag{35}$$

Using the relationship (34) to modify the bracketed term, this becomes

$$[q(x) \exp\{-i\alpha x^2\}] * \exp\{-i\beta x^2\}$$

$$= \exp\left\{\frac{-i\alpha\beta x^2}{\alpha+\beta}\right\}\left[q\left(\frac{\beta x}{\alpha+\beta}\right) * \exp\left\{\frac{-i\beta^2 x^2}{\alpha+\beta}\right\}\right]. \tag{36}$$

Similarly,

$$[Q(u) \exp\{i\alpha u^2\}] * \exp\{i\beta u^2\}$$

$$= \exp\left\{\frac{i\alpha\beta u^2}{\alpha+\beta}\right\}\left[Q\left(\frac{\beta u}{\alpha+\beta}\right) * \exp\left\{\frac{i\beta^2 u^2}{\alpha+\beta}\right\}\right], \tag{37}$$

and so on. Such relationships are of use when a transmission function is multiplied by an incident wave from a point source, $\exp\{-ikx^2/2R_0\}$, or when a transmission function $q_n(x)$ modifies the transmission function of an ideal thin lens, $\exp\{ikx^2/2f\}$.

In the limit that the number of two-dimensional objects tends to infinity and the distance between them tends to zero, expressions of the form (31) or (33) may be used to represent rigorously the scattering form any three-dimensional object (Moodie [1972]). To represent the scattering from a three-dimensional object with any desired degree of accuracy, the number of planar objects must be made sufficiently large and the distance between them sufficiently small. An example of this approach will be given in our consideration of the scattering of electrons by crystals (Chapter 11).

3.6. Partial Coherence

In this chapter we have considered the two limiting cases of coherent imaging, when the object is illuminated by a plane wave or radiation from a monochromatic point source, and incoherent imaging which applies to the imaging of a self-luminous source consisting of independently emitting point sources. Most practical situations are intermediate between these limiting, ideal cases. As has been pointed out in Section 1.3, it is always possible to derive the observed intensity in an experiment by considering separately the intensity given by each point of the source and by each different wavelength, and then adding all these intensities. However, it is often convenient to make use of the concept of a degree of coherence as a measure of partial coherence.

In its most complete form, the concept of a degree of coherence as a function of space and time provides a description of imaging or diffraction processes in terms of observable quantities so that there is no need to postulate an unobservable and hence fictitious wave function. (See, for example, Born and Wolf [1975].) in its simplest form, the use of the degree of coherence or the approximations of a "coherence width" and "coherence length" provide very convenient indications of the amount and type of the interference effects which might be observed in any experimental situation. We will confine our attention to these latter, less exacting aspects.

The degree of coherence is not a property of radiation at any one point. It is a measure of the correlation which exists between the phases of the radiation at two points. If there is a correlation between the phases at the two points, then light proceeding from those two points will give interference effects if it is allowed to overlap. The degree of coherence is defined in terms of the strength of these interference effects. Thus a practical interference experiment can be set up to measure the degree of coherence. If the two points are separated by a vector perpendicular to the direction of propagation we can measure the "lateral coherence". If they are separated by a vector parallel to the direction of propagation we measure the "longitudinal coherence", also called the "chromatic coherence" because the correlation of the phases then depends only on the spread of wavelengths or frequencies present.

To measure the lateral coherence between two points we can envisage an ideal Young's fringes experiment, in one dimension and with monochromatic radiation for simplicity. Narrow slits are placed at the two points and the interference pattern is observed on some plane beyond them. For perfect coherence, the interference pattern is given as squared cosine fringes [equations (1.30) and (2.46)]. For complete incoherence, the intensities from the slits are added separately and there are no fringes. The degree of coherence between the two points is given by the visibility of the fringes, defined as

$$\text{Visibility} = \frac{I_{max} - I_{min}}{I_{max} + I_{min}}.$$

For a fringe intensity function $A + B \cos bx$, the visibility is B/A.

As an example, we consider the coherence due to a small but finite source which may be considered as ideally incoherent, with an intensity distribution $I_0(x)$. If this is placed a distance R_0 from the two points, which are separated by a distance x_1 and the interference pattern is observed at a distance R, the observed intensity due to a point in the source at $x = X$ will be, from (1.30).

$$I_X(x) = 4 \cos^2 k\left(\frac{x}{R} + \frac{X}{R_0}\right). \tag{38}$$

The total intensity observed will then be given by integrating over the source, as

$$I(x) = \int I_0(X)\, 4\cos^2\left[k\left(\frac{x}{R} + \frac{X}{R_0}\right)x_1\right]\mathrm{d}X$$

$$= 2\int I_0(X)\mathrm{d}X + 2\cos^2\frac{kxx_1}{R}\int I_0(X)\cos k\left(\frac{2x_1}{R_0}\right)X\mathrm{d}X\,. \tag{39}$$

So that the visibility of the fringes and the degree of coherence is

$$\gamma_{12} = \left[\int I_0(X)\cos k\left(\frac{2x_1}{R_0}\right)X\,\mathrm{d}X\right]\Big/\int I_0(X)\mathrm{d}X\,. \tag{40}$$

Thus the degree of coherence is given by the normalised Fourier transform of the source intensity distribution on a scale of $2x_1/R_0\lambda$. This, in simplified form, is the Zernike–van Cittart theorem.

For a circular source of uniform intensity and diameter a, for example, the degree of coherence will take the form $J_1(\pi au)/(\pi au)$ where $u = x_1/R\lambda$. This defines the "coherence patch". As a measure of the width of this patch we take the radius to the first zero, giving $x_1 \approx \lambda/\alpha$ where $\alpha = a/R_0$, the angle subtended by the source at the points of interest. This gives a rough indication of the distances between two points for which interference effects will be observable. Conversely, if the interference fringe visibility is measured as a function of the separation of two slits, it is possible to deduce α, the angle subtended by the spurce. This is the basis for the use of the Michelson Stellar Interferometer to measure apparent diameters of stars.

In an entirely analagous fashion it is possible to define the longitudinal or chromatic coherence in terms of the visibility of fringes produced in a Michelson interferometer or any similar interferometer which takes radiation going in one direction, splits it into two paths of unequal length and recombines it to produce intensity fluctuations as a function of path length difference. For monochromatic radiation one gets fringes

$$I(x) = \cos(\pi\nu x/c)\,,$$

where ν is the frequency, x is the path length difference and c is the velocity.

For a small range of frequencies $\nu = \nu_0 + \nu^1$ around an average frequency ν_0, the observed intensity is given by adding intensities for all frequencies as

$$I(x) = \int I(\nu^1)\cos^2(\pi\nu x/c)\mathrm{d}\nu$$

and the longitudinal coherence is given by the visibility,

$$\frac{[(\int I(\nu)\cos(2\pi\nu x/c)\mathrm{d}\nu)^2 + (\int I(\nu)\sin(2\pi\nu x/c)\mathrm{d}\nu)^2]^{1/2}}{\int I(\nu)\mathrm{d}\nu}$$

or

$$\left| I(\nu) \exp (2\pi i\nu x/c) \mathrm{d}\nu \Big/ \int I(\nu) \mathrm{d}\nu \right| .$$

Thus for a spectral line which is gaussian in form of width $\nu^1 = a$, the longitudinal coherence function will be a gaussian of width $x = c/\pi a$, which may be called the coherence length.

In Chapter 4 we give estimates of the coherence width and coherence lengths commonly encountered in diffraction experiments with the usual radiations and sources.

Problems

1. A lens is used to image a coarse diffraction grating composed of slits in an opaque screen. An aperture is inserted in the back-focal plane of the lens. Find what image will be produced if the aperture lets through

(a) the central spot and the first diffracted spot on each side of it,

(b) the central spot and the first diffracted spot on one side only,

(c) the first diffracted spot on each side but not the central spot.

2. An object having a transmission function $A + B\cos(2\pi x/a)$ is imaged by an ideal thin lens as defined in Section 3.

(a) Find the variation of image intensity as a function of the position of the plane of observation.

(b) Determine how the sequence of out-of-focus images will be affected by spherical aberration of the lens, limitation of the lens aperture, and by imaging in white light rather than monochromatic light.

3.

(a) Make a simple argument to show that if a lens is used to image a perfectly incoherent source, the coherence in the image plane will depend only on the aperture of the lens and not at all on the extent of the source (provided that the source is larger than some minimum size).

(b) Using the definition of a degree of coherence between two points as the visibility of interference fringes resulting when small apertures in an opaque screen are placed at the two points, derive the result (a) mathematically.

4. An out-of-focus image of an object with transmission function $q(x) = 1 + \epsilon(x)$, where $\epsilon(x)$ may be complex but $|\epsilon(x)| \ll 1$, illuminated with plane parallel light, is recorded photographically so that the transmission function of the photographic plate is proportional to the incident intensity. The resultant "hologram" is illuminated with plane parallel light and imaged with a

lens. Show that, for different image positions, the in-focus image will be either a re-creation of $q(x)$ or a re-creation of the conjugate object $q^*(x)$, plus an unfocussed background. (This illustrates the principle of holography. Elaborations in practice are introduced to avoid confusion of the true image and its conjugate.)

5. A high-contrast (black and transparent) photographic negative of a chess board (8 squares each way with a white surround) is imaged in an optical system. The transparent parts of the negative have a uniform phase retardation. What is the form of the diffraction pattern in the back-focal plane of the imaging lens? If a screen with a slit parallel to one edge of the chess board is used to cut out all the diffraction maxima except those in a row through the central maximum, how will the image be affected? What image will result if the slit is rotated through 45°?
(*Note.* The amplitude along a line in a diffraction pattern is the Fourier transform of the projection of the object transmission function on to a parallel line.)

Section II

KINEMATICAL DIFFRACTION

CHAPTER 4

Radiations and their scattering by matter

4.1. X-rays

4.1.1. X-ray sources

For reasons of practical convenience, the X-radiation normally used in diffraction experiments is the characteristic K_α radiation from medium weight atoms, varying in wavelength from 2.28 Å for chromium to 0.71 Å for molybdenum, the most commonly used being the copper radiation of wavelength 1.54 Å (actually the K_{α_1} and K_{α_2} doublet having wavelengths approximately 1.537 and 1.541 Å). The radiation from a typical X-ray tube contains, in addition to these strong maxima, one or more K_β lines of shorter wavelength, some weak long-wavelength lines of the L series, some weak lines from impurities or contamination of the X-ray tube target, and a continuous background of white radiation starting abruptly at a short wavelength limit (for which the energy of the emitted X-ray, hc/λ, is equal to the energy of the electrons of the exciting electron beam, eE) passing through a maximum and then decreasing with increasing wavelength. Confusion of the diffraction patterns by unwanted radiation may be reduced by use of various monochromating devices, including absorbing filters, crystal monochromators and energy-discriminating detectors.

The isolated K_α lines are rather broad compared with spectral lines in the visible region, having a relative width, $\Delta\lambda/\lambda$, of the order of 10^{-4}. This limits the chromatic coherence of the radiation so that the coherence length is of the order of $10^4\lambda$ or approximately 1 μm. The more serious complication for many purposes comes from having the two adjacent K_{α_1} and K_{α_2} lines.

4.1.2. Scattering by electrons

The scattering of X-rays by matter is usually considered in terms of the scattering by a single electron, localized at the origin of coordinates by some sort of restoring force and so having some resonant frequency, ν_0, associated with it. For incident radiation

$$\boldsymbol{E} = \boldsymbol{E}_0 \exp\{\mathrm{i}(\omega t - \boldsymbol{k}\cdot\boldsymbol{r})\},$$

having a frequency $\nu \gg \nu_0$, the simple Thomson scattering theory gives a radiated wave of amplitude

$$\boldsymbol{E}_\mathrm{s} = -\boldsymbol{E}_0 \frac{e^2}{mc^2} \frac{1}{R} \sin\psi \cdot \exp\{\mathrm{i}(\omega t - \boldsymbol{k}\cdot\boldsymbol{R})\} \tag{1}$$

where R is the distance to the point of observation and ψ is the angle between the scattered beam and the direction of the acceleration of the electron.

If, as in Fig. 4.1 we consider the electron at the origin of orthogonal axes and the direction of propagation is the z-axis, then we consider the components of the incident radiation polarized in the $x-z$ and $y-z$ planes. The scattered radiation is in, say, the $y-z$ plane. Then for the component polarized in the $y-z$ plane, the direction of the electron acceleration is the y axis and $\psi = 90 - \varphi$, where φ is the scattering angle. Then

$$|\boldsymbol{E}| = E_0\left(\frac{e^2}{mc^2}\right)\frac{1}{R}\cos\varphi\,.$$

For the polarization in the $x-z$ plane, $\psi = \pi/2$ and $\cos\varphi = 1$ so that

$$|\boldsymbol{E}| = E_0\left(\frac{e^2}{mc^2}\right)\frac{1}{R}\,.$$

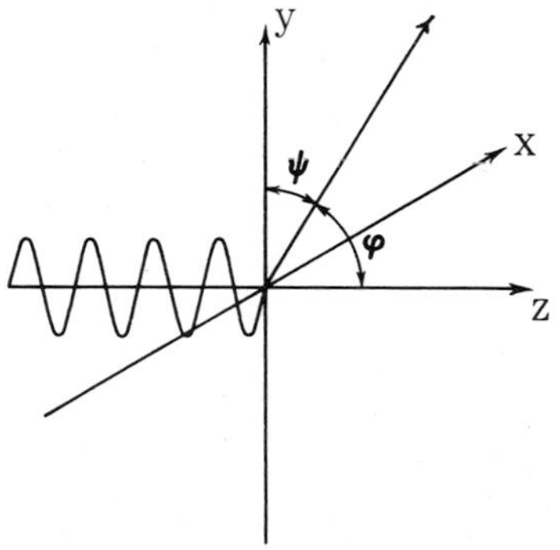

Fig. 4.1. Coordinate system for discussion of scattering of X-rays by an electron.

For unpolarized incident radiation the total scattered intensity is the sum of the intensities for the two polarizations, i.e.

$$I = I_0 \left(\frac{e^2}{mc^2}\right)^2 \frac{1}{R^2} \left(\frac{1+\cos^2\varphi}{2}\right) . \tag{2}$$

The last bracketted term here is the polarization factor. It is customary to treat the scattering as for a scalar wave function, omit the polarization factor until the final reckoning of intensities and take $(e^2/mc^2)R^{-1}$ as 1 "electron unit" of scattering.

The factor m^{-1} in (2) ensures that for scattering from atoms the scattering from the nucleus can be ignored. We consider the scattering from the cloud of electrons around the nucleus and use the nucleus only as an origin of coordinates. Then for each electron we may define a distribution function or electron-density function $\rho_n(\boldsymbol{r})$ which gives the probability that the electron should be contained in unit volume at the position defined by $\boldsymbol{r}$.

As we will see later, in Chapter 7, the purely elastic scattering is given by such a time-averaged scattering function. Hence, from equation (1.19) and (1.21), using $\rho_n(\boldsymbol{r})$ instead of $-\varphi(\boldsymbol{r})$ for the scattering strength and expressing the result in electron units in order to eliminate the constants, we find the scattering amplitude for the nth electron is

$$f_n(\boldsymbol{q}) = \int \rho_n(\boldsymbol{r}) \exp\{-\mathrm{i}\boldsymbol{q}\cdot\boldsymbol{r}\} \mathrm{d}\boldsymbol{r} , \tag{3}$$

or, putting $\boldsymbol{q} = -2\pi\boldsymbol{u}$,

$$f_n(\boldsymbol{u}) = \int \rho_n(\boldsymbol{r}) \exp\{2\pi\mathrm{i}\boldsymbol{u}\cdot\boldsymbol{r}\} \mathrm{d}\boldsymbol{r} . \tag{4}$$

The total scattering from the electron is given by (1) and is equal to one electron unit. Hence there is inelastic scattering, or Compton scattering, corresponding to the case that the X-ray photon collides with the electron, with loss of energy and change of momentum calculated in the usual way. The intensity of the inelastic scattering is then

$$I_n' = 1 - |f_n|^2 .$$

4.1.3. *Scattering by atoms*

For all the electrons associated with an atom, the elastic scattering is given by scattering from the time averaged electron density function

$$\rho(\boldsymbol{r}) = \sum_n \rho_n(\boldsymbol{r}) \,,$$

so that

$$f(\boldsymbol{u}) = \sum_n f_n(\boldsymbol{u})$$

and the so-called "atomic scattering factor" is

$$f(\boldsymbol{u}) = \int \rho(\boldsymbol{r}) \exp\{2\pi \mathrm{i} \boldsymbol{u}\cdot\boldsymbol{r}\} \mathrm{d}\boldsymbol{r} \tag{5}$$

or, if spherical symmetry of the atom is assumed, as it may always be for free atoms,

$$f(\theta) = \int_0^\infty 4\pi r^2 \rho(r) \frac{\sin 2\pi u r}{2\pi u r} \mathrm{d}r \,, \tag{6}$$

where θ is half the scattering angle φ.

For inelastic scattering, the scattering from the various electrons is incoherent, so we add intensities to give

$$I_{\text{inel.}} = \sum_n \{1 - |f_n|^2\}$$

$$= Z - \sum_n |f_n|^2 \,, \tag{7}$$

where Z is the atomic number. Thus the calculation of the Compton scattering is in general more complicated than for elastic scattering since the distribution functions for all electrons and their Fourier transforms must be calculated separately. The elastic atomic scattering factors, calculated from various approximations for the electron density function, are listed in the International Tables for X-ray Crystallography.

4.1.4. Dispersion corrections

The assumption which forms the basis for equation (1) and all the subsequent derivation, is that the frequency of the incident radiation is much greater

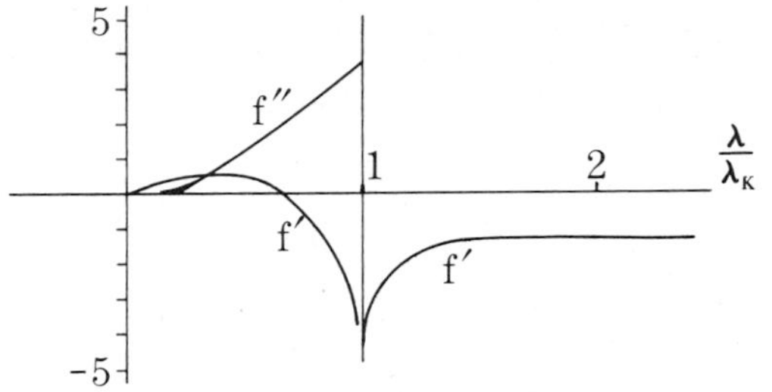

Fig. 4.2. The variation of the magnitude of the real and imaginary "anomalous scattering" components f' and f'' of the atom scattering factor for X-rays as a function of wavelength. The K absorption edge of the atoms is at λ_K.

than the frequency corresponding to any energy for excitation of the atom. This is clearly not the case if, as frequently happens, the atom has an absorption edge not far from the incident X-ray frequency. In the presence of absorption, the refractive index and also the atomic scattering factor become complex. An imaginary part and a small real part are added to $f(\boldsymbol{u})$ as defined by (5), so we write

$$f = f_0 + f' + if'' . \tag{8}$$

The variation of f' and f'' with wavelength are roughly as indicated in Fig. 4.2 with discontinuities at λ_K, the wavelength of the K absorption edge for the atom.

Since f' and f'' arise from the excitation of electrons from the inner electron shells, they may be considered as Fourier transforms of highly localized scattering distributions and so will decrease with scattering angle much more slowly than $f_c(u)$ which is given by Fourier transform of the total distribution of electrons of the atom.

4.2. Electrons

4.2.1. Sources of electrons

The electrons to be considered for electron diffraction and microscopy have energies mostly in the range 40 to 120 keV although with the advent of high voltage electron microscopes this range needs to be extended to 1 MeV or more. For electrons in the range 10–200 eV used in Low Energy Electron

Diffraction (LEED) the interactions with matter are sufficiently different to require a separate treatment.

The wavelength of the electron beam including the relativistic correction is

$$\lambda = \frac{h}{mv} = \lambda_0 \left(1 + \frac{eE}{2m_0 c^2}\right)^{-1/2} , \tag{9}$$

where λ_0 is the non-relativistic wavelength. This is 0.0548 Å for 50 keV, 0.0370 for 100 keV and 0.00867 for 1 MeV electrons.

The electron beam is produced in an electron gun in which the effective source size has a diameter of about 10 μm for a normal "hair-pin" filament, approximately 1–2 μm for heated pointed filaments and as little as 20–50 Å for field emission tips. In combination with electron lenses these sources can give well collimated beams. A divergence of 10^{-3} or 10^{-4} rads. is used for most purposes but a divergence of 10^{-6} or less can be achieved fairly readily.

In the 100 keV range, the voltage supplies may have a stability of better than 10^{-5} giving sufficient chromatic coherence for most experiments, with a coherence length of the order of 1 μm or better, often limited only by the thermal energy spread of the electron source.

4.2.2. Atom scattering amplitudes

To an approximation which is sufficiently accurate for all calculations of elastic scattering of electrons, the property of matter which is of interest is the potential distribution $\varphi(\boldsymbol{r})$. The potential distribution is related to the electron density by Poisson's equation

$$\nabla^2\varphi(\boldsymbol{r}) = -4\pi\{\rho_n(\boldsymbol{r}) - \rho_e(\boldsymbol{r})\} , \tag{10}$$

where ρ_n is the charge density due to the atomic nuclei and ρ_e is that due to the electrons.

The atom scattering amplitude for electrons is then appropriately defined as the Fourier transform of the potential distribution for the atom, $\varphi(\boldsymbol{r})$ measured in volts;

$$f_e(\boldsymbol{u}) = \int \varphi(\boldsymbol{r}) \exp\{2\pi i \boldsymbol{u}\cdot\boldsymbol{r}\} d\boldsymbol{r} . \tag{11}$$

The f values are then defined as properties of the atoms, independent of any assumption as to the scattering process or the theoretical approximation used to describe it (Dawson et al. [1974]).

The relationship of this scattering factor to that for X-rays, f_X, is derived by inserting the inverse Fourier transforms in Poisson's equation (10);

$$\nabla^2\left[\int f_e(u)\exp\{-2\pi i \boldsymbol{u}\cdot\boldsymbol{r}\}\,d\boldsymbol{u}\right]$$
$$= 4\pi\int f_X(u)\exp\{-2\pi i \boldsymbol{u}\cdot\boldsymbol{r}\}\,d\boldsymbol{u} - 4\pi\int Z\exp\{-2\pi i \boldsymbol{u}\cdot\boldsymbol{r}\}\,d\boldsymbol{u}\,,$$

where the final integral is a delta-function of weight Z, the atomic number, due to the positive charge on the nucleus. Writing the left-hand side as

$$\int(-2\pi i|u|)^2 f_e(u)\exp\{-2\pi i \boldsymbol{u}\cdot\boldsymbol{r}\}\,d\boldsymbol{u}$$

we equate integrands to obtain the Mott formula

$$f_e(\boldsymbol{u}) = \pi^{-1}(Z - f_X(\boldsymbol{u}))/u^2\,. \tag{12}$$

We note that for scattering at large angles for which f_X is small, f_e is approximately proportional to Z and decreases with angle as $(\sin^2\theta)/\lambda^2$. For $(\sin\theta)/\lambda$ tending to zero, since for neutral atoms f_X tends to Z, the value of f_e becomes indeterminate. However the limiting value is obtained from (11) as

$$f_e(0) = \int \varphi(\boldsymbol{r})\,d\boldsymbol{r}\,. \tag{13}$$

By historical accident, the atomic scattering factor for electrons has been defined as the quantity occurring in the first Born approximation for the theory of the scattering of electrons by atoms, equation (1.21); this gives

$$f_{FB}(\boldsymbol{u}) = \frac{2\pi m e}{h^2} f_e(u) = \frac{\sigma}{\lambda} f_e(u)\,,$$

where σ is the interaction constant to be defined below.

Since this quantity is not a property of the atom itself and since the first Born approximation has a very restricted range of validity for electron scattering, particularly from solids, the use of this basis for a definition is inappropriate. Its use has led to considerable confusion in the literature. However it is $f_{FB}(\boldsymbol{u})$ measured in Å for which tables of values are given, for example in the International Tables for X-ray Crystallography, Vols. 3 and 4.

4.2.3. Phase object approximation

Since appreciable scattering takes place through relatively small angles only, we may use the Fraunhofer approximation of equation (1.37) instead of the

Born approximation (1.21) to describe scattering by atoms. Regarding the atom as a quasi-two-dimensional object, we derive the transmission function $q_e(x,y)$.

From (1.6) the refractive index for the electron wave is $1 + \varphi(\boldsymbol{r})/2E$. Then the phase difference for the wave passing through the potential field $\varphi(\boldsymbol{r})$ in the z direction, relative to the wave in vacuum is $(\pi/\lambda E)\int \varphi(\boldsymbol{r})\,\mathrm{d}z$.

Putting the projection expressed by the integral equal to $\varphi(x,y)$ and $\sigma = \pi/\lambda E$, we have

$$q_e(x,y) = \exp\{-\mathrm{i}\sigma\varphi(x,y)\}\,, \tag{14}$$

and the diffraction amplitude is

$$\psi(u,v) = \mathcal{F}[\exp\{-\mathrm{i}\sigma\varphi(x,y)\}]. \tag{15}$$

With the relativistic correction, we write

$$\sigma = \frac{2\pi}{E\lambda}\{1 + (1-\beta^2)^{1/2}\}^{-1} = \frac{2\pi m_0 e\lambda}{h^2}(1 + h^2/m_0^2c^2\lambda^2)^{1/2} \tag{16}$$

where $\beta = v/c$. This relativistic interaction factor σ is seen to tend to a constant value as E increases, the limit corresponding to the non-relativistic value for $\lambda = 0.02426$, the Compton wavelength, or $E = 212$ keV. The relativistic correction for the wavelength, on the other hand, is seen from (9) to make the decrease of λ with E more rapid.

A form very close to (15), but somewhat more accurate at higher angles, is known in nuclear scattering theory as the "Molière high-energy approximation". It is derived from the general partial wave theory of scattering from a central force field by making a small angle approximation (see, for example, Wu and Ohmura [1962]).

It has been shown by Doyle [1969] by detailed calculation that this approximation and (15) are good for most atoms over the range of scattering angles normally used in diffraction experiments with solids.

If $\sigma\varphi(x,y)$ is small so that the exponential in (15) may be written

$$\exp\{-\mathrm{i}\sigma\varphi(x,y)\} \approx 1 - \mathrm{i}\sigma\varphi(x,y)\,, \tag{17}$$

then the scattering is given by

$$\psi(u,v) = \delta(u,v) - \mathrm{i}\sigma\Phi(u,v)\,. \tag{18}$$

The delta function represents the transmitted, undiffracted beam and $\Phi(u,v)$ is equal to a planar section of $f_e(\boldsymbol{u})$ defined by (11).

4.2.4. Failure of First Born Approximation

Numerical evaluation shows that, except for very light atoms, the assumption that $\sigma\varphi(x,y)$ is much smaller than unity is not justified. Then the First Born Approximation is no longer adequate. The effect of this failure may be estimated by considering the second Born Approximation (Schomaker and Glauber [1952], Glauber and Schomaker [1953]) or by invoking the partial-wave scattering theory (Hoerni and Ibers [1953]). However, the essential points are seen more immediately and graphically by use of (15), which may be written

$$\psi(u,v) = \mathcal{F}[1 + \{\cos\sigma\varphi(x,y) - 1\}] - i\mathcal{F}\{\sin\sigma\varphi(x,y)\}$$

$$= \delta(u,v) + \mathcal{F}[\cos\sigma\varphi - 1] - i\mathcal{F}[\sin\sigma\varphi] . \tag{19}$$

or, to make the analogy with (10) closer

$$\psi(u,v) = \delta(u,v) - i[\mathcal{F}\{\sin\sigma\varphi\} + i\mathcal{F}\{\cos\sigma\varphi - 1\}] , \tag{20}$$

and the atomic scattering factor $f_e(u,v)$ is the complex function in the square brackets, divided by σ.

If $\sigma\varphi$ is not too large it may be useful to expand the sine and cosine terms and take the Fourier transforms so that we get a real part of $f_e(u,v)$ equal to

$$\Phi(u,v) - \frac{\sigma^2}{3!}\{\Phi(u,v) * \Phi(u,v) * \Phi(u,v)\} + \ldots , \tag{21}$$

and an imaginary part equal to

$$\frac{\sigma}{2!}\{\Phi(u,v) * \Phi(u,v)\} - \frac{\sigma^3}{4!}\{\Phi * \Phi * \Phi * \Phi\} + \ldots . \tag{22}$$

Thus the real part is equal to $\Phi(u,v)$, the planar section of $f_e(\boldsymbol{u})$ given by (11), plus higher order terms. The imaginary term is given approximately by the two-dimensional self-convolution of $f_e(\boldsymbol{u},v)$ and so will increase more rapidly than f_e with atomic number and will fall off more slowly than the real part with $|\boldsymbol{u}|$ or $(\sin\theta)/\lambda$.

If the complex atomic scattering factor is written

$$f(\boldsymbol{u}) = |f_e(\boldsymbol{u})| \exp\{i\eta(u)\} , \tag{23}$$

then the phase angle η will increase with λ, with atomic number and with scattering angle. For the uranium atom, for example, $\eta = 0.29$ radians for $(\sin\theta)/\lambda$

= 0 and η = 2.4 rad for $(\sin\theta)/\lambda = 1.15$ Å^{-1} when E = 39.5 keV .

The effects of the failure of the first Born approximation for electron scattering from atoms were first recognized as such in the explanation by Schomaker and Glauber [1952] of the intensities of diffraction from gaseous molecules of UF_6.

It was shown later by Gjønnes [1964] and Bunyan [1963] that it may not be an adequate approximation for some gas molecules to merely replace the first Born approximation atom scattering factors by the complex scattering factors (23). In terms of the phase object approximation, (14), it is seen that if $\varphi(x,y)$ is the projected potential distribution for a molecule, the values of the real and imaginary components, (21) and (22), will depend on whether or not the atoms overlap in projection. If two atoms overlap, the contributions to $\Phi(u,v)$ will be multiplied by 2, but the contributions to the second and third order terms will be multiplied by 4 and 8. Bartell [1975] has shown that this effect can be described by an extension of Glauber's theory which provides a convenient basis for calculations.

The effects of the overlapping of atoms in projection is much more important in solids. The "pseudo-kinematical" theory proposed by Hoerni [1956], in which complex atom scattering factors replace the real ones in kinematical intensity formulas, has a very limited range of validity. It is appropriate only for "solids" consisting of single layers of atoms perpendicular to the incident beam. Coherent multiple scattering, or "dynamical scattering" must be taken into account for most experiments by use of the special theoretical treatments to be described in later chapters.

4.2.5. "Absorption" effects

As in the case of X-ray scattering, the presence of absorption will give the effect of a complex refractive index and so a complex scattering potential for electrons. In the simple formulation of (14), absorption modifies the transmission function of an atom to give

$$q(x,y) = \exp\{-\mathrm{i}\sigma\varphi(x,y) - \mu(x,y)\}$$

$$= \exp\{-\mathrm{i}[\sigma\varphi(x,y) + \mathrm{i}\mu(x,y)]\} . \qquad (24)$$

Then the atomic scattering factor for elastic scattering is convoluted by

$$\mathcal{F}[\exp\{-\mu(x,y)\}] = \delta(u,v) - \mathcal{F}[1 - \exp\{-\mu(x,y)\}]$$

$$= \delta(u,v) - M(u,v) + \tfrac{1}{2}\{M(u,v) * M(u,v)\} + \ldots \qquad (25)$$

where $M(u, v)$ is the Fourier transform of $\mu(x,y)$.

Then, in place of the bracketted part of (20), we obtain

$$f_e(u,v) = \mathcal{F}\{\sin\sigma\varphi\} + i\mathcal{F}\{\cos\sigma\varphi - 1\} - iM(u,v)$$

$$- M(u,v) * \mathcal{F}\{\sin\sigma\varphi\} - iM(u,v) * \mathcal{F}\{\cos\sigma\varphi - 1\} + \ldots . \qquad (26)$$

It is to be noted that the effect of absorption in rendering the atomic scattering factor complex is essentially different from the effect of the failure of the single-scattering approximation since for the latter the object remains a pure phase object and no energy is lost.

The absorption function $\mu(x,y)$, which may be considered as the projection of a three-dimensional function $\mu(\boldsymbol{r})$, arises from any scattering process for which the scattering is experimentally distinguished from and not interacting coherently with the scattering which is of interest.

For isolated single atoms, the only appreciable contribution to the absorption originates from the excitation of the atomic electrons. Electrons from the incident beam which have undergone an inelastic scattering process involving this excitation will have lost energy of the order of 10 eV and may be separated from the elastically scattered electrons by use of an energy analyser. Hence the inelastic scattering process may give the effect of an absorption function for elastic scattering.

For X-rays the main contribution to the absorption comes from the excitation of electrons from the inner shells of atoms and so, except in the immediate vicinity of the absorption edge, shows very little dependence on the association of atoms in molecules, liquids or solids. For electrons however, the most important inelastic scattering processes involve the outer electron shells and energy losses in the range 0 to 50 eV. The absorption coefficients are thus strongly dependent on the state of bonding or ionization of the atoms. For solids the most important contributions come from the excitation of collective electron oscillations (plasmons), the production of excited states of the crystal electrons and the excitation of lattice vibrations (phonons).

Again, for X-rays, the absorption processes subtract most of the energy from the incident photons involved so that they no longer contribute to measured intensities. This is not the case for electrons. The energy losses of the incident electrons are frequently so small that they can not be detected except by the use of special devices, so that often no distinction is made experimentally between elastic and inelastic scattering. Furthermore, if only one sort of elastic scattering is considered, e.g. only the sharp Bragg scattering, then other elastic scattering as well as the inelastic scattering may be excluded from the measurement and will then contribute to the effective absorption coefficient.

Hence the absorption coefficients for electron scattering will be strongly dependent not only on the state of aggregation of the atoms but also on the type of measurement made and the equipment used in the measurement. We therefore defer detailed discussion of the absorption coefficients for electron scattering until Chapter 12, after we have considered the elastic scattering of electrons from solids in more detail. At this stage we remark only that the imaginary part of the effective complex potential, $\mu(x,y)$ in (24) is usually smaller than the real part $\sigma\varphi(x,y)$ by a factor of from 5 to 50, and its effect may often be treated as a perturbation on the purely elastic scattering case.

4.3. Neutrons

4.3.1. Atomic scattering factors

The neutrons of interest for diffraction experiments are the thermal neutrons which are in thermal equilibrium with the atoms in a nuclear reactor and so have an average energy of about 0.025 eV for which the corresponding wavelength is about 1.5 Å. However the spread of energies or wavelengths in the beam of neutrons obtained from a reactor is quite broad and for diffraction experiments a narrow range of wavelengths is usually selected out by use of a crystal monochromator or, especially for long wavelengths, by a time-of-flight chopper device which selects a range of neutron velocities.

Neutrons are not appreciably scattered by electrons. Their main interaction is with the nucleus. We quote here the results derived initially in the context of nuclear physics and retaining something of that flavor.

Since the nucleus of an atom is very much smaller than the wavelength of the thermal neutron, the atomic scattering factor for neutrons will be isotropic, independent of scattering angle and represented by a single-valued "scattering length", b. Contributions to b include a "potential" scattering term ξ, corresponding to scattering from a hard sphere of appropriate radius and "resonance" scattering terms arising from reactions of the neutron with the nucleus. The Breit-Wigner formula for scattering from a zero-spin, isolated nucleus gives

$$b = \xi + \frac{\frac{1}{2}\Gamma_{\mathrm{n}}^{(\mathrm{r})}/k}{(E-E_{\mathrm{r}})+\frac{1}{2}\mathrm{i}(\Gamma_{\mathrm{n}}^{(\mathrm{r})}+\Gamma_{\mathrm{a}}^{(\mathrm{r})})}, \tag{27}$$

where E_{r} is the energy for resonance, $\Gamma_{\mathrm{n}}^{(\mathrm{r})}$ is the width of the resonance for re-emission of the neutron with its original energy and $\Gamma_{\mathrm{a}}^{(\mathrm{r})}$ is the width of the resonance for absorption.

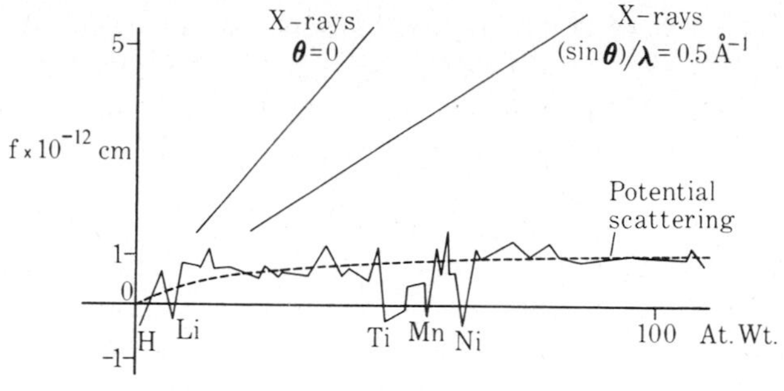

Fig. 4.3. Variation of the atomic scattering factor for neutrons, b, and the atomic scattering factor for X-rays (at particular values of $(\sin\theta)/\lambda$) with atomic weight (after Bacon [1962]).

More usually the scattering is described in terms of a scattering cross-section,

$$\sigma = 4\pi b^2 \,. \tag{28}$$

It is seen from (27) that, depending on the interactions with the nucleus, b may be positive or negative, real or complex. In practice ξ is relatively small, and being proportional to the nuclear radius, increases as the one-third power of the atomic number. The imaginary part is appreciable for only a few isotopes such as those of boron and cadmium which are strong absorbers of thermal neutrons. For a number of isotopes, including the important cases of ^{1}H, ^{7}Li, ^{48}Ti, ^{51}V, ^{53}Mn, and ^{62}Ni, b is negative. Tables of the values of b and of scattering cross-sections are given, for example by Bacon [1962] and the International Tables for X-ray Crystallography, Vol. 3. The variation with atomic number is compared with that for X-rays in Fig. 4.3.

4.3.2. Nuclear spin scattering

For neutron scattering the nuclear spin is also of importance. For a nucleus with spin, I, two scattering lengths b_+ and b_- must be defined corresponding to the formation of compound nuclei of spin $I + 1/2$ and $I - 1/2$ with relative probabilities w_+ and w_- respectively. For an assembly of atoms, or for the time average of the scattering of neutrons from a single atom, the two states will contribute at random, subject to the probabilities w_+ and w_-. Then, following the concepts to be developed in Chapter 7, the scattering will consist of "coherent" scattering from the average of the two states with a cross-section

$$\sigma_c = 4\pi(w_+ b_+ + w_- b_-)^2 \,, \tag{29}$$

and "incoherent" scattering, given by the difference between σ and the average cross-section, as

$$\sigma_{\text{inc}} = \sigma - \sigma_{\text{c}}$$

$$= 4\pi(w_+ b_+^2 + w_- b_-^2) - 4\pi(w_+ b_+ + w_- b_-)^2 . \tag{30}$$

The coherent scattering cross-section represents the scattering giving rise to diffraction effects. The incoherent scattering cross-section contributes a uniform background with no diffraction effects. This is particularly important for hydrogen for which $I = 1/2$, $b_+ = 1.04 \times 10^{-12}$ cm, $b_- = -4.7 \times 10^{-12}$ cm, so that $\sigma = 81 \times 10^{-24}$ cm^2 and $\sigma_{\text{c}} = 2 \times 10^{-24}$ cm^2. Then most of the scattering is "incoherent" background. In what follows, however, we will ignore this background scattering and consider only the b and σ_{c} values appropriate for the "coherent" scattering.

4.3.3. Isotopic disorder

For assemblies of atoms a further complication is that, for most elements the atoms present consist of a random array of the various isotopes, present either with their natural abundance or sometimes with artificial enrichment of particular isotopes. Since neutron scattering depends on the reactions of the neutrons with the nucleus, the b value may vary widely from one isotope to another. For example, b for ^{58}Ni is 1.44×10^{-12} cm, for ^{60}Ni it is 0.30×10^{-12} cm and for ^{62}Ni it is -0.87×10^{-12} cm, giving an average b for naturally occurring Ni of 1.03×10^{-12} cm.

Following the considerations of disordered variation of scattering factors given in Chapter 7 we see that for coherent diffraction effects (e.g. for giving Bragg diffraction maxima) the effective scattering factor is given by the average value

$$b = \sum_n w_n b_n \tag{31}$$

where w_n is the relative abundance of the nth isotope. The difference between scattering from this average and the total scattering is the diffuse background scattering which has intensity proportional to

$$I_{\text{diff}} = I_{\text{tot}} - |b|^2$$

$$= \sum_n w_n b_n^2 - |\sum_n w_n b_n|^2 . \tag{32}$$

4.3.4. Thermal and magnetic scattering

Two further aspects of neutron scattering are of considerable importance for practical purposes of solid-state investigations and will be treated in context in later chapters.

Firstly, because the energies of the incident neutrons are approximately the same as the thermal energies of vibration of the scattering atoms, the energy changes involved with the excitation of thermal lattice vibrations, or the creation or annihilation of phonons, are important and can be measured to provide information on phonon energies as well as momentums. Alternatively one can consider that the velocity of the thermal neutrons is so low that the diffraction process in matter is sensitive to the variation of atom positions in time so that the Fraunhofer diffraction pattern gives information concerning the relative positions of atoms in both space and time.

Secondly, because of their spin, the neutrons can interact with unpaired electron spins as well as with nuclear spins. For the near-random spin arrays of paramagnetic material this gives rise to diffuse background scattering, falling off with increasing scattering angle rather more rapidly than the X-ray scattering from an atom because the scattering is done by the electrons having unpaired spins, which are usually confined to the outer electron shells. When the electron spins are ordered in parallel or anti-parallel array as in ferromagnetic and anti-ferromagnetic materials the spin scattering gives rise to appropriate diffraction maxima. Hence neutron diffraction has become a major tool for the study of the magnetic properties of materials.

Problems

1. Estimate the magnitude of the correction to X-ray atomic scattering factors due to the inclusion of multiple-scattering terms (e.g. the second Born approximation).

2. Will Compton scattering give rise to an absorption effect for X-rays? If so, estimate its magnitude.

3. By use of the Mott formula consider the effect of ionization of an atom on its electron atomic scattering factor. What do the infinities imply? For atoms in solids, the formulas derived for isolated atoms will not apply. Suggest how scattering from ions may be treated in this case. (Then see Doyle and Turner [1968].)

CHAPTER 5

Scattering from assemblies of atoms

5.1. The kinematical approximation

The first Born approximation for scattering from a three-dimensional distribution, otherwise known as the "kinematical" or "single scattering" approximation, is given in equation (1.20). It is not limited in its application to the scattering from single atoms but may be applied to any collection of scattering matter. Normally we think in terms of assemblies of distinguishable atoms although for X-rays when $\varphi(\boldsymbol{r})$ is replaced by the electron density distribution, $\rho(\boldsymbol{r})$, the modifications of the electron distribution due to bonding may make it difficult to give a meaningful assignment of separate components of $\rho(\boldsymbol{r})$ to separate atoms. For electrons when the $\varphi(\boldsymbol{r})$ of (1.20) becomes the electrostatic potential distribution, $\varphi(\boldsymbol{r})$, the assignment to atoms may be even more difficult especially when the scattering involves, in the usual theoretical approximation, excitation of a whole crystal from one state to another i.e. the transfer from one non-localized wave function of the crystal electrons to another. However these limitations are important only for special considerations and will be treated separately when the need arises.

We will proceed under the assumption that the electron density of a collection of atoms may be written

$$\rho(\boldsymbol{r}) = \sum_i \rho_i(\boldsymbol{r}) * \delta(\boldsymbol{r} - \boldsymbol{r}_i) \tag{1}$$

where $\rho_i(\boldsymbol{r})$ is the electron density associated with the atom centered at $\boldsymbol{r} = \boldsymbol{r}_i$ and is not necessarily assumed to be the same as for a free atom. For convenience we use the notation appropriate to X-ray diffraction with the understanding that exactly the same considerations apply for electron and neutron scattering within the limits of applicability of the single-scattering approximation.

Validity of the single-scattering approximation implies that the amplitude of the single scattered radiation will be very small compared with the incident beam amplitude. Then the amplitude of the doubly- (and multiply-) scattered radiation will be very small and negligible in comparison with the singly scattered.

The amplitude of a scattered beam in a particular direction depends to a very great extent on the possibility of cooperative scattering from an ordered array of atoms. If, as in the case of X-rays and neutrons, the interaction with atoms is so weak that, in a crystal, the diffracted energy can be concentrated into one or two sharply defined directions by three dimensional diffraction before the incident beam has lost much energy, then we may consider the possibility of multiple scattering of well-defined beams. A diffracted beam will be diffracted again if it passes through another crystal region which is set at the right angle for Bragg reflection. This condition is always ensured in a large perfect single crystal, but becomes less probable in the presence of crystal defects, grain boundaries and so on. For a perfect crystal and a strong crystalline reflection, multiple scattering becomes appreciable for X-rays for path lengths of the order of 1 μm. For neutrons the necessary path lengths are several times greater. If the atoms are not sufficiently ordered to give well-defined diffracted beams, the diffracted intensity in any direction will be much less and multiple scattering effects will be correspondingly less important.

On the other hand, we have seen that for electrons the interaction with atoms is much stronger so that multiple scattering may be important in the scattering from a single heavy atom. Well-defined diffracted beams are not generated before the amplitude of scattering is appreciable. Hence multiple scattering effects become important within a distance which is of the order of one or two hundred Å for light atoms and less for heavy atoms, and are almost as strong for non-crystalline as for crystalline specimens.

It is worthwhile for any radiation to draw the distinction between "coherent" and "incoherent" multiple scattering. Here the word "coherent" is used, in accordance with a common terminology, to refer not to the incident radiation but to the correlation of the atomic positions and so to the relationships of phases of scattered radiation. It would probably be more appropriate to refer to "correlated" and "un-correlated" multiple scattering.

At one extreme, the positions of the atoms in a perfect crystal are well correlated in that they may all be related to the sites of a periodic lattice. Then the relative phases of waves scattered by any atoms may be uniquely defined and the amplitudes of the waves may be added. The correlated multiple scattering is then referred to as "dynamical" scattering.

In the other extreme, the correlation of atomic positions does not extend over the distances required to generate appreciable scattered amplitude. The relative phases of waves scattered by strongly correlated groups of atoms will vary at random from one group to another throughout the crystal. Then multiple scattering intensities are added incoherently. This case has been referred to as "multiple elastic scattering". Obviously in practice any intermediate situation, or combination of situations, may exist so that an adequate description of the scattering may be very complicated.

Fortunately, for many specimens of interest, the scattering may be described in terms of one of the two extreme cases, pure kinematical or pure dynamical diffraction, or else in terms of small perturbations of one or the other. For other specimens the physical dimensions or state of crystallinity can often be modified to simplify the scattering theory.

5.2. Real and reciprocal space

5.2.1. Reciprocal space distribution

We have seen that kinematical scattering amplitudes may be expressed in terms of the Fourier transform of a distribution in real or direct space. In real space we consider a position vector $\boldsymbol{r}$ with coordinates x, y, z. In reciprocal space we consider a position vector $\boldsymbol{u}$ with coordinates u, v, w. Then, in X-ray terminology, a distribution $\rho(\boldsymbol{r})$ in real space is related to a distribution $F(\boldsymbol{u})$ in reciprocal space by the Fourier transform,

$$F(\boldsymbol{u}) = \int \rho(\boldsymbol{r}) \exp\{2\pi \mathrm{i} \boldsymbol{u} \cdot \boldsymbol{r}\} \, \mathrm{d}\boldsymbol{r} \,, \tag{2}$$

or

$$F(u, v, w) = \iiint \rho(x, y, z) \exp\{2\pi \mathrm{i}(ux + vy + wz)\} \, \mathrm{d}x \, \mathrm{d}y \, \mathrm{d}z \,. \tag{3}$$

If an electron density distribution, $\rho(\boldsymbol{r})$, is considered to be the sum of the distributions $\rho_i(\boldsymbol{r})$ attributed to individual atoms as in equation (1) and if

$$f_i(\boldsymbol{u}) \equiv \int \rho_i(\boldsymbol{r}) \exp\{2\pi \mathrm{i} \boldsymbol{u} \cdot \boldsymbol{r}\} \, \mathrm{d}\boldsymbol{r} \,, \tag{4}$$

then Fourier transform of (1) gives

$$F(\boldsymbol{u}) = \sum_i f_i(\boldsymbol{u}) \exp\{2\pi \mathrm{i} \boldsymbol{u} \cdot \boldsymbol{r}_i\} \,. \tag{5}$$

If it is further assumed that the electron density $\rho_i(\boldsymbol{r})$ is sufficiently close to the electron density of an isolated free atom, the value of $F(\boldsymbol{u})$ can be found from the tables of atomic scattering factors given, for example, in the International Tables for X-ray Crystallography, Vol. 3.

The form of $F(\boldsymbol{u})$ corresponding to various forms of the function $\rho(\boldsymbol{r})$ may be seen by an extension to three-dimensions of the relationships and examples of Fourier transforms given in Chapter 2. For example, by extension of (2.38) and (2.42), if $\rho(\boldsymbol{r}) = 1$ inside a rectangular box of dimensions a, b, c and zero outside, then

$$F(\boldsymbol{u}) = abc \frac{\sin(\pi au)}{\pi au} \frac{\sin(\pi bv)}{\pi bv} \frac{\sin(\pi cw)}{\pi cw} . \tag{6}$$

This is the three dimensional analogue of (2.42), with a central peak of height *abc* and falling off through diminishing oscillations along each axis.

In this way we define a "shape function" and the "shape transform" often used to describe rectangular volumes of material.

5.2.2. The reciprocal lattice

By extension to three dimensions of our results for a diffraction grating, we see that, corresponding to (2.51) we have that, if

$$\rho(\boldsymbol{r}) = \sum_{n} \sum_{m=-\infty}^{\infty} \sum_{p} \delta(x - na, y - mb, z - pc) ,$$

then

$$F(\boldsymbol{u}) = \sum_{h} \sum_{k} \sum_{l} \delta\left(u - \frac{h}{a}, v - \frac{k}{b}, w - \frac{l}{c}\right) . \tag{7}$$

Thus for a periodic lattice in real space with periodicities a, b, c the corresponding distribution in reciprocal space is a lattice of points with periodicities a^{-1}, b^{-1}, c^{-1}. This is the "reciprocal lattice" for the special case of rectangular axes.

Corresponding to (2.50) and (2.51) we have the relationship, that, if $\rho(\boldsymbol{r})$ is a periodic function expressed by the Fourier series

$$\rho(\boldsymbol{r}) = \sum_{h} \sum_{k} \sum_{l} F_{hkl} \exp\left\{-2\pi i\left(\frac{hx}{a} + \frac{ky}{b} + \frac{lz}{c}\right)\right\} , \tag{8}$$

then

$$F(\boldsymbol{u}) = \sum_{h} \sum_{k} \sum_{l} F_{hkl}\, \delta\left(u - \frac{h}{a}, v - \frac{k}{b}, w - \frac{l}{c}\right) , \tag{9}$$

which represents a reciprocal lattice with each reciprocal lattice point weighted by the Fourier coefficient.

For a periodic function in real space chopped off by rectangular shapefunction having dimensions A, B, C, we have from (2.59), (6) and (9),

$$F(\boldsymbol{u}) = \sum_h \sum_k \sum_l F_{hkl}\, \delta\left(u - \frac{h}{a}, v - \frac{k}{b}, w - \frac{l}{c}\right)$$

$$* ABC \frac{\sin(\pi A u)}{\pi A u} \frac{\sin(\pi B v)}{\pi B v} \frac{\sin(\pi C w)}{\pi C w}, \qquad (10)$$

which implies that each point of the weighted reciprocal lattice, (9), is spread out into a continuous distribution given by a shape transform of the form (6).

5.2.3. Friedel's law and the phase problem

There are several relations of importance which follow from the definitions of Chapter 2.

Provided that no absorption effect is important, $\rho(\boldsymbol{r})$ may be assumed to be a real function. Then

$$F(-\boldsymbol{u}) = \int \rho(\boldsymbol{r}) \exp\{2\pi i(-\boldsymbol{u})\cdot\boldsymbol{r}\}\, d\boldsymbol{r}$$

$$= \int \rho(\boldsymbol{r}) \exp\{-2\pi i \boldsymbol{u}\cdot\boldsymbol{r}\}\, d\boldsymbol{r}$$

$$= F^*(\boldsymbol{u}), \qquad (11)$$

so that

$$|F(-\boldsymbol{u})|^2 = |F(\boldsymbol{u})|^2 .$$

Then it follows from (2.22) that

$$|\mathcal{F}\rho(-\boldsymbol{r})|^2 = |\mathcal{F}\rho(\boldsymbol{r})|^2 . \qquad (12)$$

Since the intensity of diffraction is proportional to $|F(\boldsymbol{u})|^2$, this is a convenient statement of Friedel's law which implies that inversion of a crystal through a center of symmetry does not change the diffraction intensities in a kinematical approximation.

The inversion of the relationship (2) gives

$$\rho(\boldsymbol{r}) = \int F(\boldsymbol{u}) \exp\{-2\pi i \boldsymbol{u}\cdot\boldsymbol{r}\}\, d\boldsymbol{u} . \qquad (13)$$

If the diffraction amplitudes $\psi(l, m, n)$ could be measured so that $F(\boldsymbol{u})$ could be derived, then the distribution $\rho(\boldsymbol{r})$ could be deduced by numerical evaluation of this integral. However the measurement of wave amplitudes is not possible for the radiations we consider. Only the intensities given by $\psi\psi^*$ can be recorded. Thus information on relative phases of the diffracted beams is lost and the function $\rho(\boldsymbol{r})$ cannot be deduced directly.

For most experimental situations complete knowledge of $\rho(\boldsymbol{r})$ is neither necessary nor desirable since it would involve knowledge of the relative positions and bonding of all the atoms of the sample and represent a quantity of data which would be difficult to handle. We therefore proceed to investigate the type of information which may be derived directly from the observable intensities.

5.3. The generalized Patterson function

The observable intensity of diffracted radiation is related in a known way to $\psi\psi^*$ and so to $|F(\boldsymbol{u})|^2$. This function, $|F(\boldsymbol{u})|^2$, sometimes written $J(\boldsymbol{u})$, is a function in reciprocal space. It is a real positive function and, from (11), is centro-symmetric about the reciprocal space origin. The form of the function for particular cases is readily derived from the relations for $F(\boldsymbol{u})$ given, for example, in equations (6) to (9).

By inverse Fourier transform, we see that the corresponding function in real space is

$$\mathcal{F}^{-1}\{|F(\boldsymbol{u})|^2\} \equiv P(\boldsymbol{r})$$
$$= \rho(\boldsymbol{r}) * \rho(-\boldsymbol{r})$$
$$= \int \rho(\boldsymbol{R})\,\rho(\boldsymbol{r}+\boldsymbol{R})\,\mathrm{d}\boldsymbol{R}\,. \tag{14}$$

This function of real space, $P(\boldsymbol{r})$, we call the generalized Patterson function to distinguish it from the Patterson function employed for crystal structure analysis and usually referring to periodic structures only. When no chance of confusion exist, we will call it simply the "Patterson". It is similar to the Q-function of Hosemann and Baggchi [1962]. It may also be called a "self-correlation" function, as we shall see. It is the real space function directly derivable from the observable intensities.

The greatest value of $P(\boldsymbol{r})$ will occur at the origin, $\boldsymbol{r} = 0$, since

$$P(0) = \int \rho^2(\boldsymbol{R})\,\mathrm{d}\boldsymbol{R} = \int |F(\boldsymbol{u})|^2\,\mathrm{d}\boldsymbol{u}\ , \tag{15}$$

from (2.66). It will tend to have a relatively large value when $\boldsymbol{r}$ is a vector which connects points having high $\rho(\boldsymbol{r})$ values.

When $\rho(\boldsymbol{r})$ represents the electron density for a collection of atoms, we write, as before,

$$\rho(\boldsymbol{r}) = \sum_i \rho_i(\boldsymbol{r}) * \delta(\boldsymbol{r} - \boldsymbol{r}_i) . \qquad (16)$$

Then

$$\rho(\boldsymbol{r}) * \rho(-\boldsymbol{r}) = \sum_i \sum_j \rho_i(\boldsymbol{r}) * \rho_j(-\boldsymbol{r}) * \delta(\boldsymbol{r} - \boldsymbol{r}_i) * \delta(\boldsymbol{r} + \boldsymbol{r}_j)$$

$$= \sum_i \sum_j [\rho_i(\boldsymbol{r}) * \rho_j(-\boldsymbol{r})] * \delta\{\boldsymbol{r} - (\boldsymbol{r}_i - \boldsymbol{r}_j)\} . \qquad (17)$$

Since $\rho_i(\boldsymbol{r})$ is a peaked function, roughly represented by a gaussian, such that $\int \rho_i(\boldsymbol{r})\,\mathrm{d}\boldsymbol{r} = Z_i$, the convolution $\rho_i * \rho_j$, will represent a slightly broader gaussian-like peak of integrated weight Z_iZ_j. This will be placed at the end of the interatomic vector $\boldsymbol{r}_i - \boldsymbol{r}_j$ relative to the origin of $P(\boldsymbol{r})$. There will be an identical peak at $\boldsymbol{r}_j - \boldsymbol{r}_i$. Thus $P(\boldsymbol{r})$ is made up of peaks corresponding to all inter-atomic vectors present. The weight of each peak corresponds to the sum of all products Z_iZ_j for the pairs of atoms having that interatomic vector. This is illustrated for simple two- and three- and four-atom objects in Fig. 5.1.

The peak at the origin of $P(\boldsymbol{r})$ corresponds to all vectors between centers of atoms of length zero and so is

$$\delta(\boldsymbol{r}) * \sum_i [\rho_i(\boldsymbol{r}) * \rho_i(-\boldsymbol{r})] ,$$

and has weight $\Sigma_i Z_i^2$.

If there are two vectors of the same length and direction in the object, as in Fig. 5.1(c), the corresponding Patterson peaks will be twice as large. Thus the Patterson function may be regarded as a mapping of the weighted probability that two atoms will be found a distance $\boldsymbol{r}$ apart, the weighting being according to the atomic numbers. Or we may say that $P(\boldsymbol{r})$ gives the probability that if any one atom is taken as origin, there will be another atom at a vector distance $\boldsymbol{r}$ from it. Thus it is, in essence, a correlation function giving the spatial correlation of electron densities.

Another way of looking at $P(\boldsymbol{r})$ is illustrated in Fig. 5.2, where we have redrawn the Patterson of the three-atom structure of Fig. 5.1(b) in two ways to emphasize that it may be regarded as a superposition of "images" of $\rho(\boldsymbol{r})$ or its inverse, $\rho(-\boldsymbol{r})$. We may consider each of the atoms in turn to be placed at the origin and the resulting "images" added with a weighting factor given by the Z for the atom at the origin. Alternatively, we may consider the distribution

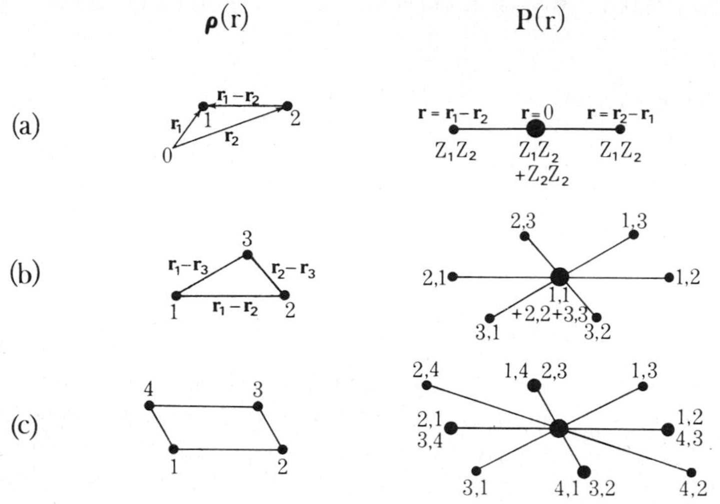

Fig. 5.1. Diagrams to suggest the form of the electron density distribution, $\rho(\boldsymbol{r})$, and the Patterson function, $P(\boldsymbol{r})$ for (a) two atoms, at positions defined by vectors $\boldsymbol{r}_1$ and $\boldsymbol{r}_2$; (b) three atoms, and (c) four atoms forming a parallelogram.

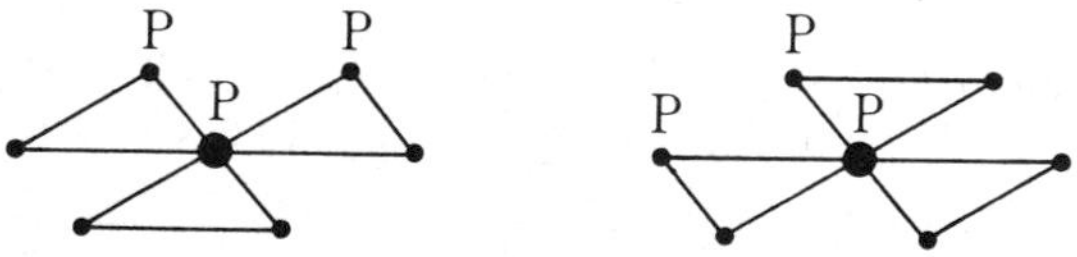

Fig. 5.2. Representation of the Patterson function for three atoms, Fig. 5.1(b), as a superposition of images of the object or its inverse.

$\rho(\boldsymbol{r})$ to be translated so that one of its atoms is placed, in turn, at each of the atom positions of $\rho(-\boldsymbol{r})$, the points marked by P in the diagram.

For systems composed of a very large number, N, of atoms, the Patterson function will have $N^2 - N + 1$ peaks (since N peaks are superimposed at the origin), and so will be too complicated to be interpreted except in a statistical sense as a probability-density or self-correlation function. We now consider a few representative cases to illustrate the application of these concepts.

5.4. Examples of correlation functions

5.4.1. Finite volume limitations

A distribution of atoms over a finite volume is often described in terms of a function representing an infinite distribution $\rho'(\boldsymbol{r})$ multiplied by a shape-function $s(\boldsymbol{r})$ i.e.

$$\rho(\boldsymbol{r}) = \rho'(\boldsymbol{r})\, s(\boldsymbol{r}) .$$

In reciprocal space,

$$F(\boldsymbol{u}) = F'(\boldsymbol{u}) * S(\boldsymbol{u}) .$$

If $s(\boldsymbol{r})$ has dimensions which are very large compared to atomic dimensions so that $S(\boldsymbol{u})$ is a very narrow function compared with the variations of $F'(\boldsymbol{u})$ which are of interest, then it is a reasonable approximation to put

$$|F(\boldsymbol{u})|^2 = |F'(\boldsymbol{u})|^2 * |S(\boldsymbol{u})|^2 . \tag{18}$$

The most obvious error in this assumption comes where $F'(\boldsymbol{u})$ changes sign, when $|F'(\boldsymbol{u}) * S(\boldsymbol{u})|^2$ will have zero values but $|F'(\boldsymbol{u})|^2 * |S(\boldsymbol{u})|^2$ will not. Also, $|F'(\boldsymbol{u}) * S(\boldsymbol{u})|^2$ will have a greater region of near-zero values, but the region affected will have approximately the half-width of $S(\boldsymbol{u})$.

Fourier transform of (18) gives the corresponding assumption for real space, that

$$P(\boldsymbol{r}) = P'(\boldsymbol{r})\, [s(\boldsymbol{r}) * s(-\boldsymbol{r})] , \tag{19}$$

i.e. the effect of spatial limitation of the sample is to multiply the Patterson for the infinite crystal (the Patterson "per unit volume") by the self-convolution of the shape transform. In one dimension, if $s(x)$ is the slit function given by

$$s(x) = \begin{cases} 1 & \text{if } |x| \leqslant A/2 , \\ 0 & \text{if } |x| > A/2 . \end{cases}$$

Then

$$s(x) * s(x) = \begin{cases} x + A & \text{if } -A < x < 0 , \\ A - x & \text{if } 0 < x < A , \\ 0 & \text{elsewhere} , \end{cases}$$

which is the Fourier transform of $|S(\boldsymbol{u})|^2$; see Fig. 5.3.

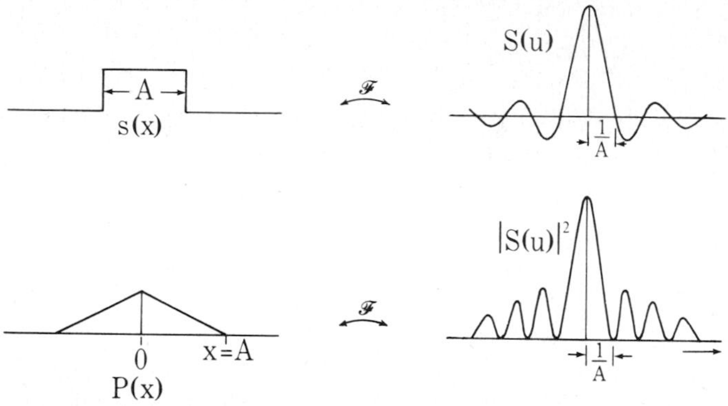

Fig. 5.3. Diagram of a shape function $s(x)$ and its Patterson function $P(x)$, and the corresponding Fourier transforms.

Similarly in three dimensions, if the sample is a rectangular block limited by planes $x = \pm A/2, y = \pm B/2, z = \pm C/2$, where the dimensions A, B, C are much greater than the dimensions of the groups of atoms of interest, the Patterson per unit volume is multiplied by a factor $ABC = V$ at the origin and decreases linearly to zero in a distance A in the x direction, B in the y direction, C in the z direction, corresponding to the fact that no interatomic vector can have x, y, z components greater than A, B, C.

Usually, on the assumption that A, B and C are very much greater than any interatomic vectors of interest, this "shape-convolution" function is ignored or omitted.

5.4.2. Finite crystals

As a simple model, we consider a crystal having one atom in a rectangular unit cell which has axes $\boldsymbol{a}, \boldsymbol{b}, \boldsymbol{c}$. The crystal has dimensions $A = N_1 a, B = N_2 b$, $C = N_3 c$ and so can be represented by

$$[\rho_0(\boldsymbol{r}) * \sum_n \delta(\boldsymbol{r} - \boldsymbol{R}_n)]\ s(\boldsymbol{r})\,, \tag{20}$$

where $s(\boldsymbol{r})$ is the shape function having dimensions A, B, C and $\boldsymbol{R}_n = n_1 \boldsymbol{a} + n_2 \boldsymbol{b} + n_3 \boldsymbol{c}$.

Then, following the considerations of (a), the Patterson function may be written,

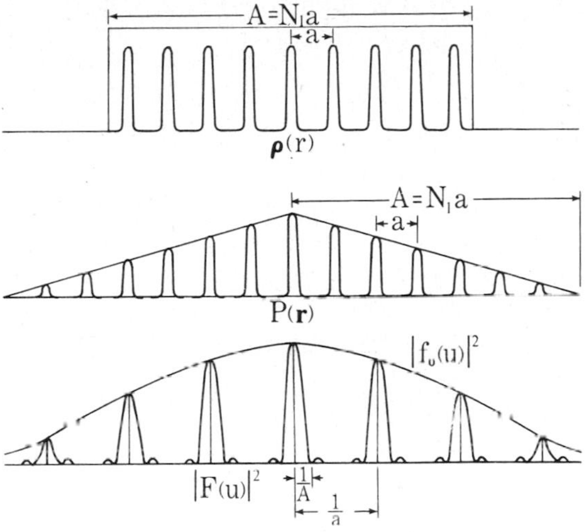

Fig. 5.4. Representation of a finite crystal as the product of a periodic object and a shape function, the corresponding Patterson function and its Fourier transform.

$$\rho_0(\boldsymbol{r}) * \rho_0(-\boldsymbol{r}) * \left[\sum_n \sum_m \delta(\boldsymbol{r} - \boldsymbol{R}_n + \boldsymbol{R}_m)\right] [s(\boldsymbol{r}) * s(-\boldsymbol{r})] \,, \tag{21}$$

$$= N\rho_0(\boldsymbol{r}) * \rho_0(-\boldsymbol{r}) * \left[\sum_n \delta(\boldsymbol{r} - \boldsymbol{R}_n)\right] [s(\boldsymbol{r}) * s(-\boldsymbol{r})] \,, \tag{22}$$

since the convolution of the two sets of N delta functions gives N times the set of all vectors between lattice points multiplied by the shape convolution, as suggested in the one-dimensional example of Fig. 5.4.

Then the Fourier transform of (25) gives the corresponding distribution in reciprocal space as

$$|F(\boldsymbol{u})|^2 = N|f_0(\boldsymbol{u})|^2 \sum_{h,k,l} \delta\left(u - \frac{h}{a}, v - \frac{k}{b}, w - \frac{l}{c}\right) * |S(\boldsymbol{u})|^2 \,, \tag{23}$$

which is the reciprocal lattice having spacings a^{-1}, b^{-1}, c^{-1} with reciprocal lattice points weighted by the function $|f_0(\boldsymbol{u})|^2$, and with each reciprocal lattice point spread out by convolution with the function

$$|S(\boldsymbol{u})|^2 = A^2B^2C^2 \frac{\sin^2(\pi Au)}{(\pi Au)^2} \frac{\sin^2(\pi Bv)}{(\pi Bv)^2} \frac{\sin^2(\pi Cw)}{(\pi Cw)^2}, \tag{24}$$

which is the square of the function (6).

This case represents the most perfect ordering of atoms within a finite volume with the correlation function of atomic positions limited only by the shape-correlation. There are many stages of ordering intermediate between this and the case of an ideal gas representing the minimum of ordering. Several examples of intermediate degrees of order, particularly those approximating to the order of the perfect crystal, will be considered in later chapters.

5.5. Correlation in space and time

5.5.1. Four-dimensional Patterson

The concept of a Patterson function, which is a correlation function for the distribution of electron density in space, can be extended to include correlations in time. In space we ask; what is the probability that an atom will be separated from a given atom by a vector $\boldsymbol{r}$? We may ask, similarly: if an atom is at the point $\boldsymbol{r}$ at time 0, what is the probability that there will be an atom at that point (or at any other point) after a time t?

The function which includes such correlations is the four-dimensional Patterson function

$$P(\boldsymbol{r}, t) \equiv \int \rho(\boldsymbol{R}, T)\, \rho(\boldsymbol{r} + \boldsymbol{R}, t + T)\, \mathrm{d}\boldsymbol{R}\, \mathrm{d}T$$

$$= \rho(\boldsymbol{r}, t) * \rho(-\boldsymbol{r}, -t)\,. \tag{25}$$

Fourier transformation gives a function in four-dimensional reciprocal space. The transform with respect to time, following equation (2.31), gives a function of frequency so we write

$$|F(\boldsymbol{u}, \nu)|^2 = \iint P(\boldsymbol{r}, t) \exp\{2\pi \mathrm{i}(\boldsymbol{u} \cdot \boldsymbol{r} + \nu t)\}\, \mathrm{d}\boldsymbol{r}\, \mathrm{d}t\,. \tag{26}$$

Just as in the spatial relationship, $\boldsymbol{u}$ is related to the vector $\boldsymbol{q}$ of a diffraction experiment, giving the change in the wave vector $\boldsymbol{k}$, or in the momentum $h\boldsymbol{k}$, so ν is related to the change in frequency from the incident frequency, or the change in energy $h\nu$ of the incident photons (or particles).

This type of formalism, introduced by van Hove [1954], is of special value for neutron diffraction for which changes in both momentum and energy of the incident neutrons may be measured. For scattering from a crystal having thermal vibration of the atoms or, more appropriately, a distribution of phonons, both the momentums and energies of the phonons may be deduced.

5.5.2. *Special cases*

A few special cases will serve to illustrate the properties and relationships of the function $P(\boldsymbol{r}, t)$. For the point $\boldsymbol{r} = 0$ we have

$$P(0, t) = \iint \rho(\boldsymbol{R}, T)\, \rho(\boldsymbol{R}, t + T)\, \mathrm{d}T\, \mathrm{d}\boldsymbol{R}\,. \tag{27}$$

The integral over T gives the probability that if there is an atom at $\boldsymbol{R}$ at time 0 there will be an atom at $\boldsymbol{R}$ at time t, and this is averaged over all points $\boldsymbol{R}$ This will give an indication of the rate of diffusion of atoms away from their initial positions.

For purely elastic scattering there is no change of frequency of the incident radiation so that $\nu = 0$. Then the reciprocal space function of interest is

$$|F(\boldsymbol{u}, 0)|^2 = \iint P(\boldsymbol{r}, t) \exp\{2\pi i(\boldsymbol{u} \cdot \boldsymbol{r})\}\, \mathrm{d}\boldsymbol{r}\, \mathrm{d}t$$

$$= \int \left[\int P(\boldsymbol{r}, t)\, \mathrm{d}t\right] \exp\{2\pi i \boldsymbol{u} \cdot \boldsymbol{r}\}\, \mathrm{d}\boldsymbol{r}\,. \tag{28}$$

This is also equal to the Fourier transform of the three-dimensional Patterson function of the time average of the electron density.

Hence the section of the four-dimensional reciprocal space distribution, $\nu = 0$, corresponds to the Fourier transform of the time-average of the four-dimensional Patterson function.

If a measurement is made in which the scattered radiation intensities for all energies are added together, the reciprocal space function of interest is

$$\int |F(\boldsymbol{u}, \nu)|^2\, \mathrm{d}\nu = \iint P(\boldsymbol{r}, t) \exp\{2\pi i(\boldsymbol{u} \cdot \boldsymbol{r})\}\, \delta(t)\, \mathrm{d}\boldsymbol{r}\, \mathrm{d}t$$

$$= \int P(\boldsymbol{r}, 0) \exp\{2\pi i \boldsymbol{u} \cdot \boldsymbol{r}\}\, \mathrm{d}\boldsymbol{r}\,. \tag{29}$$

Hence the total scattered intensity, elastic plus inelastic, is related to $P(\boldsymbol{r}, 0)$, which gives the correlations of atomic positions for no difference in times; corresponding to the sum of all correlation functions for instantaneous "pictures" of the atomic configuration.

Expressions such as (28), however, are unsatisfactory as they stand because they lead to infinities. If $P(\boldsymbol{r}, t)$ refers to a system which is a constant, or repeats periodically in time, the integral $\int P(\boldsymbol{r}, t)\, \mathrm{d}t$ will be infinite. This unsatisfactory result must arise because we have not described the experimental system adequately.

Measurements of intensities are always made with instruments of finite resolution. The finite energy resolution may be taken into account by replacing $|F(\boldsymbol{u}, \nu)|^2$ by, say,

$$|F(\boldsymbol{u}, \nu)|^2 * (\pi M^2)^{-1/2} \exp\{-\nu^2/M^2\},$$

where M is a measure of the range of frequencies included. The Fourier transform of this function is then

$$P(\boldsymbol{r}, t) \exp\{-2\pi^2 M^2 t^2\}.$$

Then the measurement of intensities for all ν values is given in the limit $M \to \infty$, giving, by Fourier transform, $P(\boldsymbol{r}, 0)$.

In the limit that M tends to zero we return to the ideal case of measuring $|F(\boldsymbol{u}, \nu)|^2$ and obtaining $P(\boldsymbol{r}, t)$ by Fourier transform.

The case where near-elastic scattering only is measured is represented by selecting only a small range of frequency around the origin:

$$\int |F(u, V)|^2 \exp\{-V^2/M^2\}\, \mathrm{d}V.$$

The Fourier transform of this is

$$P(\boldsymbol{r}, t) * (\pi^{1/2} M) \exp\{-2\pi^2 M^2 t^2\}.$$

In the limiting case $M \to 0$, corresponding to purely elastic scattering this integral tends to $\langle P(\boldsymbol{r}, t)\rangle_t$, i.e. the time average of $P(\boldsymbol{r}, t)$, since $P(\boldsymbol{r}, t)$ is convoluted with a function which has a width tending to infinity but an integrated value always equal to unity.

In what follows we will not complicate the mathematics by using these more complete formulations, but will bear in mind that the difficulties arising from the over-simplified expressions may be avoided in this way.

5.5.3. Ideal monatomic gas or liquid

We consider the idealised case of N non-interacting atoms contained in a volume V such that the distribution of atom positions is completely random, i.e. there is no correlation of atom center positions except for the limitations of the finite volume. Each atom has an electron density distribution $\rho_0(\boldsymbol{r})$. We ignore the usual practical limitation that the electron distributions of separate atoms cannot overlap appreciably.

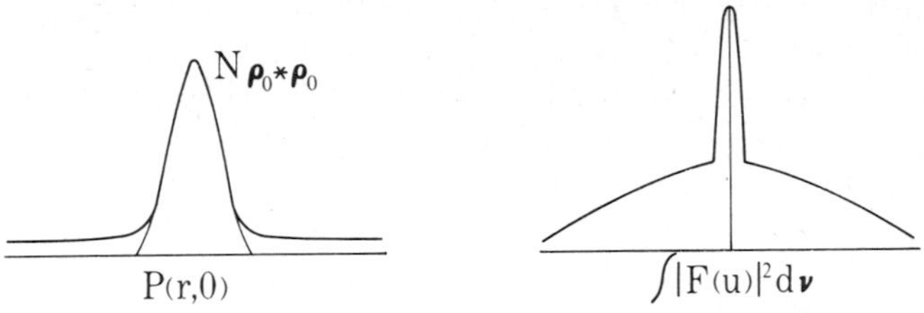

Fig. 5.5. Instantaneous correlation function, $P(\boldsymbol{r}, 0)$, for an ideal monatomic gas and the corresponding distribution in reciprocal space.

The form of the Patterson function $P(\boldsymbol{r}, t)$ can be written down immediately. The instantaneous correlation function $P(\boldsymbol{r}, 0)$ will have a peak at the origin given by $N\rho_0(\boldsymbol{r}) * \rho_0(\boldsymbol{r})$. For any non-zero length vector the probability of finding an atom at that vector distance from a given atom has the same value which depends only on the gas or liquid density, apart from the finite volume effect discussed in 5.4.1. Hence we can write, as a reasonable approximation except near $\boldsymbol{r} = 0$,

$$P(\boldsymbol{r}, 0) = \frac{N}{V} \int \{\rho_0(\boldsymbol{r}) * \rho_0(\boldsymbol{r})\} \,\mathrm{d}\boldsymbol{r}\, [s(\boldsymbol{r}) * s(\boldsymbol{r})]$$

$$= Z^2 \frac{N}{V} [s(\boldsymbol{r}) * s(-\boldsymbol{r})] \,. \tag{30}$$

The form of $P(\boldsymbol{r}, 0)$ is then as shown in Fig. 5.5. This function could be derived from the scattering power distribution for all energies, $\int |F(\boldsymbol{u}, \nu)|^2 \,\mathrm{d}\nu$, which will consist of a central peak coming from the Fourier transform of (30) plus $N|f_0(\boldsymbol{u})|^2$ from the origin peak of $P(\boldsymbol{r}, 0)$. Thus the scattering will be N times that for a single atom.

The correlation $P(\boldsymbol{r}, t)$ for $\boldsymbol{r}, t$ not equal to zero will be the same. For $\boldsymbol{r} = 0$, the value of $P(0, t)$ will give the probability that if an atom is at a point at time 0, there will be an atom there at time t. For large times this will tend to the random value of (30). The rate at which $P(0, t)$ approaches this random value will depend on the average velocity of the atom or the self-diffusion coefficient.

More formally and generally we may describe the electron density function as

$$\rho(\boldsymbol{r}, t) = \rho_0(\boldsymbol{r}) * g(\boldsymbol{r}, t)$$

where

$$g(\boldsymbol{r}, t) = \sum_n \delta\{\boldsymbol{r} - \boldsymbol{r}_n(t)\} . \tag{31}$$

Thus $g(\boldsymbol{r}, t)$ is a distribution function: a set of delta functions describing the positions of the atom centers in space and time. Then

$$P(\boldsymbol{r}, t) = \rho_0(\boldsymbol{r}) * \rho_0(-\boldsymbol{r}) * g(\boldsymbol{r}, t) * g(-\boldsymbol{r}, -t) . \tag{32}$$

The convolutions are over space and time. Since $\rho_0(\boldsymbol{r})$ does not depend on time the first convolution is

$$\iint \rho_0(\boldsymbol{R})\, \rho_0(\boldsymbol{r} + \boldsymbol{R})\, \mathrm{d}\boldsymbol{R}\, \mathrm{d}T \ ,$$

where the normalizing factor must be introduced as before so that the integral $\int \mathrm{d}T$ does not give an infinity. The distribution function convolution

$$\mathcal{G}(\boldsymbol{r}, t) = g(\boldsymbol{r}, t) * g(-\boldsymbol{r}, -t) ,$$

gives the probability that, given an atom centered $\boldsymbol{r} = 0$ at time $t = 0$, there will be an atom centered at the position $\boldsymbol{r}$ at time t. Provided that we are dealing with a very large number of atoms, $\mathcal{G}(\boldsymbol{r}, t)$ can be considered to be a continuous function.

Then in a diffraction experiment in which all frequencies are recorded with equal efficiency, the observation measures $\int |F(\boldsymbol{u}, \nu)|^2\, \mathrm{d}\nu$ from which one may deduce $P(\boldsymbol{r}, 0)$ which depends on

$$\begin{aligned}\mathcal{G}(\boldsymbol{r}, 0) &= [g(\boldsymbol{r}, t) * g(-\boldsymbol{r}, -t)]_{t=0} \\ &= \int g(\boldsymbol{R}, T)\, g(\boldsymbol{r} + \boldsymbol{R}, T)\, \mathrm{d}\boldsymbol{R}\, \mathrm{d}T \\ &= \langle g(\boldsymbol{r}, t) \underset{r}{*} g(-\boldsymbol{r}, -t)\rangle_t \end{aligned} \tag{33}$$

and this is the time average of the instantaneous distribution Patterson.

The purely elastic intensity, $|F(\boldsymbol{u}, 0)|^2$, gives $\int P(\boldsymbol{r}, t)\, \mathrm{d}t$, which depends on

$$\begin{aligned}\int \mathcal{G}(\boldsymbol{r}, t)\, \mathrm{d}t &= \iint g(\boldsymbol{R}, T)\, g(\boldsymbol{r} + \boldsymbol{R}, t + T)\, \mathrm{d}t\, \mathrm{d}T\, \mathrm{d}\boldsymbol{R} \\ &= \int \langle g(\boldsymbol{R}, t)\rangle_t\, \langle g(\boldsymbol{r} + \boldsymbol{R}, t)\rangle_t\, \mathrm{d}\boldsymbol{R} \\ &= \langle g(\boldsymbol{r}, t)\rangle_t \underset{r}{*} \langle g(-\boldsymbol{r}, -t)\rangle_t \ , \end{aligned} \tag{34}$$

and this is the self convolution of the time average of the distribution function.

5.5.4. *Real monatomic gases and liquids*

An important difference between (33) and (34) appears when we introduce the restriction that atoms cannot overlap appreciably, i.e. that at a particular time t the distribution Patterson $g(\boldsymbol{r}, t) * g(-\boldsymbol{r}, -t)$ and therefore in general $\mathcal{G}(\boldsymbol{r}, 0)$ cannot contain any points between $|\boldsymbol{r}| = 0$ and $|\boldsymbol{r}| = 2r_0$ where r_0 is the effective radius of an atom. For $|\boldsymbol{r}| > 2r_0$ some fluctuations in $\mathcal{G}(\boldsymbol{r}, 0)$ will occur because atoms usually have an attractive interaction. There will tend to be an excess of atoms at the smallest possible, "nearest neighbor" distance; then especially for liquids, an appreciable but less marked tendency for atoms to occur also at a second nearest, third nearest, etc. neighbor distance, so that $\mathcal{G}(\boldsymbol{r}, 0)$ will have a form as suggested, in one dimension, in Fig. 5.6(a), with a delta-function at the origin. The form of $P(\boldsymbol{r}, 0)$ is then as suggested in Fig. 5.6(b) and the distribution of scattering power will oscillate with $|\boldsymbol{u}|$ as suggested in Fig. 5.6(c).

On the other hand the purely elastic scattering $|F(\boldsymbol{u}, 0)|^2$ derives from the self convolution of the time average of the distribution function, $\langle g(\boldsymbol{r}, t)\rangle_t$. But,

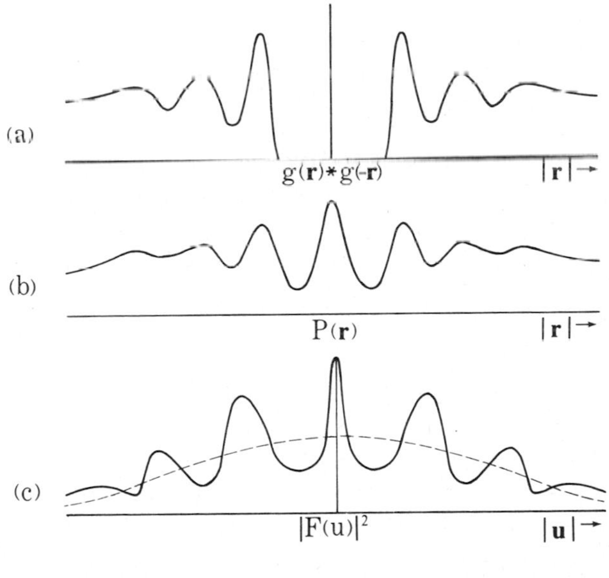

Fig. 5.6 Diagrams suggesting the form of, (a) the distribution Patterson, (b) the actual Patterson function and (c) the scattering power distribution for a real monatomic gas or liquid.

since all atoms are moving and all positions for an atom are equally probable, this averaged distribution function is a constant within the volume of the system. The convolution (34) and $\int P(\boldsymbol{r}, t)\,dt$ will also then be constants, multiplied by the convolution of the shape function $[s(\boldsymbol{r}) * s(-\boldsymbol{r})]$. Then the purely elastic scattering consists of the single sharp peak $|S(\boldsymbol{u})|^2$.

Thus we deduce that, apart from the sharp central, zero angle peak, all diffracted intensity is inelastic. This is a somewhat academic point for X-ray or electron diffraction because the rate of decay of $P(0, t)$ with time is commensurate with the Boltzmann distribution of atom velocities for normal temperatures and average energies of the atoms of kT ($\approx$0.02 eV). Energy changes of the incident radiation of this magnitude are not detectable and the normal measurements will reflect the Patterson $P(\boldsymbol{r}, 0)$ as given by (33). For neutron diffraction however such changes in energy, or frequency, can be detected and the situation represented by (34) may be approached.

A further stage of complication is added if we consider gases or liquids composed of molecules rather than independent atoms. Then instead of $\rho_0(\boldsymbol{r})$ for a single atom we must consider $\rho_0(\boldsymbol{r}, \theta_n, \varphi_n)$ where the angles θ_n, φ_n specify the orientation of the molecule and are functions of time. The total scattered intensity then corresponds to a time average of the instantaneous Pattersons,

$$P(\boldsymbol{r}, 0) = \left\langle \sum_n \sum_m \rho_0(\boldsymbol{r}, \theta_n, \varphi_n) * \rho_0(-\boldsymbol{r}, \theta_m, \varphi_m) * \delta(\boldsymbol{r} - \boldsymbol{r}_n + \boldsymbol{r}_m) \right\rangle_t . \qquad (35)$$

If it can be assumed as a first approximation that the relative positions of the centers of all molecules are completely random, all these terms for $n \neq m$ give only a continuous smooth background and the Patterson will be the same for a monatomic gas except that the origin peak is replaced by

$$\left\langle \sum_n \rho_0(\boldsymbol{r}, \theta_n, \varphi_n) * \rho_0(-\boldsymbol{r}, \theta_n, \varphi_n) \right\rangle_t .$$

For each orientation $\theta_n, \varphi_n, P(\boldsymbol{r}, 0)$ will contain a set of peaks corresponding to the interatomic vectors within the molecule. The summation over n and the averaging over time will reproduce this set of peaks in all possible orientations about the origin with equal probability. Then the radial distribution $P(r, 0)$ will contain a set of peaks corresponding to the interatomic vector as suggested in Fig. 5.7.(a) and the intensity function will correspondingly show fluctuations about the smooth $N|f_0(\boldsymbol{u})|^2$ curve, Fig. 5.7(b).

A more detailed and complete discussion of diffraction from liquids along lines related to the above is given by Guinier [1963].

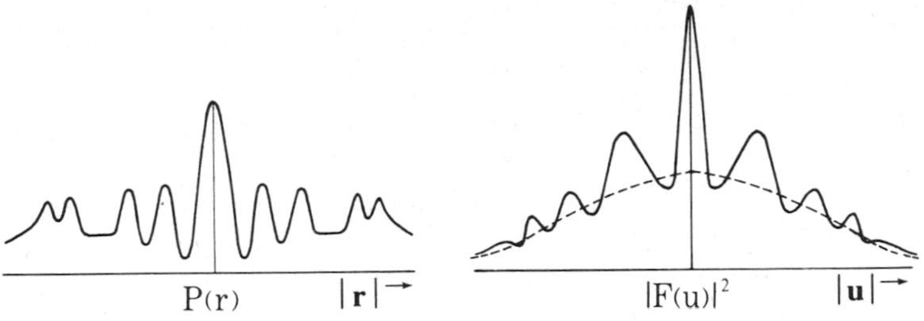

Fig. 5.7. Diagrams suggesting the Patterson function for a molecular gas, and the corresponding variation of scattering power in reciprocal space.

5.5.5. *The hydrogen atom*

When dealing with quantum processes of excitation of the scattering object considerable caution is required and it cannot necessarily be assumed that the simple classical considerations we have used can be carried over (van Hove [1954]). However a direct application of the above theory gives an adequate treatment for one case; the scattering of X-rays by a hydrogen atom, i.e. by the electron of a hydrogen atom. For this electron $P(\boldsymbol{r}, 0)$ is a delta function at the origin, since all instantaneous pictures show a single point, one electron. Then the total scattering, in electron units squared, is given by

$$\int |F(\boldsymbol{u}, \nu)|^2 \, d\nu = 1 .$$

For the elastic scattering we consider

$$\int P(\boldsymbol{r}, t) \, dt = \rho_e(\boldsymbol{r}) * \rho_e(-\boldsymbol{r}) ,$$

where $\rho_e(\boldsymbol{r})$ is the time average distribution of the electron position or the electron density function for the electron in a hydrogen atom. Then in reciprocal space

$$|F(\boldsymbol{u}, 0)|^2 = f_e^2(\boldsymbol{u}) . \tag{33}$$

The inelastic (Compton) scattering is then given by the difference between the total and the elastic scattering as,

$$1 - f_e^2(\boldsymbol{u}) , \tag{34}$$

in accordance with our result of equation (4.7).

The application of this formalism to the scattering of neutrons by phonons will be given in Chapter 12.

5.6. Diffraction geometry and intensities

We have established that kinematical, elastic diffraction amplitudes and intensities, obtained by the scattering of X-rays from electron density distributions, may be related to the distributions in reciprocal space given by Fourier transform of $\rho(\boldsymbol{r})$ or $P(\boldsymbol{r})$. The next step is to show how the amplitudes or intensities for particular experimental arrangements may be derived from the reciprocal space distributions. The argument may be carried through either for diffraction amplitudes in terms of the reciprocal space distribution $F(\boldsymbol{u})$ or for diffraction intensities in terms of the distribution $|F(\boldsymbol{u})|^2$, which may be called the distribution of the scattering power. For the moment we restrict ourselves to the latter as being more appropriate to general diffraction experiments in which intensities or energy fluxes are measured.

From equations (1.20) and (1.21), the scattered amplitude is given in the assymptotic limit of large R, as a function of $\boldsymbol{q} = \boldsymbol{k} - \boldsymbol{k}_0$;

$$\psi(\boldsymbol{q}) = -\frac{\mu}{4\pi} \int \varphi(\boldsymbol{r}) \exp\{-\mathrm{i}\boldsymbol{q}\cdot\boldsymbol{r}\}\, \mathrm{d}\boldsymbol{r} \;. \tag{35}$$

In the appropriate units and putting $\boldsymbol{q} = -2\pi\boldsymbol{u}$, we have $\psi(\boldsymbol{q}) = F(\boldsymbol{u})$, given by equation (2) and the intensity is

$$I(\boldsymbol{q}) = \psi\psi^*(\boldsymbol{q}) = \int P(\boldsymbol{r}) \exp\{2\pi\mathrm{i}\boldsymbol{u}\cdot\boldsymbol{r}\}\, \mathrm{d}\boldsymbol{r} \;. \tag{36}$$

Thus for an incident monochromatic beam in a direction defined by the wave vector $\boldsymbol{k}_0$, the intensity diffracted elastically in a particular direction defined by wave vector $\boldsymbol{k}$ will be equal to the value of the function $|F(\boldsymbol{u})|^2$ at the position in reciprocal space defined by $\boldsymbol{u} = (\boldsymbol{k} - \boldsymbol{k}_0)/2\pi$.

This relationship is expressed by the Ewald sphere construction in reciprocal space, Fig. 5.8. A vector of length λ^{-1} $(= |\boldsymbol{k}_0|/2\pi)$ is drawn to the origin, O, of reciprocal space, in the direction of $\boldsymbol{k}_0$ from the point P. A sphere of radius λ^{-1} is drawn around P as center. Then for any point on the sphere, $\boldsymbol{u}$, the radial vector (length λ^{-1}) from P represents the direction of the diffracted beam $\boldsymbol{k}$ such that $\boldsymbol{u} = (\boldsymbol{k} - \boldsymbol{k}_0)/2\pi$. The intensity of this diffracted beam will be $|F(\boldsymbol{u})|^2$. Thus this Ewald sphere construction gives the directions and intensities for all diffracted beams produced for a given incident beam direction.

Unless $|F(\boldsymbol{u})|^2$ is isotropic, its orientation in reciprocal space is defined in

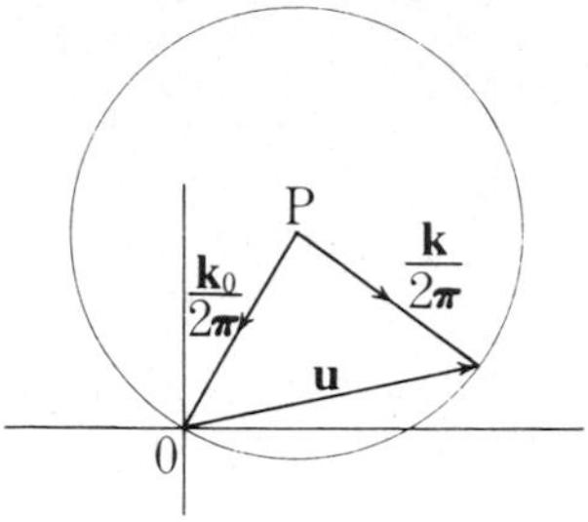

Fig. 5.8. The Ewald sphere construction.

terms of the orientation of $P(\boldsymbol{r})$ in real space. Then a rotation of the sample in real space produces the corresponding rotation of the scattering power distribution $|F(\boldsymbol{u})|^2$ in the reciprocal space. For a constant incident beam direction the diffracted intensities then vary as regions of different scattering power are rotated through the Ewald sphere. Naturally, exactly the same sequence of intensities will be produced if the distribution $|F(\boldsymbol{u})|^2$ is kept stationary and the Ewald sphere is rotated in the opposite direction, i.e. if the sample is kept stationary and the direction of the incident beam is rotated.

The form of the diffraction pattern recorded in practice will depend on the geometry of the recording system and the wavelength of the radiation. The function $|F(\boldsymbol{u})|^2$ will decrease, on the average, with the square of the atomic scattering factor for an average atom, $|f(\boldsymbol{u})|^2$. If the mean radius of an atom is taken to be about 0.5 Å, the half-width of the distribution $|f(\boldsymbol{u})|^2$ will be of the order of 2 Å^{-1} and the range of $|\boldsymbol{u}|$ which will normally be of interest will be several times this: say 5 Å^{-1}.

For X-ray and neutron diffraction, the wavelengths are of the order of 1 Å so that the diameter of the Ewald sphere will be 2 Å^{-1}. Thus the whole of the intersection of the sphere with the function $|F(\boldsymbol{u})|^2$ will be of interest and scattering through all angles from 0 to π will normally be recorded as suggested by Fig. 5.9(a). Thus for photographic recording it is customary to use a cylindrical film with the sample on the axis. For electronic recording with a photon- or particle-counting detector, a goniometer stage is used which allows the detector to be swung through diffraction angles which are as large as is convenient.

For the diffraction of electron having a wavelength of about 0.04 Å, the diameter of the Ewald sphere will be 50 Å^{-1}. On this sphere, only the small region of radius about 5 Å^{-1} around the reciprocal space origin will be of interest and the scattering will be predominantly through small angles as suggested

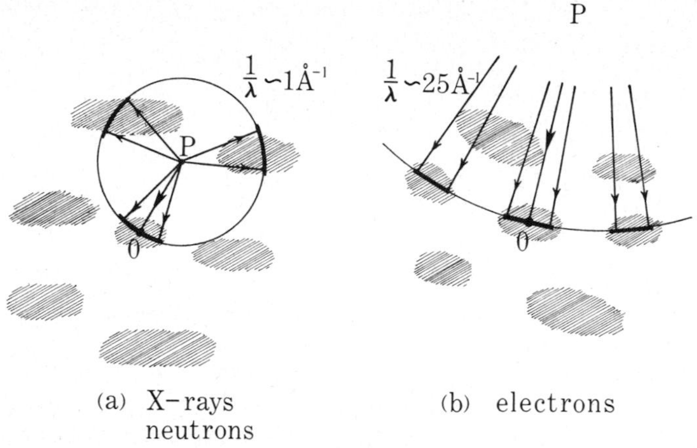

Fig. 5.9. Comparison of the scales of the Ewald spheres for X-rays, neutrons and electrons, in relation to normal scattering power distributions.

by Fig. 5.9(b). The diffraction pattern may be recorded on a flat plate or film placed perpendicular to the incident beam at some distance behind the specimen and will represent an almost plane section of the distribution of scattering power in reciprocal space.

In this way it would seem that the intensities to be observed for a given radiation and for a particular geometry of the experiment may be deduced for scattering from any system for which the Patterson function can be derived or postulated. However our discussion so far has been for the idealized case of perfectly plane and monochromatic waves. These considerations must be broadened in order to make contact with experimental situations.

5.7. Practical considerations

5.7.1. Finite sources and detectors

In X-ray diffraction experiments, the finite source size results in an angle of convergence of the incident beam at any point of the sample of something like 10^{-4} to 10^{-3} radians. The intensity is not necessarily uniform over this range of angles. In neutron diffraction the angle of convergence is often made greater than this since the source intensity is low. For electron diffraction the angle of

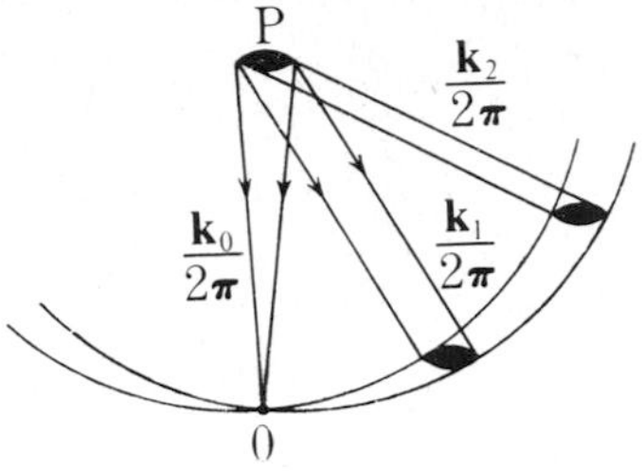

Fig. 5.10. The effect of a finite source size in spreading the Ewald sphere into a spherical shell of varying thickness.

convergence may be made much less, although it may be as high as 10^{-3} radians when the specimen is immersed in the magnetic field of the objective lens in an electron microscope and the "selected-area electron diffraction" technique is used (see Chapter 13).

Such a convergence of the incident beam modifies the picture of the Ewald sphere in reciprocal space in a way suggested in Fig. 5.10. Drawing the incident beam directions as vectors $\boldsymbol{k}_0$ to the reciprocal lattice origin, O, the origin points P are distributed over a disc having the shape of the source and a relative weighting of the points given by the intensity distribution of the source. Corresponding to each point of this disc there will be a differently-oriented Ewald sphere, so that we may think of an Ewald sphere thickened into a spherical shell of thickness which varies with distance from O.

For a given diffracted beam direction, the vectors $\boldsymbol{k}/2\pi$ from the source disc at P, each drawn to the corresponding Ewald sphere, define a disc-shaped section of the "Ewald shell". This we may call the "scattering disc". The total intensity scattered in the direction $\boldsymbol{k}$ will then be given by integrating $|F(\boldsymbol{u})|^2$ over this scattering disc with a weighting factor corresponding to the intensity of the source disc at P. Clearly, the size and shape of the scattering disc will vary with the angle of scattering, so the effect on the intensities cannot be represented by a simple convolution of some shape function with $|F(\boldsymbol{u})|^2$ unless all scattering angles are small as in the case of electron diffraction.

When the recording of the diffraction pattern is photographic, the resolution of the photographic plate or film is normally sufficient to allow each diffracted beam direction of interest to be recorded separately. For electronic recording, however, the detector usually accepts a finite angular range of diffracted beams from each point of the sample. Then for each incident beam direction the diffracted intensity will depend on the value of not just for one point on the Ewald

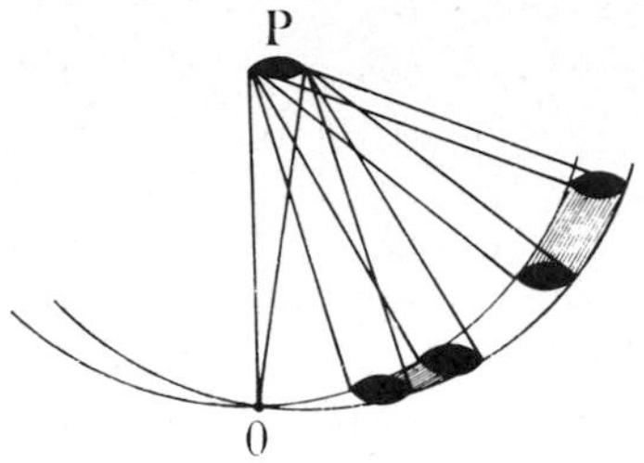

Fig. 5.11. The combined effects of finite source size and finite detector aperture size to define "scattering volumes" in reciprocal space over which the scattering power is integrated.

sphere but for a finite area of the Ewald sphere. Combining this with the effect of the finite source size, as suggested by Fig. 5.11, the recorded intensity for any setting of the crystal and the detector will come from a volume of reciprocal space. Both the size and shape of this "scattering volume" will vary with scattering angle.

5.7.2. Wavelength spread

A further complication comes from the finite range of wavelengths present for any real source. For X-rays the natural half-widths of the characteristic emission lines are of the order of $10^{-4}\ \lambda$ or more. For neutron diffraction, since the radiation used is selected out of a continuous "white radiation" distribution, the range of wavelengths used may be made greater in order to increase the total intensity of incident radiation. For electrons the radiation is usually much more nearly monochromatic with a width of about $10^{-6}\ \lambda$.

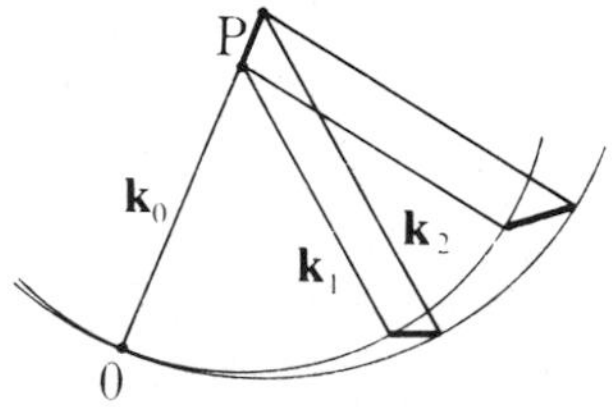

Fig. 5.12. The effect of a spread of wavelength in producing a spread in values of the Ewald sphere radius.

The spread of wavelengths produces a spread in the values of the Ewald sphere radius, as suggested in Fig. 5.12 for the case of very small incident and diffracted beam divergence. This results in a broadening of the scattering region into a line rather than a disc since the origin points P lie on a line. There is an essential difference from the case illustrated in Fig. 5.10 in that the wavelength spread gives a scattering line which varies in its orientation as well as length, being small and parallel to $\boldsymbol{k}_0$ in the limit of small scattering angles; of medium length and roughly perpendicular to $\boldsymbol{k}_0$ for a scattering angle like the larger one shown in the figure, and of maximum length and oppositely directed to $\boldsymbol{k}_0$ for a scattering angle of π.

Adding this effect to the effects of the divergence of the incident and scattered beams suggested in Fig. 5.11 gives a scattering volume having a shape which is complicated and varies with scattering angle in a complicated way. Thus, in general, the relationship of the observed intensity to the function $|F(\boldsymbol{u})|^2$ is only to be derived by laborious calculation from a detailed knowledge of the parameters of the experimental arrangement.

5.7.3. Integrated intensities

The usual procedure to simplify matters for the measurement of continuous distributions of scattering power is to ensure that the scattering volume is much smaller than the region over which $|F(\boldsymbol{u})|^2$ varies significantly. In the case of diffraction from well-crystallized materials, when the object is to measure the integrated scattering power contained in sharp maxima separated by a low flat background, the usual scheme is to generate a scattering volume in the shape of a thin disc, e.g. by having a well-collimated, nearly-monochromatic beam and a relatively wide angle of acceptance of the detector so that the disc is a portion of the Ewald sphere. Then the disc is swept through the sharp maximum of scattering power, e.g. by rotating the crystal, and the observed intensity is integrated over time.

For each experimental situation involving these simplifying special cases, the variation of the shape and size of the scattering volume and the rate at which it is scanned through reciprocal space are taken into account by multiplying the observed intensities by an appropriate factor, known as the Lorentz Factor, which may usually be taken as a function of the magnitude of the scattering angle only. Detailed derivations are given, for example, by Warren [1969].

For the diffraction of electrons with medium or high energies (greater than 50 keV say) all these considerations of scattering volumes are greatly simplified by the small scattering angles involved. The effect of the wavelength spread is negligible. The effect of the finite source size is to convolute the two-dimensional scattering power distribution, $|F(u, v)|^2$, with a source function, say $S(u, v)$ since this may be assumed independent of scattering angle. Then

$$I(u, v) = |F(u, v)|^2 * S(u, v) . \tag{37}$$

Also if this intensity distribution is observed with a detector of finite aperture or photographic plate of finite resolution with sensitivity represented by a function $D(u, v)$, the measured intensity is given by

$$\begin{aligned} I_{\text{obs}} &= I(u, v) * D(u, v) \\ &= |F(u, v)|^2 * S(u, v) * D(u, v) . \end{aligned} \tag{38}$$

The Lorentz factor then derives solely from the way in which intensities are integrated over time when a sample is rotated or, more commonly, integrated over crystal orientation when the sample contains a range of crystal orientations.

In neutron diffraction involving measurements of intensity as a function of energy loss as well as diffraction angles, the complications may be correspondingly greater than for X-ray diffraction. We prefer not to deal with them here.

5.8. Sections and projections

For a number of purposes it is convenient to deal with two- or one-dimensional sections or projections of the three-dimensional functions $\rho(\boldsymbol{r})$, $P(\boldsymbol{r})$, $F(\boldsymbol{u})$ and $|F(\boldsymbol{u})|^2$. It is much easier to represent and appreciate functions plotted in one or two dimensions and the amount of data required to define them is often very much less. For the case of a radially symmetric function, such as $P(\boldsymbol{r})$ for a gas or liquid, no new information is gained by going to more dimensions than one.

Experimentally, electron diffraction patterns represent, to a first approximation, planar sections of reciprocal space. In X-ray diffraction a planar section of reciprocal space may be obtained by use of the Buerger Precession Camera or similar devices, and the simplest form of operation of the X-ray diffractometer, the $\theta - 2\theta$ scan, gives a section along a radial straight line.

It is therefore appropriate to summarize now the general relationships between sections and projections in real and reciprocal space. We do so with reference to the functions $\rho(\boldsymbol{r})$ and $F(\boldsymbol{u})$ for convenience but draw our examples from our broader range of interest.

The projection of the function $\rho(\boldsymbol{r})$ in the direction of the z-axis is

$$\rho(x,y) = \int_{-\infty}^{\infty} \rho(\boldsymbol{r})\, \mathrm{d}z\,. \tag{39}$$

In terms of $F(\boldsymbol{u})$ this is

$$\rho(x,y) = \int\!\!\int F(\boldsymbol{u}) \exp\{-2\pi\mathrm{i}(ux+vy+wz)\}\, \mathrm{d}z\, \mathrm{d}\boldsymbol{u}\,. \tag{40}$$

The integral over z is the delta function $\delta(w)$ so that $\rho(x,y)$ is the inverse Fourier transform of $F(u,v,0)$, the section of $F(u,v,w)$ on the plane $w=0$.

Similarly, a planar section of $\rho(\boldsymbol{r})$ on the $z=0$ plane is

$$\begin{aligned}\rho(x,y,0) &= \int F(\boldsymbol{u}) \exp\{-2\pi\mathrm{i}(ux+vy)\}\, \mathrm{d}u\, \mathrm{d}v\, \mathrm{d}w \\ &= \int [\int F(u,v,w)\, \mathrm{d}w] \exp\{-2\pi\mathrm{i}(ux+vy)\}\, \mathrm{d}u\, \mathrm{d}v\,. \end{aligned} \tag{41}$$

Thus we have the general relationship, that a planar section through the origin in real space corresponds to a projection of the reciprocal space distribution on a parallel plane, and vice versa.

For example, the electron diffraction pattern, to a first approximation is a planar section of reciprocal space so that Fourier transform of the intensity distribution of the diffraction pattern gives the projection of the Patterson function $P(\boldsymbol{r})$ in the beam direction. This is the approximation that the object may be treated as a two-dimensional phase- and amplitude-object.

We have seen examples in Section (5) above for the case of the four-dimensional distributions in space and time when intensity is measured as a function of scattering angles and frequency change. Thus the section of reciprocal space on the $\nu=0$ plane, appropriate for purely elastic scattering, equation (31), gives the projection of the Patterson function in the time direction or the time-average of the correlation function. The projection of the four-dimensional distribution scattering power in reciprocal space in the ν direction, given by the integral over ν in equation (32), is the Fourier transform of the section of the Patterson function $P(\boldsymbol{r},0)$ which is the sum of the instantaneous spatial correlations of the object.

If the section in real space is not through the origin, a corresponding phase factor is introduced in reciprocal space. Thus the Fourier transform of the section $\rho(x,y,c)$ gives the modulated projection

$$\int F(u,v,w) \exp\{-2\pi\mathrm{i}wc\}\, \mathrm{d}w\,. \tag{42}$$

Extending the considerations to one-dimensional sections and projections, we have

$$\mathcal{F}\rho(x, 0, 0) = \int\int F(u, v, w)\, dv\, dw\,, \tag{43}$$

and

$$\mathcal{F}\int\int \rho(x, y, z)\, dy\, dz = F(u, 0, 0)\,. \tag{44}$$

Thus the value of $\rho(\boldsymbol{r})$ along the line of the x axis is given by Fourier transform of the projection of $F(\boldsymbol{u})$ along both v and w directions on to the u-axis.

An example from the four-dimensional situation is that of equation (30), where the function of interest is $P(0, t)$, the correlation in time for the position at the origin. This function would be given by projecting the reciprocal space function $|F(\boldsymbol{u}, \nu)|^2$ in the u, v and w directions, i.e. by integrating over all scattering directions for each frequency change, ν.

Problems

1. Sketch the Patterson function in two dimensions for a stationary molecule in the form of an equilateral triangle of atoms. Is there a centro-symmetrical object or a group of centro-symmetric objects which will give the same Patterson peaks (apart from the origin peak)? Derive an expression for the corresponding distribution $|F(\boldsymbol{u})|^2$ in reciprocal space. Repeat the considerations for an ideal gas of such molecules.

2. A diatomic molecule of fixed orientation (along x axis) vibrates about a fixed center of mass with frequency ν_0. Sketch the correlation function $P(x, t)$ and the corresponding reciprocal space distribution $|F(\boldsymbol{u}, \nu)|^2$.

3. Derive the Lorentz factor to be applied to X-ray diffraction from a gas and for the intensities of rings in a powder pattern (from crystallites in all possible orientations, each giving the same set of sharp peaks in reciprocal space).

CHAPTER 6

Diffraction from crystals

6.1. Ideal crystals

Although, as is well known, real crystals contain many different types of imperfections and faults, including point defects, impurities, dislocations, stacking faults and so on, it is often possible to consider the main diffraction effects as coming from an ideally periodic average crystal, as we shall see in Chapter 7. The discussion of the kinematical diffraction from ideal periodic crystals forms the basis for the important field of crystal structure analysis and so merits some special attention here.

An ideal crystal is made up by the repetition in three dimensions of a unit cell containing one or more atoms. In general the unit cell is not rectangular. It is defined by three vectors $\boldsymbol{a}, \boldsymbol{b}, \boldsymbol{c}$ which have length a, b, c and angles between the axes α, β, γ. Hence, writing $\rho_0(\boldsymbol{r})$ for the contents of the unit cell and $s(\boldsymbol{r})$ for the shape function,

$$\rho(\boldsymbol{r}) = \left[\rho_0(\boldsymbol{r}) * \sum_l \sum_m \sum_n \delta\{\boldsymbol{r} - (l\boldsymbol{a}+m\boldsymbol{b}+n\boldsymbol{c})\}\right] s(\boldsymbol{r})$$

or

$$\rho(\boldsymbol{r}) = \left[\rho_0(\boldsymbol{r}) * \sum_l \sum_m \sum_n \delta(x-la, y-mb, z-nc)\right] s(\boldsymbol{r}) , \qquad (1)$$

where x, y, z are coordinates with respect to the axes, in the directions of the unit vectors $\boldsymbol{x}, \boldsymbol{y}, \boldsymbol{z}$ parallel to $\boldsymbol{a}, \boldsymbol{b}, \boldsymbol{c}$.

This is a generalization of equation (2.56) following (5.8). For the special case for which $\boldsymbol{a}, \boldsymbol{b}, \boldsymbol{c}$ are at right angles, we saw (equation (5.9)) that Fourier transform gives a reciprocal lattice of points with spacings a^{-1}, b^{-1}, c^{-1} i.e. that the reciprocal lattice is defined by vectors, $\boldsymbol{a}^*, \boldsymbol{b}^*, \boldsymbol{c}^*$ such that $|\boldsymbol{a}^*| = a^{-1}$, $|\boldsymbol{b}^*| = b^{-1}$, $|\boldsymbol{c}^*| = c^{-1}$. For the more general case of non-rectangular axes, we must redefine the reciprocal lattice and put

$$\boldsymbol{a}^* = \frac{\boldsymbol{b} \times \boldsymbol{c}}{\boldsymbol{a}(\boldsymbol{b} \times \boldsymbol{c})} = \frac{\boldsymbol{b} \times \boldsymbol{c}}{V},$$

$$\boldsymbol{b}^* = \frac{\boldsymbol{c} \times \boldsymbol{a}}{V},$$

$$\boldsymbol{c}^* = \frac{\boldsymbol{a} \times \boldsymbol{b}}{V}, \tag{2}$$

where V is the unit cell volume, so that

$$\boldsymbol{a}^* \cdot \boldsymbol{b} = \boldsymbol{a}^* \cdot \boldsymbol{c} = \boldsymbol{b}^* \cdot \boldsymbol{c} = \ldots = 0\,, \tag{3}$$

$$\boldsymbol{a} \cdot \boldsymbol{a}^* = \boldsymbol{b} \cdot \boldsymbol{b}^* = \boldsymbol{c} \cdot \boldsymbol{c}^* = 1\,. \tag{4}$$

Then Fourier transform of (1) gives

$$F(\boldsymbol{u}) = F_0(\boldsymbol{u}) \sum_h \sum_k \sum_l \delta\{\boldsymbol{u} - (h\boldsymbol{a}^* + k\boldsymbol{b}^* + l\boldsymbol{c}^*)\} * S(\boldsymbol{u})$$

or

$$F(u, v, w) = F_0(\boldsymbol{u}) \sum_h \sum_k \sum_l \delta(u - ha^*, v - kb^*, w - lc^*) * S(\boldsymbol{u})\,. \tag{5}$$

For a crystal having the dimensions A, B, C in the directions of the three axes, the function $S(\boldsymbol{u})$ will have the same form as in equation (5.10). Alternatively, since the only values of $F_0(\boldsymbol{u})$ of interest are those at the reciprocal lattice points we may write

$$F(\boldsymbol{u}) = \sum_h \sum_k \sum_l F_{hkl}\, \delta\{\boldsymbol{u} - (h\boldsymbol{a}^* + k\boldsymbol{b}^* + l\boldsymbol{c}^*)\} * S(\boldsymbol{u})\,. \tag{6}$$

Then F_{hkl} is the "structure factor" or preferably, the "structure amplitude" for the h, k, l reciprocal lattice point and is given by

$$F_{hkl} = \int_0^a \int_0^b \int_0^c \rho(x, y, z) \exp\left\{2\pi i\left(\frac{hx}{a} + \frac{ky}{b} + \frac{lz}{c}\right)\right\} \mathrm{d}x\, \mathrm{d}y\, \mathrm{d}z$$

or

$$F_{\boldsymbol{h}} \equiv F_{hkl} = \int \rho(\boldsymbol{r}) \exp\{2\pi i \boldsymbol{h} \cdot \boldsymbol{r}\}\, \mathrm{d}\boldsymbol{r} \tag{7}$$

where the integration is over the unit cell and $\boldsymbol{h}$ is the vector $h\boldsymbol{a}^* + k\boldsymbol{b}^* + l\boldsymbol{c}^*$.

An alternative and often useful convention is to use fractional coordinates which we signify for the moment as X, Y, Z so that distances are measured in terms of unit cell parameters;

$$X = x/a\ , \quad Y = y/b\ , \quad Z = z/c\ . \tag{8}$$

Then (7) becomes,

$$F_{hkl} = V \iiint_0^1 \rho(X, Y, Z) \exp\{2\pi i(hX + kY + lZ)\} \mathrm{d}X \mathrm{d}Y \mathrm{d}Z\ . \tag{9}$$

By inverse transform of (6) we derive the alternatives to (1);

$$\rho(x, y, z) = V^{-1} \sum_h \sum_k \sum_l F_{hkl} \exp\left\{-2\pi i\left(\frac{hx}{a} + \frac{ky}{b} + \frac{lz}{c}\right)\right\} s(\boldsymbol{r}) \tag{10}$$

$$\rho(\boldsymbol{r}) = \sum_{\boldsymbol{h}} F_{\boldsymbol{h}} \exp\{-2\pi i \boldsymbol{h}\cdot\boldsymbol{r}\}\, s(\boldsymbol{r})\ , \tag{11}$$

or

$$\rho(X, Y, Z) = \sum_h \sum_k \sum_l F_{hkl} \exp\{-2\pi i(hX + kY + lZ)\}\, s(\boldsymbol{r})\ . \tag{12}$$

If the electron density in the unit cell is assumed to be the summation of the electron densities for individual atoms, we write

$$\rho_0(\boldsymbol{r}) = \sum_i \rho_i(\boldsymbol{r}) * \delta(\boldsymbol{r} - \boldsymbol{r}_i)\ , \tag{13}$$

so that

$$F_{hkl} = \sum_i f_i \exp\{2\pi i(hX_i + kY_i + lZ_i)\ , \tag{14}$$

where X_i, Y_i, Z_i are the fractional coordinates of the atom at $\boldsymbol{r}_i$ and f_i is the atomic scattering factor.

The distribution of scattering power in reciprocal space consists of a sharp peak around each reciprocal lattice point of the form $|S(\boldsymbol{u})|^2$ so that we may write, to a good approximation

$$|F(\boldsymbol{u})|^2 = \sum_h \sum_k \sum_l |F_{hkl}|^2 \delta\{\boldsymbol{u} - (h\boldsymbol{a}^* + k\boldsymbol{b}^* + l\boldsymbol{c}^*)\} * |S(\boldsymbol{u})|^2\ . \tag{15}$$

Inverse transformation of this then gives, following (12),

$$P(X,Y,Z) = V^{-1} \sum_h \sum_k \sum_l |F_{hkl}|^2 \exp\left\{-2\pi i\left(\frac{hx}{a}+\frac{ky}{b}+\frac{lz}{c}\right)\right\} [s(\boldsymbol{r}) * s(-\boldsymbol{r})] , \tag{16}$$

$$P(\boldsymbol{r}) = \left[\rho_0(\boldsymbol{r}) * \rho_0(-\boldsymbol{r}) * \sum_l \sum_m \sum_n \delta\{\boldsymbol{r}-(l\boldsymbol{a}+m\boldsymbol{b}+n\boldsymbol{c})\}\right] [s(\boldsymbol{r}) * s(-\boldsymbol{r})] . \tag{17}$$

Thus, apart from the gradual fall-off due to the shape convolution, the Patterson function is periodic, as foreseen in equation (5.17), and is made up by repetition of the self-convolution of the contents of the unit cell.

For convenience the shape convolution and its transform are often omitted. Then it must be understood that the periodic functions in real space and the delta functions in reciprocal space are operational abstractions which can be expanded into the more realistic descriptions of equations (15) to (17) when necessary.

6.2. Diffraction geometry

6.2.1. Laue and Bragg diffraction conditions

From equation (15) we see that for the ideal finite crystal the distribution of scattering power is a sharp peak, $|S(\boldsymbol{u})|^2$, around each point of the reciprocal lattice defined by the vectors $\boldsymbol{a}^*$, $\boldsymbol{b}^*$, $\boldsymbol{c}^*$. Then a sharply defined diffracted beam will be generated when the Ewald sphere cuts through one of these sharp peaks of scattering power. From our previous considerations of the Ewald sphere construction we see that the geometric condition to be satisfied is

$$\boldsymbol{q} \equiv \boldsymbol{k} - \boldsymbol{k}_0 = 2\pi \boldsymbol{h} \equiv 2\pi(h\boldsymbol{a}^* + k\boldsymbol{b}^* + l\boldsymbol{c}^*) . \tag{18}$$

We may write this in terms of the projections of $\boldsymbol{q}$ on the real space axes as

$$\boldsymbol{q}\cdot\boldsymbol{a} = 2\pi h , \quad \boldsymbol{q}\cdot\boldsymbol{b} = 2\pi k , \quad \boldsymbol{q}\cdot\boldsymbol{c} = 2\pi l , \tag{19}$$

which represent the well-known "Laue conditions" for diffraction.

The condition for diffraction may also be written in terms of the concept of planes of atoms in the crystal. The periodicity of the crystal ensures that sets of parallel planes may be drawn to pass through atom centers at regular intervals. These sets of planes are denoted by the Miller indices *hkl* if, when one plane is drawn through an atom at the unit cell origin, the intercepts of the next plane of the set on the axes are a/h, b/k, c/l.

It is easy to see in the case of rectangular axes, that the perpendicular distance between planes of the *hkl* set is d_{hkl} where

$$1/d_{hkl}^2 = \left(\frac{h^2}{a^2} + \frac{k^2}{b^2} + \frac{l^2}{c^2}\right). \tag{20}$$

But this is exactly the square of the distance from the reciprocal lattice origin to the *hkl* reciprocal lattice point and the direction of the perpendicular to the *hkl* lattice planes is the direction from the origin to the *hkl* reciprocal lattice point. This relationship holds also for non-orthogonal axes, although then the relationship (20) becomes more complicated.

Hence the condition that a sharp diffraction beam should be generated is, from (19) and (20),

$$|\boldsymbol{q}| = 2\pi/d_{hkl} ,$$

or, since $|\boldsymbol{k}_0| = |\boldsymbol{k}| = 2\pi/\lambda$, and the angle between $\boldsymbol{k}$ and $\boldsymbol{k}_0$ is 2θ, we have

$$2d_{hkl} \sin\theta_{hkl} = \lambda , \tag{21}$$

which is Bragg's law. The condition that $\boldsymbol{q}$ should be perpendicular to the lattice planes is equivalent to Bragg's concept of a "reflection" in the optical sense from the planes, subject to (21). We refer for convenience to the strong well-defined diffracted beams given under these conditions as "Bragg reflections".

6.2.2. *Shape transforms*

The Bragg or Laue conditions refer to the Ewald sphere passing exactly through a reciprocal lattice point. Our more detailed discussion of diffraction conditions suggests that, for monochromatic incident plane waves, the diffracted intensity for this case would be given by a near-planar section of the distribution of scattering power, $|S(\boldsymbol{u})|^2$, given a weighting $|F_{\boldsymbol{h}}|^2$. If, in practice, the divergence of the incident beam, finite acceptance angle of the detector and wavelength spread of the incident radiation together define a scattering volume, as described in the last chapter, which is very much larger in size than $|S(\boldsymbol{u})|^2$, then the observed intensity may correspond to a three-dimensional integration over the peak of scattering power and so be proportional to $|F_{\boldsymbol{h}}|^2$. Alternatively a thin "scattering disc" may be swept through the peak at a constant rate so that the observed intensity integrated over time will be proportional to the integral over $|S(\boldsymbol{u})|^2 |F_{\boldsymbol{h}}|^2$ and so will be proportional to $|F_{\boldsymbol{h}}|^2$.

For X-rays and neutrons the peaks of scattering power are very sharp. The kinematical scattering approximation may apply for perfect crystal regions several thousand times the dimensions of the unit cell so that the dimensions of the $|S(\boldsymbol{u})|^2$ distribution will be several thousand times smaller than the reciprocal lattice unit cell dimensions. Let us take as a representative value a half-width of $|S(\boldsymbol{u})|^2$ of $1/2000$ Å^{-1}. An incident beam divergence of 10^{-3} radians will give the Ewald sphere a thickness of about $1/2000$ Å^{-1} for a reciprocal lattice point with $1/d_{\boldsymbol{h}} = 0.5$ Å^{-1} and the wavelength spread of the incident beam will add to this. Hence even for a single crystal region of this size and for favorable diffraction conditions it would not be feasible to see anything of the details of the function $|S(\boldsymbol{u})|^2$. Only an integrated intensity could be recorded.

However with readily available sources of radiation the intensity scattered by such a small region is too small to be observed. The crystals normally used for diffraction under near-kinematical conditions are imperfect and may be thought of as very large numbers of small regions of this order of size, or a "mosaic" of small crystals, having a spread of orientations of perhaps 10^{-2} to 10^{-3} radians. Then we may consider the Ewald sphere to be cutting an aggregate of a very large number of slightly rotated reciprocal lattice configurations or, equivalently, we may consider the reciprocal space distribution for a single average crystal cut by a further-broadened Ewald sphere. This makes it even more nearly certain that only an integration over the scattering power peak will be seen and no information on individual $|S(\boldsymbol{u})|^2$ functions is attainable (Chapter 16).

6.2.3. Special cases for electron diffraction

For electron diffraction the situation is quite different. Crystals giving near-kinematical intensities are normally a few hundred Å in size in at least the one direction parallel to the incident beam. Sources are sufficiently bright to allow diffraction from single crystals of this size to be observed readily and the monochromatization and collimation give a broadening of the Ewald sphere with an angular spread of as little as 10^{-5} radians. Thus for a reflection with $1/d_{\boldsymbol{h}} = 0.5$ Å^{-1} the extent of the shape transform function may be 10^{-2} Å^{-1} or more while the thickness of the Ewald sphere may be as little as 5×10^{-6} Å^{-1}. Thus near-planar sections of the scattering power peak are frequently observed. Fig. 6.1 is a portion of a diffraction pattern from small

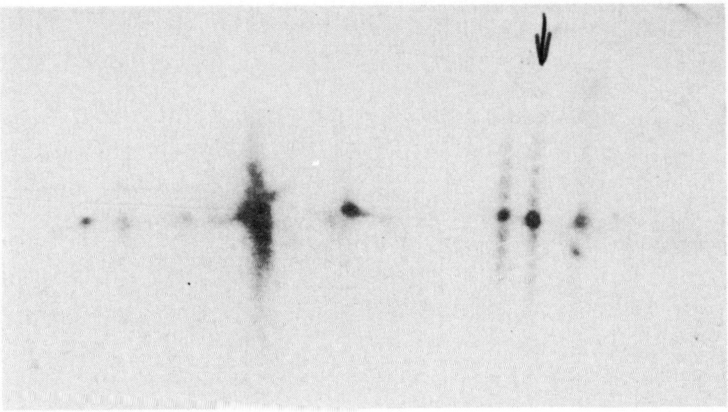

Fig. 6.1. Portion of a diffraction pattern (enlargement of spots on one ring) from small needle-like crystals of ZnO, showing shape-transform modulation of the scattering power (Rees and Spink [1950]).

needle-like crystals of ZnO (Rees and Spink [1950]). The limitation of crystal size in the direction perpendicular to the beam gives rise to an extension of the peak of scattering power in the plane of the Ewald sphere. The modulation of the intensity corresponding to the $(\sin^2 x)/x^2$ form of $|S(\boldsymbol{u})|^2$ is clearly seen in the spots from several individual needle crystals. (The intensity variation is usually modified by dynamical effects but for these particular cases, this is not very obvious.)

It follows that in electron diffraction special techniques of specimen preparation or of intensity recording are required in order to obtain integrated intensities proportional, in the kinematical approximation, to $|F(\boldsymbol{h})|^2$. For example, the techniques for crystal structure analysis by electron diffraction developed in the USSR (Pinsker [1953], Vainshtein [1964]) depend mostly on the use of oriented polycrystalline specimens having a random distribution of orientations about one axis so that each reciprocal lattice spot is spread into an annular ring and the section of this by the Ewald sphere gives an integrated intensity. Single crystal patterns are often obtained from extended thin sheets of crystal, of the order of 100 Å thick but possibly microns in diameter. Inevitably these thin sheets are often bent. This again provides an integration over the scattering power maximum, although the observed intensities will be proportional to $|F(\boldsymbol{h})|^2$ only if the bending is sufficiently uniform or sufficiently random to ensure that all crystal orientations are equally represented.

A further consequence of the difference in geometry between the X-ray

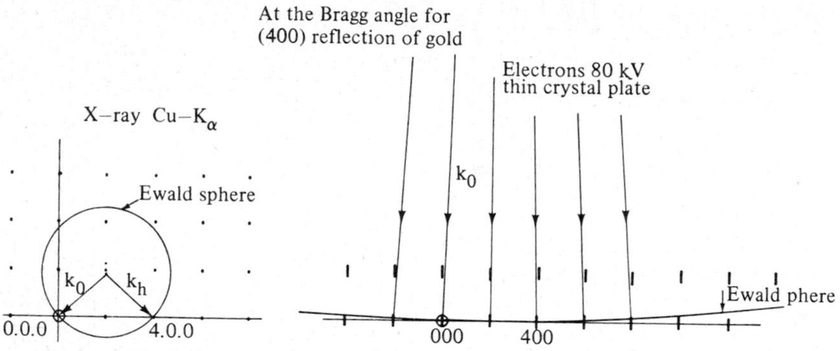

Fig. 6.2. Ewald sphere diagram for CuK_α X-rays and for 80 keV electrons for a crystal of gold or aluminum when the Bragg condition is satisfied for the 400 reflection. The crystal dimensions in the beam direction are assumed to be 1000 Å for the X-rays and 50 Å for the electrons.

and electron cases is the difference in the number of diffracted beams produced at any one time. For X-rays, even with the spread of the maxima of scattering power or the Ewald sphere which we have discussed, the probability that a strong reflection should occur for any particular incident beam orientation is low for crystals with small unit cells. If a strong reflection does occur, it is unlikely that a second one will be generated. For electrons on the other hand, the Ewald sphere normally cuts a number of the extended scattering-power regions and for particular orientations the number of diffracted beams may be large. This is illustrated in Fig. 6.2 drawn approximately to scale for the diffraction of CuK_α X-rays and 80 keV electrons from crystals of gold or aluminum with the Bragg reflection condition satisfied for the 400 reciprocal lattice point in each case. For the X-ray scattering the perfect crystal regions are assumed to be 1000 Å or more in size. For electrons the crystal is assumed to be a thin film 50 Å thick.

It is seen that for electrons, especially when the third dimension is taken into account, the number of simultaneously diffracted beams is quite large. If the voltage of the electron beam is increased and the wavelength is correspondingly reduced, the Ewald sphere will become more nearly planar and the number of reflections for such an orientation will increase rapidly, especially for the voltage range greater than about 200 keV for which relativistic effects become important.

If the thin crystals used in electron diffraction are bent, the rotation of the

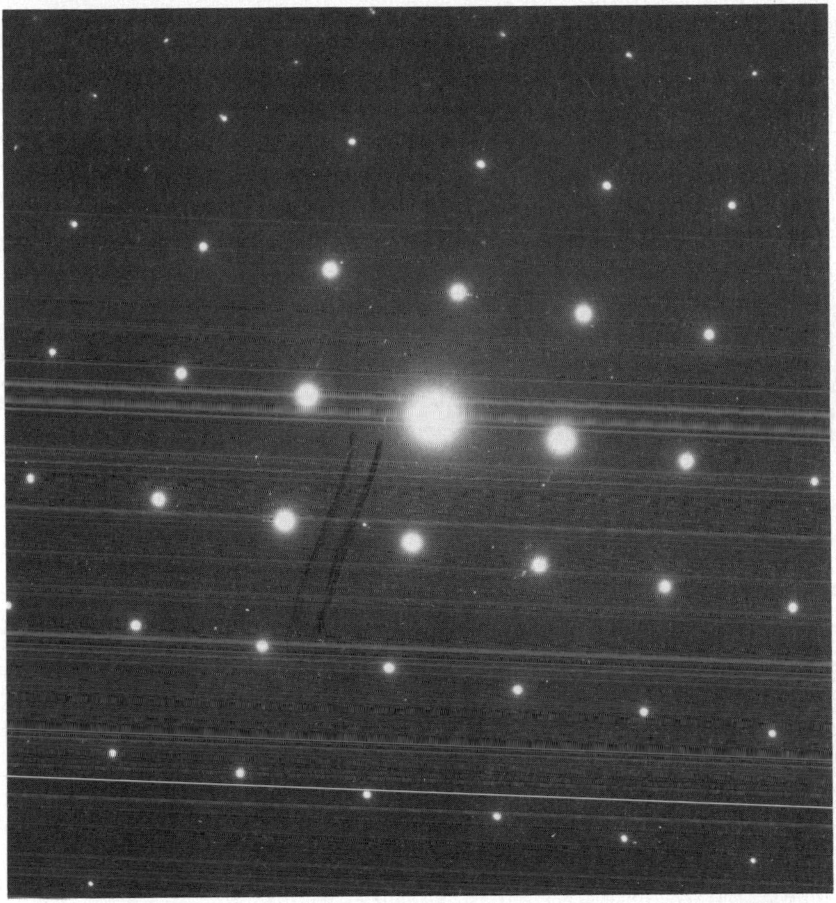

Fig. 6.3. Electron diffraction pattern from a thin, bent crystal of BiOCl with the incident beam nearly parallel to the *c*-axis.

reciprocal lattice relative to the Ewald sphere will ensure that many more diffracted beams are produced, giving the diffraction pattern the appearance of a complete section of the reciprocal lattice as in Fig. 6.3, a diffraction pattern obtained from a thin bent crystal with the incident beam approximately perpendicular to the $hk0$ reciprocal lattice plane.

Obviously this diffraction pattern contains information on the unit cell geometry and symmetry of the crystal. Also, provided that a kinematical approximation can be used, the relative intensities of the spots should allow values of $|F_{\boldsymbol{h}}|^2$ to be derived as a basis for the study of crystal structures. The possibilities in this direction have been reviewed by Cowley [1967].

6.3. Crystal structure analysis

6.3.1. The phase problem

The primary object of the analysis of the crystal structure of a substance not previously studied is to find the positions of the centers of the atoms, as given by the maxima in the electron density function $\rho(\boldsymbol{r})$ and to identify the atoms present from the relative weights of the maxima. As secondary objectives to be achieved by use of more accurate data and more extended analysis, one may hope to find the distortions of the free-atom electron densities of the atoms due to ionization and bonding and also the mean-square displacements of the atoms from their average positions due to thermal vibration. Thus the aim is to determine $\rho(\boldsymbol{r})$ within the unit cell of the average, periodic lattice.

As we have seen in Chapter 5, we may write

$$\rho(x,y,z) = V^{-1} \sum_h \sum_k \sum_l F_{hkl} \exp\left\{-2\pi\mathrm{i}\left(\frac{hx}{a} + \frac{ky}{b} + \frac{lz}{c}\right)\right\}, \tag{22}$$

or, from (12), in fractional coordinates,

$$\rho(X,Y,Z) = \sum_h \sum_k \sum_l F_{hkl} \exp\{-2\pi\mathrm{i}(hX+kY+lZ)\}\,. \tag{23}$$

However, the observable intensities give us $|F_{hkl}|^2$, not F_{hkl}. The phases of F_{hkl} (or $F_{\boldsymbol{h}}$) can not influence the intensities for kinematical scattering without absorption. In general $F_{\boldsymbol{h}}$ is complex and can be written $|F_{\boldsymbol{h}}| \exp\{\mathrm{i}\eta_{\boldsymbol{h}}\}$ and it is the phase factor $\eta_{\boldsymbol{h}}$ which is lost. For a centrosymmetrical crystal $F_{\boldsymbol{h}}$ is real so that it is the choice between positive and negative signs which can not be made.

The question of overcoming this lack of the information essential for the derivation of $\rho(\boldsymbol{r})$ constitutes the phase problem of crystal structure analysis. In principle this phase problem may be solved in many ways since, for example, either dynamical diffraction effect or scattering with absorption may give scattering sensitive to the relative phases of the reflections and these effects are never completely absent. In practice, however, the phase problem remains as a serious hindrance to the derivation of electron density distributions and a great deal of ingenuity has been expended on the development of procedures to overcome this limitation.

The observable quantity in reciprocal space is $|F_{\boldsymbol{h}}|^2$ and from this the Patterson function $P(\boldsymbol{r})$ for the crystal (equation (17)) is directly derived. The de-convolution of $P(\boldsymbol{r})$ to give $\rho(\boldsymbol{r})$ is in general not possible. In order to proceed it is necessary to make use of the available knowledge of the form of $\rho(\boldsymbol{r})$ and any information about the structure which may be gained from other sources.

6.3.2. Supplementary information

It is known that $\rho(\boldsymbol{r})$ is a positive, real function having a small, near-constant value except for isolated peaks at the atom positions. The relative weights of the peaks depend on the atomic numbers of the atoms and the separations of the peaks are not less than some fairly well defined minimum value.

The number of each kind of atom contained in the unit cell may usually be deduced from the chemical analysis of the material, the unit cell dimensions (given by the geometry of the diffraction patterns) and the density of the crystal.

Information on the possible arrangements of atoms within the unit cell may be provided by consideration of the symmetry of the crystal structure. For each crystal, the atomic arrangement must conform with the symmetry elements of one of the 230 possible space groups. From our previous considerations it can be seen that a symmetry operation in real space involving a rotation of the crystal about an axis or reflection in a plane must be accompanied by the same symmetry operation in reciprocal space. Operations of a screw axis or glide plane involving translations in real space must involve multiplication by a phase factor in reciprocal space which may make the amplitudes of some points of reciprocal space equal to zero, giving "systematic absences" of some reflections. Thus a great deal of information on real space symmetry may be derived from observations on intensity distributions in reciprocal space. The important exception, as we have seen, is that the presence or absence of a center of symmetry can not be deduced directly from observation of diffraction intensities since $|F(\boldsymbol{u})|^2 = |F(-\boldsymbol{u})|^2$. As a consequence, only 58 of the space groups can be identified uniquely from kinematical diffraction data and altogether it is possible to distinguish only 122 sets of one or more space groups. In some cases the presence or absence of a center of symmetry may be indicated by non-diffraction measurements such as the observation of piezoelectricity.

For simple structures the available information on symmetry may be sufficient to locate all atoms in the unit cell. For more complicated structures it may limit the possible positions considerably. For example, if there are one or two atoms of a particular kind in a unit cell which contains a 3- or 4-fold rotation axis, those one or two atoms must lie on that axis. For a detailed discussion of such considerations see, for example Lipson and Cochran [1966].

The problem of determining the remaining parameters needed to define the crystal structure may be considered in real space as the problem of finding the positions $\boldsymbol{r}_i$ of the individual atoms: or it may be considered in reciprocal space as the problem of assigning phases to the amplitudes $|F_{\boldsymbol{h}}|$. The main techniques which have been developed to assist with these problems have been described in detail in a number of books (e.g. Lipson and Cochran [1966], Woolfson [1970], Ladd and Palmer [1977]) and so will not be treated extensively here. We mention only a few of the key ideas and methods related to our more general considerations.

6.4. Structure analysis methods

6.4.1. Trial and error

The earliest and most direct approach to finding the parameters not determined by symmetry is to postulate a structure and calculate intensities for comparison with observed intensities. From a set of atomic positions with fractional coordinates x_i, y_i, z_i and atomic scattering factors f_i assumed initially to be those of isolated atoms, the structure amplitude is calculated as

$$F_{hkl} = \sum_i f_i \exp\{2\pi i(hx_i + ky_i + lz_i)\}\,. \tag{24}$$

Then $|F_{\boldsymbol{h}}|^2_{\text{calc}}$ is compared with $|F_{\boldsymbol{h}}|^2_{\text{obs}}$ derived from the intensities.

As a measure of agreement for the often large number of values involved it is usual to calculate the R-factor,

$$R = \frac{\sum_{\boldsymbol{h}} \big|\,|F_{\boldsymbol{h}}|_{\text{obs}} - |F_{\boldsymbol{h}}|_{\text{calc}}\big|}{\sum_{\boldsymbol{h}} |F_{\boldsymbol{h}}|_{\text{obs}}}\,, \tag{25}$$

Then changes in the postulated atomic coordinates are made in an effort to minimize R. For some purposes it is considered to be more appropriate to minimize R_1 defined by

$$R_1 = \sum_{h} W_{h}(|F_{\text{obs}}| - |F_{\text{calc}}|)^2 , \tag{26}$$

where W_h is a weighting factor which is determined from estimates of the relative accuracy of the various experimental measurements.

Once the parameters of the structure have been determined with moderate accuracy systematic least-squares minimization procedures may be applied to refine them.

An alternative to this refinement in terms of the reciprocal space quantities is the real-space procedure of calculating successive Fourier maps, or contoured maps of approximations to the electron density found by summation of the series

$$\rho_n(\boldsymbol{r}) = \sum_{h} |F_{h}|_{\text{obs}} S_{h \cdot \text{calc}} \exp\{-2\pi i(\boldsymbol{h}\cdot\boldsymbol{r})\} , \tag{27}$$

where $S_{h \cdot \text{calc}}$ is the sign of the structure amplitude calculated from (24) for a trial structure. With all the calculated signs correct $\rho_n(\boldsymbol{r})$ will show well-shaped symmetrical peaks at the atom positions and a flat back-ground. Deviations from this suggest changes to be made in the trial structure.

6.4.2. *Patterson function*

The Patterson function for the periodic crystal structure may be calculated from the observed intensities as

$$P(\boldsymbol{r}) = \tfrac{1}{2} \sum_{h} |F_{h}|^2 \cos\{2\pi \boldsymbol{h}\cdot\boldsymbol{r}\} , \tag{28}$$

where the center of symmetry allows us to replace the exponential by the cosine. As we have seen in Chapter 5 in the case of the Generalized Patterson function, $P(\boldsymbol{r})$ gives an appropriately weighted mapping of the inter-atomic vectors. As indicated by equation (17), $P(\boldsymbol{r})$ may be considered as the periodic repetition of the Patterson function for the contents of one unit cell, $P_0(\boldsymbol{r})$. However this repetition may create some complication because it gives some overlapping of different parts of $P_0(\boldsymbol{r})$.

For relatively simple structures the recognition of particular interatomic vectors may allow a determination of the structure on a trial- and error-basis, but in general the Patterson contains too many poorly resolved peaks for this to be useful.

One approach to the interpretation of the Patterson is the so-called "image seeking" procedure introduced by Buerger [1959] based on the interpretation of the Patterson which is illustrated in Fig. 5.2. One seeks systematically for images of the structure which are repeated in the Patterson with the vector separations of the inverse structure.

6.4.3. Optical analogues

The possibility of using optical diffraction observations as an aid to crystal structure analysis (Taylor and Lipson [1964]) follows directly from the considerations of Chapters 1 and 3. In an optical diffractometer a well-collimated beam of light passes through a specimen: then a long focal-length lens is used to produce a diffraction pattern which is viewed through an eye-piece or photographed. If the transmission function of the object is $q(x,y)$ the intensity distribution in the diffraction pattern will be

$$\left|Q\left(\frac{x}{f\lambda},\frac{y}{f\lambda}\right)\right|^2 .$$

If $q(x,y)$ is made to represent a projection of the structure of a crystal, $\int \rho(\boldsymbol{r})\mathrm{d}z = \rho(x,y)$, then the optical diffraction pattern will have intensity distribution $|F(u, v, 0)|^2$ with $u = x/f\lambda$, $v = y/f\lambda$. In practice it is not convenient to reproduce the smoothly varying function $\rho(x,y)$ as the transmission function of an object. Instead small holes are punched in an opaque mask at the atom positions and the diameter of the holes is varied to represent atoms of different atomic number.

Also it is not necessary to punch holes for all atoms in a large number of unit cells in order to get the $|F(u, v, 0)|^2$ values at the reciprocal lattice points. Instead we may write

$$\rho(x,y) = \rho_0(x,y) * \sum_m \sum_n \delta(\boldsymbol{r} - m\boldsymbol{a} - n\boldsymbol{b}) , \tag{29}$$

and the diffraction pattern is then

$$|F(u, v, 0)|^2 = \left|F_0(u, v, 0) \sum_h \sum_k \delta(u - h\boldsymbol{a}^*, v - k\boldsymbol{b}^*)\right|^2 , \tag{30}$$

where the spread due to shape transform is understood. Then if the object inserted represents only the atoms of $\rho_0(x,y)$, the diffraction pattern intensity will be $|F_0(u, v, 0)|^2$ and this may be sampled at the points on the net defined by $\boldsymbol{a}^*, \boldsymbol{b}^*$ to give the $|F_{hk0}|^2$ values.

In this way diffraction patterns for trial structures may be viewed rapidly and tested for general agreement with the reciprocal space distributions deduced from X-ray diffraction data. More importantly an appreciation may be generated for the type of diffraction effects to be expected for various molecules or groups of atoms in various orientations, leading to the rapid development of a powerful intuitive ability to relate reciprocal space and real space distributions.

There are a number of refinements, modifications and developments of this scheme. For example, the sign of $F_0(u, v, 0)$ for a centro-symmetric $\rho_0(x,y)$ can be readily determined. If an extra hole is punched at the origin of $\rho_0(x,y)$ this adds a constant term to $F_0(u, v, 0)$. Then for positive areas of $F_0(u, v, 0)$ the intensity $|F(u, v, 0)|^2$ will increase and for negative areas of $F_0(u, v, 0)$ it will decrease.

For the reverse transformation, a two-dimensional reciprocal lattice section is simulated by making holes of area proportional to $|F_{hk0}|$. For a centro-symmetric structure the negative signs on some reflections may be produced by covering the holes with phase plates having a retardation of π. Then the intensity in the diffraction plane is $\rho^2(x,y)$.

6.4.4. *Other methods*

Of the many other techniques used, one class involves the use of specially favorable or specially prepared crystals. If one or more heavy atoms are incorporated in the unit cell, their contributions will tend to have a dominating influence on the signs of the structure factors. Hence if the heavy-atom positions are found (for example by recognizing the heavy atom to heavy atom vector peaks in the Patterson) this will allow sufficient of the signs of the structure factors to be determined to allow a first model of the structure to be derived.

For some crystals it is possible to make an isomorphous replacement, which involves the substitution of one kind of an atom for another at some sites in the unit cell without appreciable displacement of the other atoms on other sites. Suppose that we write the scattering power for a centro-symmetric crystal containing atom type 1 as

$$I_1(\boldsymbol{u}) = (F_1(\boldsymbol{u})+F_r(\boldsymbol{u}))^2 = F_1^2 + 2F_1 F_r + F_r^2 \tag{31}$$

where F_1 is the contribution from the type 1 atoms and F_r is from the rest.

Similarly when type 2 atoms are substituted

$$I_2(\boldsymbol{u}) = (F_2(\boldsymbol{u})+F_r(\boldsymbol{u}))^2 = F_2^2 + 2F_2 F_r + F_r^2 \,. \tag{32}$$

Subtracting the two sets of intensities

$$I_1 - I_2 = F_1^2 - F_2^2 + 2(F_1 - F_2) F_r \tag{33}$$

so that, if the positions of the replacement atoms are known, F_1 and F_2 are known and F_r is determined in magnitude and sign. If the structure has no center of symmetry more than one substitution is required.

In a variation of this method no replacement of atoms is made, but the atomic scattering factor for one type of atom is changed by changing the wavelength of the incident radiation to the vicinity of the absorption edge for the element.

Another class of techniques of growing importance involves the so-called "direct methods" in which algebraic equalities or inequalities between the structure amplitudes are derived from the known properties of the electron density function; for example the fact that $\rho(\boldsymbol{r})$ is a real positive function and that it consists of peaks of an approximately known shape and size. An important feature of these methods is that they involve only the computerized manipulation of numerical data. Coupled with least-squares refinement procedures they offer the possibility of almost automatic, fully computerized structure analysis (see Hauptman [1972], Woolfson [1961], Ladd and Palmer [1977], Hauptman [1978]).

6.5. Neutron diffraction structure analysis

6.5.1. Nuclear scattering

Compared with the X-ray case the main differences in the application of neutron diffraction for structure analysis from non-magnetic materials arise from the erratic variation of the scattering length in magnitude and sign with varying atomic number.

The light elements such as carbon, oxygen and even hydrogen scatter just as strongly as the heavy elements and so can be located with equal ease whereas for X-ray diffraction they are often undetectable. Particularly for hydrogen the X-ray scattering comes from a single, rather diffusely spread electron, or less if the hydrogen is partially ionized. For neutrons the scattering is from the nucleus and is quite large and negative; hence the application of neutron diffraction for the important study of hydrogen positions in organic and biologically significant crystals.

The hydrogen positions are distinguished as negative peaks in the Fourier maps because of the negative b values. Similarly, negative peaks appear in the Patterson maps corresponding to vectors between atoms having positive and negative b values.

In practice hydrogen is not favored because it gives a high background of "incoherent" scattering (see Chapter 4). It is replaced when possible by deuterium which has a high positive b value and little background scattering.

A further limitation of X-ray diffraction is the difficulty of distinguishing between elements which are close together in the periodic table since the f values increase smoothly with atomic number. For a number of important cases the b values for neutrons differ greatly from one atom to the next and so allow them to be distinguished. For example for the transition elements we find $b_{Mn} = -0.36$, $b_{Fe} = 0.96$, $b_{Co} = 0.25$ and $b_{Ni} = 1.03$.

6.5.2. Magnetic scattering

For magnetic materials we must distinguish two atomic scattering factors for neutrons, the nuclear scattering amplitude b, independent of scattering angle, and the magnetic scattering amplitude p, depending on the distribution of unpaired electron spins and given by

$$p = \frac{e^2\gamma}{2mc^2}\, gJf\,, \tag{34}$$

where γ is the magnetic moment of the neutron, g is the Lande splitting factor, J is the spin-orbit quantum number and f is the form factor given by Fourier transform of the distribution of electrons having unpaired spins. But since the spin scattering involves the interactions of vector quantities we must also define a magnetic interaction vector $\boldsymbol{q}$ given by

$$\boldsymbol{q} = \boldsymbol{\varepsilon}(\boldsymbol{\varepsilon}\cdot\boldsymbol{K}) - \boldsymbol{K}\,,$$

where $\boldsymbol{\varepsilon}$ is the unit vector perpendicular to the diffracting plane and $\boldsymbol{K}$ is the unit vector in the direction of the magnetic moment. Then $\boldsymbol{q} = 0$ for $\boldsymbol{\varepsilon}$ parallel to $\boldsymbol{K}$ and $|\boldsymbol{q}| = 1$ for $\boldsymbol{\varepsilon}$ perpendicular to $\boldsymbol{K}$.

For an ordered array of spins in the lattice, the nuclear and magnetic scattering contribute independently to the diffracted intensity provided that the neutron beam is unpolarized (Bacon [1962]) so that

$$|F(\boldsymbol{u})|^2 = |F_n(\boldsymbol{u})|^2 + |F_{mag}(\boldsymbol{u})|^2 \tag{35}$$

where

$$|F_n(\boldsymbol{u})|^2 = \Big|\sum_i b_i \exp\{-M_i\} \exp\{2\pi i \boldsymbol{b}\cdot\boldsymbol{r}_i\}\Big|^2 \tag{36}$$

$$|F_{\text{mag}}(\boldsymbol{u})|^2 = \sum_i \sum_j p_i \boldsymbol{q}_i \cdot p_j \boldsymbol{q}_j \exp\{-M_i - M_j\} \exp\{2\pi i \boldsymbol{b}\cdot(\boldsymbol{r}_i - \boldsymbol{r}_j)\}\,, \tag{37}$$

where we have inserted the Debye-Waller factors, $\exp\{-M_i\}$, arising from the thermal vibrations of the atoms (Chapters 7 and 12).

For a ferromagnetic material with all spins lined up parallel,

$$\begin{aligned} \boldsymbol{q}_i \cdot \boldsymbol{q}_j &= q^2 \\ &= \sin^2\alpha\,, \end{aligned} \tag{38}$$

where α is the angle between $\boldsymbol{K}$ and $\boldsymbol{\varepsilon}$.

For a simple anti-ferromagnetic case with spins on different sublattices of the structure aligned anti-parallel we have for some atom pairs that $\boldsymbol{K}_j = -\boldsymbol{K}_i$ so that $\boldsymbol{q}_i \cdot \boldsymbol{q}_j = -\sin^2\alpha$. Then (37) may be written

$$|F^2_{\text{mag}}(\boldsymbol{u})|^2 = \sin^2\alpha \sum_i \sum_j (\pm) p_i p_j \exp\{-M_i - M_j\} \exp\{2\pi i \boldsymbol{b}\cdot(\boldsymbol{r}_i - \boldsymbol{r}_j)\} \tag{39}$$

where the plus and minus signs refer to pairs of atoms with parallel and anti-parallel spins. Correspondingly if the $\sin^2\alpha$ factor is ignored, a Patterson function can be drawn with positive peaks for parallel pairs of spins and negative peaks for anti-parallel pairs.

For anti-ferromagnetic crystals the separation of like atoms, in the simplest case, into two sets, spin up and spin down, will almost inevitably lower the symmetry of the crystal and in many cases produces a unit cell size which is some multiple of that seen by X-rays. Then new, purely magnetic, "superlattice" reflections appear in the diffraction pattern. Gradually more and more complicated magnetic superlattice structures are being found with large repeat distances and spins inclined to each other at a variety of angles. For these, the simple formulation of (39) cannot be used.

6.6. Electron diffraction structure analysis

The atomic scattering factors for electrons increase smoothly with atomic number except at low scattering angles but not quite as rapidly as for X-rays.

The difference is most apparent for hydrogen. The electron scattering depends on the potential field of the nucleus which is partially screened by the orbital electron. Partial ionization decreases the screening and increases the scattering factor. It has been estimated by Vainshtein [1964] that the ratio of the scattering by carbon and hydrogen is about 10 for X-rays but only 3 or 4 for electrons. However, in view of the ease of detection of hydrogen atoms by neutron diffraction the use of electron diffraction for these purposes is restricted to special cases for which neutron diffraction methods are not applicable.

One factor of interest is that the detection of hydrogen atoms is probably the one case in which the results should differ appreciably with the technique. X-ray diffraction should indicate the position of maximum electron density. Neutron diffraction should indicate the mean position of the nucleus. These positions may well be different when the atom is polarized as in an asymmetric bond. Then the peak of potential deduced from electron diffraction should be near to the position of the positive nucleus, but since the contribution of the electron cloud is negative, displacement of the electron distribution in one direction should displace the potential peak slightly in the opposite direction.

Determinations of the N–H bond length in NH_4Cl crystals give 0.94 ± 0.03 Å by X-ray diffraction, 1.03 ± 0.02 Å by neutron diffraction and 1.02 ± 0.02 Å (Kuwabara [1959] or 1.04 ± 0.02 Å (Avilov et al. [1973]) by electron diffraction.

Undoubtedly the most important possibility for electron diffraction structure analysis is in the examination of crystals too small to be studied by any other method. In electron microscopes the selected-area diffraction technique can be used to obtain single crystal patterns from crystals a few hundred Å in thickness and of lateral extent which is some reasonable fraction of the minimum practical area which is about 1 μm for 100 keV electrons and as little as 500 Å for 500 to 1000 keV electrons. By the use of electron microbeams such as are produced now in scanning electron microscopes, the diameter of the specimen area may be reduced to 20 Å or less. Thus single crystal patterns can be obtained from clay minerals, aerosols, colloids and many inorganic materials which give at best very diffuse powder patterns with X-ray diffraction.

Because of the strong dynamical scattering effects occurring in electron diffraction most of the structure analyses that have been attempted using electrons have made use of very thin bent crystals, disordered crystals or polycrystalline aggregates since for each of these cases the dynamical effects are minimized by an averaging process. A summary of the current position has been given by Cowley [1967].

Problems

1. A zinc single crystal has a hexagonal unit cell ($a = b \neq c$, $\alpha = \beta = 90°$, $\gamma = 120°$) with 3-fold symmetry about the c-axis. The zinc atoms have coordinates 0, 0, 0 and 2/3, 1/3, 1/2.

Find the axes and angles of the reciprocal lattice. Describe the form of the X-ray diffraction pattern recorded on a cylindrical film when the crystal is rotated about the c-axis which coincides with the axis of the cylindrical film. Find expressions for the intensities of the reflections in terms of the atomic scattering factors. Which reflections are "forbidden" i.e. have zero intensity as a result of space-group symmetries?

2. For a certain problem of structure analysis it is desirable to have a projection on the $x-y$ plane of the limited region of the unit cell lying between $z = 0.4$ and $z = 0.6$. What series should be summed to give this?

3. MnF_2 crystallizes with a tetragonal unit cell (SnO_2-type) with $a = 4.865$, $c = 3.284$ Å: Space Group $D_{4h}^{14} - P_{42/mnm}$, $Z = 2$, with

Mn at 0, 0, 0; 1/2, 1/2, 1/2
F at x, x, 0 etc. with $x = 0.31$.

(a) What are the symmetry elements of the unit cell?
(b) What reflections are forbidden by symmetry?
(c) What series should be summed to give the electron density along a line through an Mn atom and a neighboring F atom in the $x-y$ plane?
(d) Sketch the form of the projection of the Patterson function on to the $x-z$ plane (z-axis is direction of c-axis of unit cell).

Note: Make use of International Tables of X-ray Crystallography, Vol. 1.

4. Below a certain critical temperature MnF_2 is anti-ferromagnetic with the spin of the Mn atom at 0, 0, 0, directed along the positive c-axis and the spin of the atom at 1/2, 1/2, 1/2, directed along the negative c-axis (see Bacon, p. 311).

(a) How will the symmetry and forbidden reflections for neutron diffraction differ from those for X-ray diffraction (c.f. problem 3)?
(b) Sketch the Patterson function for the projection on to the $x-z$ plane derived from neutron diffraction intensities including both nuclear and magnetic scattering.

5. Equalities used as a basis for some of the "direct" methods for structure analysis (Sayre's equation) can be derived by considering the Fourier transforms of the electron density distribution and the square of the electron density distributions for a structure composed of equal atoms. From the fact that these two distributions have maxima at the same positions but of different

form, derive the relationship

$$F_h = g(\theta) \sum_{h'} F_{h'} F_{h-h'} ,$$

where $g(\theta)$ is a known function of scattering angle.

6. For use in an optical diffractometer a mask is made to represent a section of reciprocal space for a perfect crystal by drilling small holes in an opaque screen at positions corresponding to the reciprocal lattice points. The areas of the holes are made proportional to the $|F_{hk0}|^2$ values. To what extent will the diffraction pattern of this mask represent the Patterson function of the crystal (a) near the center of the diffraction pattern, and (b) far from the center where the intensity is relatively small?

7. Two crystal structures may be assumed to be identical except that one atom per unit cell (which may be assumed to be at the origin) is different for the two cases. How will the two corresponding Patterson functions differ? How may the differences be used to determine the crystal structures unambiguously?

CHAPTER 7

Diffraction from imperfect crystals

7.1. Formulation of the diffraction problem

7.1.1. Types of defects

The great variety of defects, imperfections, faults and other irregularities which may occur in real crystals provides a great variety of diffraction effects, including modifications of the sharp maxima about the reciprocal lattice points and also continuous distributions of intensity in the "background" regions between these maxima. These effects are of interest not only as an indication of the limitations of the ideal crystal model used for structure analysis purposes but also as one of the most powerful means available for the study of the nature of defects in crystals.

We distinguish two main classes of crystal defects: those for which it is possible to define an average periodic lattice and those for which this is not possible. This is an artificial and incomplete classification, but provides a useful starting point. The first class of defects includes mostly localized defects, including the single-atom or "point" defects such as vacancies, interstitial atoms, substitutions of atoms; also small clusters of point defects and the localized strain fields associated with point defects or clusters. For these cases the defect is surrounded by the three-dimensional bulk of crystal which defines an average periodicity, acting as a frame of reference and defining a basis from which deviations can be measured. The most commonly occurring example of a deviation of this type from an ideally periodic crystal lattice is provided by the thermal vibrations of atoms about their mean lattice positions.

In the second class of defects the simplest type of defect to appreciate is the planar fault or twin plane which cuts right across the crystal, giving the unit cells on one side a shift or change of orientation with respect to those on the other side. Then, while the shift or change of orientation between the two parts of the crystal may be readily defined, it is not feasible to refer both parts to one average lattice. For a progression of faults or twins occuring at more-or-less random intervals, certain definite geometric relationships may be maintained between various parts of the crystal, even though the full lattice periodicity is lost. Some well-defined sharp maxima of scattering power will occur in reciprocal space, but an average lattice is not relavant in that, however it is chosen, displacements from it equal to large fractions of the unit cell dimensions will occur over a large proportion of the crystal.

In more complicated cases, faults may give progressive deviations from any locally defined regular lattice in two or three dimensions. As the frequency and variety of faults increases the distance over which any correlation between atom sites may be traced decreases and the crystallinity of the material decreases. In the limit, a liquid-like structure may be described in terms of a fault in the periodicity occuring every few unit cells in every direction. However, such extreme cases fall within the ill-defined boundary region between amporhous and polycrystalline materials. We confine our attention here to cases in which axial directions and principal periodicities may still be defined.

7.1.2. General diffraction formulation

Since for crystals with defects or faults it is not possible to use the special formulas for diffraction intensities developed in the last chapter for ideal crystals, it is necessary to revert to the earlier more general formulations of the diffraction problem. A common starting point is that of equation (5.5), from which the reciprocal space scattering power function is written in terms of the atomic scattering factors and positions; thus

$$F(\boldsymbol{u}) = \sum_i f_i \exp\{2\pi i \boldsymbol{u}\cdot\boldsymbol{r}_i\}\,. \tag{1}$$

$$J(\boldsymbol{u}) \equiv |F(\boldsymbol{u})|^2 = \sum_i \sum_j f_i f_j^* \exp\{2\pi i \boldsymbol{u}\cdot(\boldsymbol{r}_i - \boldsymbol{r}_j)\}\,. \tag{2}$$

It is not possible to insert into this expression the atomic scattering factors and positions of all atoms of the specimen. The intensity must be evaluated in terms of statistical relationships between these quantities.

One approach is to consider the average surroundings of an atom or of a particular type of atom. For example, if one atom of a statistically equivalent set is indicated by the subscript n, and there are N_n such atoms, we may take each such atom in turn as the origin and find its average surroundings. Then

$$J(\boldsymbol{u}) = \sum_n N_n f_n \left\langle \sum_j f_{j-n} \exp\{2\pi i \boldsymbol{u}\cdot(\boldsymbol{r}_j - \boldsymbol{r}_n)\}\right\rangle\,. \tag{3}$$

where the brackets $\langle\,\rangle$ represent an averaging over all configurations of atoms about an atom of type n. For a monatomic solid, this reduces to

$$J(\boldsymbol{u}) = Nf^2 \sum_j \langle \exp\{2\pi i \boldsymbol{u} \cdot \boldsymbol{r}_j\} \rangle \,, \tag{4}$$

where $\boldsymbol{r}_j$ is the position of an atom referred to an origin taken on another atom.

To be more general we should admit the possibility that the atomic positions should vary with time. Then, as we have seen before, for purely elastic scattering the diffracted amplitude is given by the time average of the structure so that

$$J_{\text{el}}(\boldsymbol{u}) = |\sum_i f_i \langle \exp\{2\pi i \boldsymbol{u} \cdot \boldsymbol{r}_i(t)\} \rangle_t|^2 \,. \tag{5}$$

Usually for X-ray or electron diffraction the intensity measurement does not distinguish between purely elastic and inelastic scattering. The observed intensity is then a time average of the instantaneous intensities, or

$$\langle J_{\text{tot}}(\boldsymbol{u}, t) \rangle_t = \left\langle \sum_i \sum_j f_i f_j \exp\{2\pi i \boldsymbol{u} \cdot (\boldsymbol{r}_i - \boldsymbol{r}_j)\} \right\rangle_t$$

$$= \sum_i \sum_j f_i f_j \langle \exp 2\pi i \boldsymbol{u} \cdot (\boldsymbol{r}_i - \boldsymbol{r}_j) \rangle_t \,. \tag{6}$$

In cases of practical significance the averaging over time can be considered completely equivalent to an averaging over space. Usually the region of the sample which can give coherent diffraction effects is limited in size to be very much smaller than the total illuminated size because of the limitations on coherence of the incident beam. The lateral coherence of the beam is usually limited to a few hundred Å by the convergence of the beam from a finite incoherent source. The coherence of the beam in the direction of propagation is limited by the spread in wavelengths to about 1 μm for X-rays and less for neutrons. Thus the total diffracted intensity from a sample may be regarded as the sum of intensities from a very large number of independent but statistically equivalent regions. This will be the same as the sum of the intensities obtained from any one region at different times. In time any one region will take on all the possible configurations of the atoms present and for a large number of independent regions all possible configurations will be represented at any one time.

7.2. Patterson function approach

7.2.1. Patterson with an average periodic structure

The discussion of the intensity expression in terms of the average surroundings of a given atom, as in equations (3) and (4) suggests the use of the Patterson function description and, in fact, this forms the basis for an alternative and very powerful method for studying the diffraction by imperfect crystals. We may apply directly the considerations of Chapter 5.

The two main classes of crystal imperfection which we have distinguished require somewhat different approaches and will be considered separately.

When an average periodic lattice may be defined we write

$$\rho(\boldsymbol{r}) = \langle \rho(\boldsymbol{r}) \rangle + \Delta\rho(\boldsymbol{r}) , \tag{7}$$

where $\langle \rho(\boldsymbol{r}) \rangle$ is the electron density distribution for the average lattice, defined to be time independent and periodic (when we ignore shape-function limitations) and $\Delta\rho(\boldsymbol{r})$ represents the deviation from the average lattice, which is essentially non-periodic. This deviation from the average lattice may be time-dependent or not and the averaging process may be an averaging over time or over space. As suggested above, the two cases may be considered equivalent. By definition, for the averaging process involved, $\langle \Delta\rho \rangle = 0$.

From (7) we may write the Patterson function as

$$\begin{aligned} P(\boldsymbol{r}) &= \{\langle \rho(\boldsymbol{r}) \rangle + \Delta\rho(\boldsymbol{r})\} * \{\langle \rho(-\boldsymbol{r}) \rangle + \Delta\rho(-\boldsymbol{r})\} \\ &= \{\langle \rho(\boldsymbol{r}) \rangle * \langle \rho(-\boldsymbol{r}) \rangle\} + \{\langle \rho(\boldsymbol{r}) \rangle * \Delta\rho(-\boldsymbol{r})\} \\ &\quad + \{\Delta\rho(\boldsymbol{r}) * \Delta\rho(-\boldsymbol{r})\} + \{\Delta\rho(\boldsymbol{r}) * \langle \rho(-\boldsymbol{r}) \rangle\} . \end{aligned} \tag{8}$$

In terms of spatial averaging of time-independent functions, the second term may be written

$$\langle \rho_0(\boldsymbol{r}) \rangle * \sum_r \delta(\boldsymbol{r} - \boldsymbol{R}_n) * \Delta\rho(-\boldsymbol{r}) ,$$

where $\boldsymbol{R}_n$ is a lattice vector of the average lattice. Then the convolution of $\Delta\rho$ with $\Sigma\delta(\boldsymbol{r} - \boldsymbol{R}_n)$ represents the superposition of the function $\Delta\rho$ with all lattice vector shifts. There is thus an averaging of $\Delta\rho$ values at every point of the unit cell and by definition this is zero. The same applies to the convolution in the fourth term of (8). Hence

$$P(\boldsymbol{r}) = \{\langle \rho(\boldsymbol{r}) \rangle * \langle \rho(-\boldsymbol{r}) \rangle\} + \{\Delta\rho(\boldsymbol{r}) * \Delta\rho(-\boldsymbol{r})\} . \tag{9}$$

The same argument holds if the averaging is over time rather than space to give the time-average of the instantaneous Patterson function $P(\boldsymbol{r}, 0)$. Fourier transform of (9) gives

$$J(\boldsymbol{u}) = |\bar{F}|^2 + |\Delta F|^2 , \tag{10}$$

where $\bar{F}$ and ΔF are the Fourier transforms of $\langle \rho(\boldsymbol{r}) \rangle$ and $\Delta\rho$.

Thus we have the general result that the total scattering power distribution is the sum of that for the average lattice and that for the deviations from the average lattice considered separately. Since $\langle\rho(\boldsymbol{r})\rangle$ is periodic, $|\bar{F}|^2$ will consist of sharp peaks at the reciprocal lattice points only, giving the sharp Bragg reflections in diffraction patterns. Since $\Delta\rho$ is non-periodic, $\Delta\rho * \Delta\rho$ will be non-periodic and will decrease rapidly with distance from the origin. Hence $|\Delta F|^2$ will represent a continuous distribution of scattering power between the reciprocal lattice points and so will give rise to diffuse scattering in diffraction patterns. It may be noted that for time dependent perturbations of an average periodic structure, the first terms of (9) and (10) represent scattering from the time average structure and so correspond to the purely elastic scattering while the second term represents the inelastic scattering.

7.2.2. Patterson function with no average structure

For the second class of crystal imperfections, an average periodic lattice can not be meaningfully defined and there can not be the same separation into sharp Bragg reflections and diffuse scattering. Then, in general, the Patterson function must be evaluated for the whole structure or the intensity expression (2) must be evaluated using any simplifications which seem appropriate for the particular case in hand.

For a large number of situations the structure may be described in terms of the repetition of one or more definite units of structure with irregular translations. For example the structure may be made up of identical planes of atoms which have an irregular inter-planar spacing: or identical lines of atoms which are lined up parallel but have irregular spacings in the two directions at right angles to their lengths: or identical blocks of atoms separated from their neighbors by variable spacings in three dimensions. Then the electron density distribution may be written

$$\rho(\boldsymbol{r}) = \rho_0(\boldsymbol{r}) * d(\boldsymbol{r})\,, \tag{11}$$

where the "distribution function", $d(\boldsymbol{r})$, is a set of delta functions representing the positions of reference points for each reproduction of the unit of structure $\rho_0(\boldsymbol{r})$. Then the Patterson is

$$P(\boldsymbol{r}) = \rho_0(\boldsymbol{r}) * \rho_0(-\boldsymbol{r}) * d(\boldsymbol{r}) * d(-\boldsymbol{r})\,, \tag{12}$$

and the reciprocal space distribution of scattering pow'

$$J(\boldsymbol{u}) = |F_0(\boldsymbol{u})|^2\,|D(\boldsymbol{u})|^2\,. \tag{13}$$

Thus the scattering power function for the individual unit of structure may be calculated first and then multiplied by the function corresponding to the distribution function.

If there is more than one type of structural unit involved, we may write

$$\rho(\boldsymbol{r}) = \sum_n \rho_n(\boldsymbol{r}) * d_n(\boldsymbol{r}) \,. \tag{14}$$

Then the $P(\boldsymbol{r})$ and $J(\boldsymbol{u})$ expressions involve cross-product terms and are correspondingly more complicated.

In order to illustrate the treatment of diffraction problems for the two classes of crystal imperfections we have discussed, we proceed to work out a few simple examples using both the direct calculation of intensities from equation (2) and the calculation via the Patterson. This will then serve as a basis for the discussion of the more complicated problems to be met in the later chapters of Section IV.

7.3. Deviations from an average lattice

7.3.1. Random vacancies: no relaxation

As a first simple example we consider a monatomic crystal of simple structure. There are N atom sites but a number n of them, distributed at random, are vacant. We ignore any displacement of the atoms around the vacancies from their equilibrium lattice sites (relaxation).

Then the scattering power $J(\boldsymbol{u})$ can be evaluated from equation (2) by considering in turn the various possible vectors $\boldsymbol{r}_i - \boldsymbol{r}_j$.

For $i = j$ and $\boldsymbol{r}_i - \boldsymbol{r}_j = 0$, there will be $N - n$ vectors for which $f_i = f_j = f$ and n vectors for which $f_i = f_j = 0$, so that the contribution is $(N - n)f^2$.

For any other vector length there will be a probability $(N - n)/N$ of having an atom at i and $(N - n)/N$ of having an atom at j so that the contribution to $J(\boldsymbol{u})$ is

$$\left(\frac{N-n}{N}\right)^2 f^2 \exp\{2\pi \mathrm{i}\boldsymbol{u}\cdot(\boldsymbol{r}_i - \boldsymbol{r}_j)\} \,.$$

Then

$$J(\boldsymbol{u}) = (N-n)f^2 + \left(\frac{N-n}{N}\right)^2 f^2 \sum_{i \neq j}\sum \exp\{2\pi \mathrm{i}\boldsymbol{u}\cdot(\boldsymbol{r}_i - \boldsymbol{r}_j)\}$$

$$= \left(\frac{N-n}{N}\right)^2 f^2 \sum_i \sum_j \exp\{2\pi \mathrm{i}\boldsymbol{u}\cdot(\boldsymbol{r}_i - \boldsymbol{r}_j)\} + (N-n)f^2 - \frac{(N-n)^2}{N} f^2 \,.$$

Here, we have removed the restriction $i \neq j$ on the summation by adding and subtracting a zero term. Then

$$J(\boldsymbol{u}) = \left(\frac{N-n}{N}\right)^2 f^2 \sum_i \sum_j \exp\{2\pi i \boldsymbol{u} \cdot (\boldsymbol{r}_i - \boldsymbol{r}_j)\} + \frac{n(N-n)}{N} f^2 . \qquad (15)$$

The first term of (15) represents the scattering power for a perfectly ordered lattice with no defects but with the average scattering factor of the atoms reduced in the ratio $(N-n)/N$. The second term is a continous distribution of scattering power in reciprocal space, falling off with $|\boldsymbol{u}|$ as f^2 and of strength proportional to the number of defects (if $n \ll N$). This is equal to the scattering from n independent, isolated atoms.

Alternatively we can follow the formulation represented by equations (7), (9) and (10). The average structure will be periodic with a fraction $(N-n)/N$ of $\rho_0(\boldsymbol{r})$ at each lattice point. The corresponding contribution to the distribution $|\bar{F}|^2$ is that of the crystal with no vacancies, but with each sharp peak at a

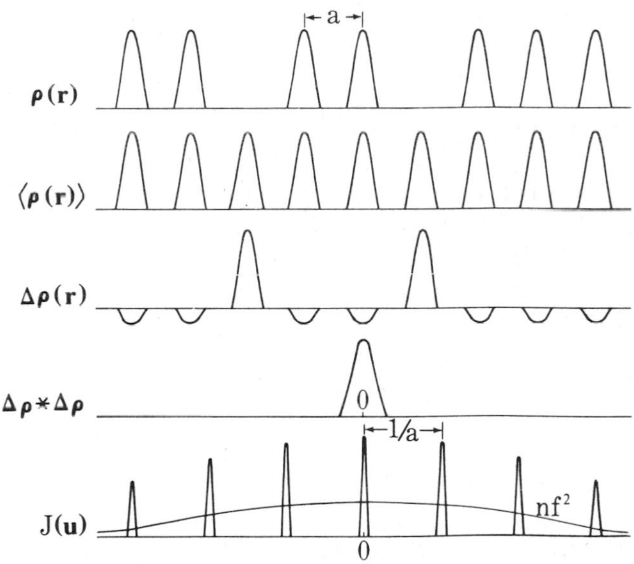

Fig. 7.1. One-dimensional diagrams representing the electron density distribution, $\rho(\boldsymbol{r})$ for a structure having random vacancies, the periodic average structure $\langle\rho(\boldsymbol{r})\rangle$, the deviations from the average structure $\Delta\rho(\boldsymbol{r})$, the Patterson function for this deviation function, and the scattering power distribution in reciprocal space showing sharp peaks plus diffuse scattering.

reciprocal lattice point reduced in weight by a factor $(N - n)^2/N^2$. The deviation from the average lattice will be $(N - n)\,\rho_0(\boldsymbol{r})/N$ for each site where there is a vacancy and $-n\rho_0(\boldsymbol{r})/N$ for each site where there is no vacancy, as suggested in Fig. 7.1. Then the origin peak of the Patterson of $\Delta\rho$ is

$$\rho_0(\boldsymbol{r}) * \rho_0(\boldsymbol{r}) \left[n\left(\frac{N-n}{N}\right)^2 + (N-n)\left(\frac{n}{N}\right)^2 \right]$$

$$= \frac{n(N-n)}{N}\,\rho_0(\boldsymbol{r}) * \rho_0(\boldsymbol{r})\,. \tag{16}$$

For any other Patterson peaks there will be n contributions of weight

$$\frac{n}{N}\left\{\left(\frac{N-n}{N}\right)\left(-\frac{n}{N}\right) + \left(\frac{n}{N}\right)\left(\frac{N-n}{N}\right)\right\},$$

when vacancy sites are taken as origin; and $N - n$ contributions of weight

$$\left(\frac{N-n}{N}\right)\left\{\left(\frac{N-n}{N}\right)\left(-\frac{n}{N}\right) + \left(\frac{n}{N}\right)\left(\frac{N-n}{N}\right)\right\},$$

when an atom is taken at the origin. The factor in curly brackets in each case, representing a random weighting of peak heights according to the occupancies, is equal to zero.

Hence the defect Patterson consists of the origin peak only. The total scattering power distribution is then $(N - n)^2/N^2$ times that for a crystal with no vacancies plus, from (16),

$$\mathcal{F}\{\Delta\rho * \Delta\rho\} = \frac{n(N-n)}{N}\,f^2\,,$$

as in equation (15).

7.3.2. *Clustered vacancies*

The particular advantage of the formulation of scattering via the Patterson is the possibility of obtaining an immediate appreciation of the form of the diffuse scattering for any particular type of defect and, in many cases, rapid quantitative estimates of the magnitude of the scattering power function.

We consider, for example, a case similar to (a) except that the distribution of vacancies is not random but shows some preference for the vacancies to clump together. To be specific we consider that the vacancies occur in pairs

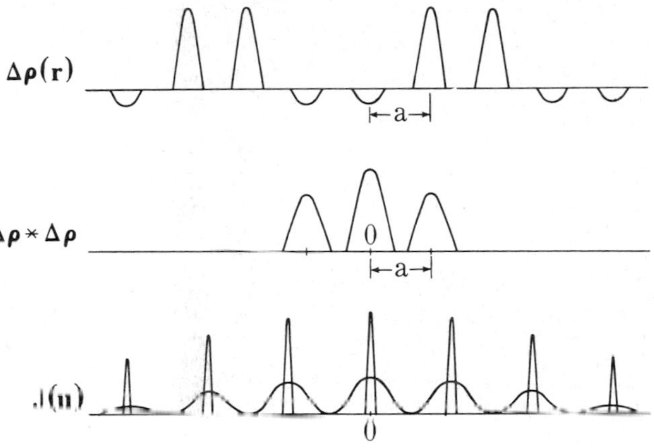

Fig. 7.2. The deviation from the average structure, the corresponding Patterson function and the scattering power distribution for a case such as in Fig. 7.1 when the vacancies occur in pairs.

parallel to one axis. For the same total number of vacancies, n, the average structure will be exactly the same as for (a) so that $|\bar{F}|^2$ will be the same.

The origin peak of $\Delta\rho * \Delta\rho$ will be the same as for (a) and there will be no correlation for interatomic vectors of greater than the nearest-neighbor distance but, as suggested in Fig. 7.2, there will be peaks at nearest neighbor distances of weight $n(N - n)/2N$ since the probability of finding a vacancy at distance a from a given vacancy is half that of finding a vacancy at distance zero. The diffuse scattering will then be modulated with a periodicity a^{-1}, thus;

$$|\Delta F(\boldsymbol{u})|^2 = \frac{n(N-n)}{N} f^2(1 + \cos\{2\pi a u\}) . \tag{17}$$

If lines of more than two vacancies may occur in the one direction, the correlation will give peaks in the defect Patterson extending further from the origin and the modulation of the diffuse scattering function will be more sharply peaked around the reciprocal lattice positions. Similarly for a tendency for the formation of three-dimensional clumps of vacancies the Patterson will show correlations in three dimensions and the modulation of the diffuse part of the intensity will appear in all directions.

With very little change, this simple theory may be applied to determine the scattering from crystals with random or clustered interstitial atoms or with occasional impurity atoms substituting for regular lattice atoms.

7.3.3. *Lattice relaxation*

We have neglected so far the effect of point defects on the surrounding atoms. For interstitials and, to a lesser extent, for vacancies, the neighboring atoms may be displaced from their average lattice sites by an appreciable fraction of the unit cell dimensions. This relaxation of the surrounding lattice may affect a large number of atoms and give rise to diffuse scattering effects which are more pronounced than the scattering from the defects themselves which we have considered above.

For simplicity we consider an idealized one-dimensional case in which n small interstitial atoms of negligible scattering factor are inserted at random in the lattice half-way between two atoms. These nearest-neighbor atoms are displaced a distance $a/4$ but no other atoms have appreciable displacement, as suggested by Fig. 7.3.

Then the average lattice has peaks of $(N - n)\, \rho_0(\boldsymbol{r})/N$ at each lattice site and subsidiary peaks of $n\rho_0(\boldsymbol{r})/N$ at distances $\pm a/4$ from the lattice sites. Hence

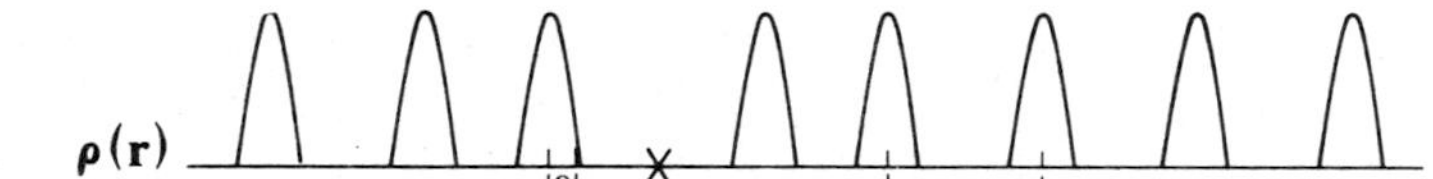

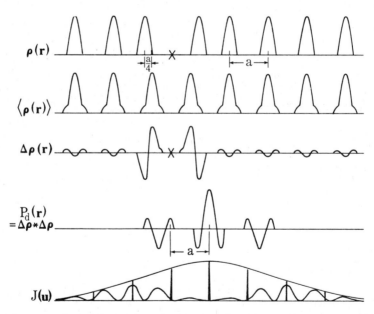

Fig. 7.3. The same set of functions as in Fig. 7.1 for the case of a one-dimensional structure in which a light interstitial atom at X displaces its nearest neighbors by a distance $a/4$.

$F(\boldsymbol{u})$ will be modulated by a factor

$$\frac{N-n}{N} + 2\frac{n}{N}\cos 2\pi(au/4)\,,$$

and the sharp peaks of scattering power in $|F(\boldsymbol{u})|^2$ will be multiplied by

$$\left\{1 + 2\frac{n}{N-n}\cos 2\pi(au/4)\right\}^2 \left(\frac{N-n}{N}\right)^2.$$

By considering the vectors between positive and negative peaks of $\Delta(\boldsymbol{r})$ it is seen that the defect Patterson will have a positive peak of relative weight 4 at the origin and negative peaks of relative weight 2 at $x = \pm a/4$. Then there will be the same grouping of peaks, inverted, around $x = \pm 5a/4$. Thus the diffuse scattering distribution will be of the form

$$\frac{n(N-n)}{N} f^2 \left(1 - \cos 2\pi \frac{au}{4}\right)\left\{1 - \cos 2\pi\left(\frac{5au}{4}\right)\right\}. \tag{18}$$

It is noted that, since the integrated value of $\Delta\rho * \Delta\rho$ is zero, the diffuse scattering function is zero at the reciprocal space origin. Apart from the f^2 fall-off, the maximum diffuse scattering comes at a position corresponding to the inverse of the displacements of the atoms.

This illustrates the general principle that diffuse scattering power arising from the displacement of atoms, without any change of scattering factors, is zero at the reciprocal lattice origin and increases with scattering angle. At the same time the sharp Bragg reflections are reduced by a factor which is unity for $|\boldsymbol{u}| = 0$ and decreases as $|\boldsymbol{u}|$ increases.

In more realistic models, the relaxation of atoms around a point defect is not limited to the nearest neighbor atoms but there are displacements of atoms decreasing gradually in three dimensions with distance from the center of dilation or contraction. Then the correlations in the defect Patterson extend to greater distances. The diffuse scattering power shows a steady overall increase with $|\boldsymbol{u}|$, apart from the fall-off with f^2, and tends to have local maxima near to the reciprocal lattice point positions. The decrease of the sharp peaks with angle which is added to the f^2 decrease, can be expressed to a first approximation as $\exp\{-\beta u^2\}$ and so has roughly the same form as the Debye–Waller factor due to thermal vibrations (see also Chapter 12). This results because, when all the atom displacements are taken into account, the peaks of the average lattice $\langle\rho(\boldsymbol{r})\rangle$ are spread out as if by convolution with something like a gaussian function.

7.3.4. Thermal vibrations – Einstein model

Finally, as an example involving a time-dependent perturbation of the lattice, we consider the case of a simple monatomic lattice in which all atoms independently vibrate about their mean lattice positions. In the harmonic approximation we may assume that the time average of the electron density function around each lattice point is given by spreading out the atom with a gaussian spread function.

In one dimension, for simplicity, we write

$$\langle\rho(x)\rangle_t = \sum_n \delta(x - na) * \rho_0(x) * (\pi b^2)^{-1/2} \exp\{-x^2/b^2\} . \tag{19}$$

Then

$$\bar{F} = F_0(\boldsymbol{u}) \sum_h \delta(u - h/a) \exp\{-\pi^2 b^2 u^2\} .$$

and

$$|\bar{F}|^2 = |F_0(\boldsymbol{u})|^2 \exp\{-2\pi^2 b^2 u^2\} \sum_h \delta(u - h/a) . \tag{20}$$

Thus if the root mean-square deviation of an atom from its lattice site is b, the intensities of the sharp reflections are decreased by the Debye–Waller factor $\exp\{-2\pi^2 b^2 u^2\}$.

As a slight variation of our former procedure, we now find the diffuse, inelastic scattering as the difference between the total scattering and the elastic scattering given by the average structure.

If all atoms are spread out by convolution with the gaussian function used in (19) and there is no correlation between the movements of different atoms, then the distribution of lengths of the interatomic vector between two atoms will depend on the spread of positions of both atoms. Then the Patterson peaks will be spread out by convolution by

$$(\pi b^2)^{-1/2} \exp\{-x^2/b^2\} * (\pi b^2)^{-1/2} \exp\{-x^2/b^2\}$$

$$= (2\pi b^2)^{-1/2} \exp\{-x^2/2b^2\} . \tag{21}$$

This is true of all Patterson peaks except the peak at the origin since the vector from the center of an atom to itself is always zero (each atom "sees" itself at rest). Thus the total time-average of the instantaneous Patterson function may be written

$$P(x, 0) = N\rho_0(x) * \rho_0(x)$$
$$* \left[\delta(x) + \sum_{n\neq 0} \delta(x - na) * (2\pi b^2)^{-1/2} \exp\{-x^2/2b^2\}\right], \qquad (22)$$

and the scattering power is

$$\int J(u, \nu)\, d\nu = N|F_0|^2 \exp\{-2\pi^2 b^2 u^2\} \sum_h \delta(u - h/a)$$
$$+ N|F_0|^2 \,[1 - \exp\{-2\pi^2 b^2 u^2\}] \,. \qquad (23)$$

Here the first term is the elastic scattering from the average lattice as we found in equation (20). The second term is the diffuse inelastic scattering. The form of this is illustrated in Fig. 7.4.

As we have seen in the case of static displacements of atoms from their mean lattice positions, the effect of correlation between atom displacements will be to modulate the diffuse background scattering. If neighboring atoms tend to move in the same direction at the same time, as in acoustic-mode lattice vibrations, the diffuse scattering will tend to peak at the positions of the reciprocal lattice points.

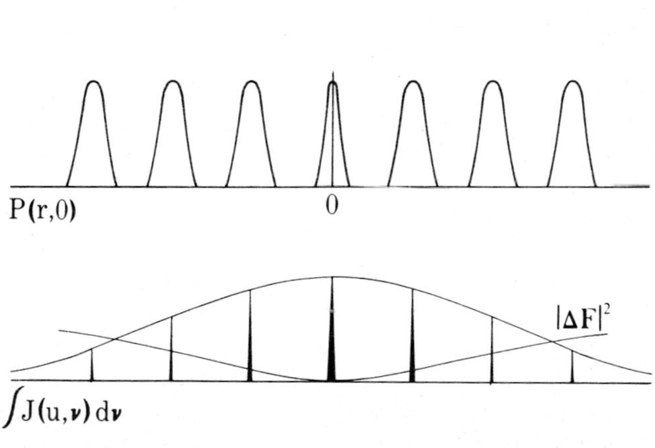

Fig. 7.4. One-dimensional diagrams illustrating the Patterson function and corresponding scattering distributions for uncorrelated atomic vibrations (Einstein model).

7.4. Imperfect crystals with no average lattice

7.4.1. Uneven separation of lattice planes

One of the few cases for which a non-repetitive structure can be simply defined and allows a simple analytical solution for the diffraction problem is that of parallel identical planes of atoms (or layers composed of several planar arrays of atoms) stacked in such a way that the spacing between them is not regular. This has some relevance in practice in that it may be considered a reasonable model for some crystals of clay minerals, for example, in which the two-dimensionally infinite sheets of atoms, consisting of tightly-bonded close-packed oxygen arrays with metal atoms in octahedral or tetrahedral sites, have only weak bonding between them. Then a variation of the number or nature of the ions or molecules lying in the spaces between these sheets may modify the distances between the sheets without affecting their relative orientations.

Then, following (11), the structure may be described in terms of the electron density distribution of the individual sheets of atoms, $\rho_0(\boldsymbol{r})$ and a one-dimensional distribution function, $d(z)$, a set of delta functions giving the positions of the equivalent reference points within the sheets:

$$\rho(\boldsymbol{r}) = \rho_0(\boldsymbol{r}) * d(0, 0, z) . \tag{24}$$

If we can assume $\rho_0(\boldsymbol{r})$ to be known, the problem is then to define $d(z)$ or $d(z) * d(z)$ and deduce $|D(u)|^2$ for use in (13).

The function $\rho_0(\boldsymbol{r})$ is considered to have periodicities a, b in the x and y directions but to be non-periodic and of limited extent (say 5 to 15 Å) in the z-direction. Then $|F_0(\boldsymbol{u})|^2$ will consist of a set of continuous lines parallel to the w-axis of reciprocal space with regular spacings a^*, b^* in the u and v directions. The variations of scattering power along these lines will depend on the relative positions of the atoms within the sheets.

To find a model for $d(z) * d(-z)$, we assume that the spacing between any two sheets of atoms has no influence on the spacing of any other sheets. Further we assume that the spacings between adjacent planes show a gaussian distribution about some mean value, c. Then the one-dimensional correlation function $d(z) * d(-z)$ has a form illustrated in Fig. 7.5.

For $z = 0$ we have a delta function of weight N corresponding to the zero distance of each reference point from itself. Around $z = \pm c$ there is a gaussian peak of half-width say, γ, corresponding to the distribution of nearest neighbor distance. Around $z = \pm 2c$ there will be a broader gaussian peak since for each position of the sheet which is nearest neighbor to a given sheet there will be a gaussian distribution of distances to the next, or second nearest neighbor sheet.

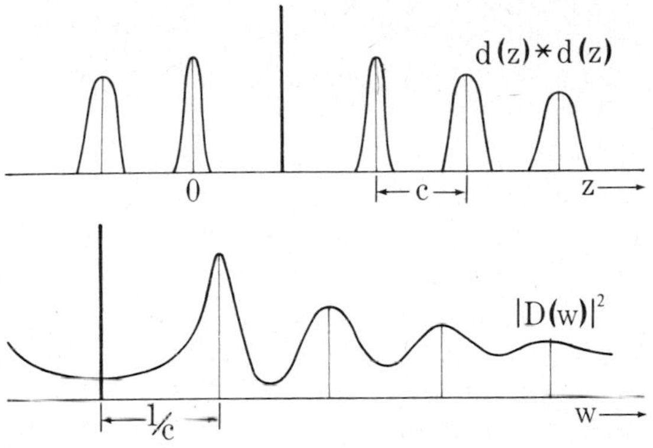

Fig. 7.5. The distribution function Patterson and its Fourier transform for the case of parallel planes of atoms having random variations of the inter-planar spacing.

The second nearest neighbor then has a distribution of positions, relative to the origin position, of

$$(\pi\gamma^2)^{-1/2}\exp\{-z^2/\gamma^2\} * (\pi\gamma^2)^{-1/2}\exp\{-z^2/\gamma^2\}$$

$$= (2\pi\gamma^2)^{-1/2}\exp\{-z^2/2\gamma^2\}\,. \qquad (25)$$

Similarly for the nth-nearest neighbor sheets to a given sheet, the spread of positions will be given by an n-fold convolution of the nearest-neighbor distribution. Thus the Patterson of $d(z)$ may be written

$$d(z) * d(-z) = \sum_{n=-\infty}^{\infty} \delta(z - nc) * \frac{1}{(\pi|n|\gamma^2)^{1/2}}\exp\{-z^2/|n|\gamma^2\}\,. \qquad (26)$$

We note the need to insert $|n|$ in order to make the distribution symmetrical. Fourier transform of this gives

$$|D(w)|^2 = \sum_n \exp\{-\pi^2 w^2 |n| \gamma^2\}\exp\{2\pi i w n c\}$$

$$= \sum_{n=-\infty}^{\infty} \exp\{-(\pi^2 w^2 \gamma^2 \mp 2\pi i w c)|n|\}\,, \qquad (27)$$

where the minus sign refers to positive n.

Using the relationship $\Sigma_0^\infty x^n = (1-x)^{-1}$, this becomes

$$|D(w)|^2 = [1 - \exp\{-\pi^2 w^2 \gamma^2 + 2\pi i w c\}]^{-1}$$

$$+ [1 - \exp\{-\pi^2 w^2 \gamma^2 - 2\pi i w c\}]^{-1} - 1$$

$$= \frac{1 - \exp\{-2\pi^2 w^2 \gamma^2\}}{1 + \exp\{-2\pi^2 w^2 \gamma^2\} - 2\exp\{-\pi^2 w^2 \gamma^2\} \cos 2\pi w c} . \quad (28)$$

Fig. 7.5 shows the form of this function. There is a delta function at $w = 0$, then maxima at $w = l/c$ for integral l. The heights of the maxima are given by

$$\left|D\left(\frac{l}{c}\right)\right|^2 = \frac{1}{l^2} \frac{2c^2}{\pi^2 \gamma^2} \quad \text{if} \quad l^2 \gamma^2 \ll c^2 . \quad (29)$$

In between the maxima, for $w = (2l+1)/2c$, the minima have values which increase initially with w^2. The widths of the maxima increase approximately as l^2.

The total distribution of scattering power is then given by multiplying this function by the values of $|F_0(\boldsymbol{u})|^2$. It therefore consists of a set of regularly-spaced lines parallel to the w direction having sharp points on the $u - v$ plane and increasingly diffuse maxima as the distance from this plane increases.

This result gives an indication of what might be expected for more complicated cases for which the irregularity of spacing occurs in more than one dimension. A two-dimensional equivalent would be the case of long rod-like molecules packed together so that the ordering is near-perfect in the direction of the rods, with equivalent atoms in all rods lying in the same plane, but with an irregular spacing between the rods introduced by a random variation of side-groups of atoms attached to the molecules. An example in three dimensions would be provided by the packing of large molecules together into a lattice which is irregular because the presence of disordered side-groups or absorbed atoms gives a variation of the distance between molecules in all directions. In each case the maxima of scattering power in reciprocal space become progressively broader in the directions of the real-space irregularities.

7.4.2. Disordered orientations

When the disorder of the units of structure involves a relative rotation instead of, or in addition to, a relative displacement between neighbors, the use of a distribution function to simplify the Patterson is no longer possible and either the intensity expression (2) or the Patterson function must be derived from first principles.

We give as an example of such situations the case of the "turbostratic"

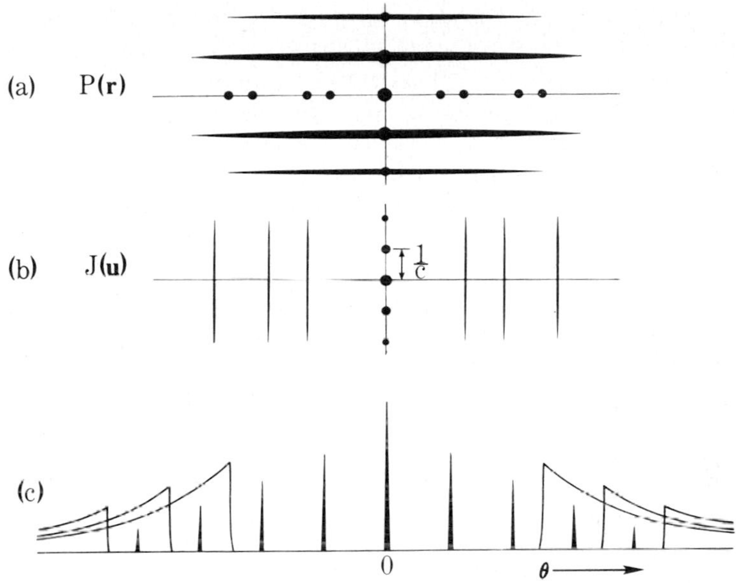

Fig. 7.6. (a) The Patterson function, and (b) the corresponding distribution in reciprocal space for a turbo-stratic structure of parallel, equally-spaced planes of atoms, having arbitrary relative orientations of the axes within the planes. The diagram (c) suggests the form of the power pattern given by such a structure, consisting of sharp rings corresponding to the $(0\ 0\ l)$ reflections and broad asymmetrical peaks corresponding to the $(h\ k\ l)$ reflections with intensity a continuous function of the variable l, but with integral h, k values.

structure which has been used as a model for considering the diffraction from poorly-crystalline carbon black. Planar sheets of carbon atoms such as occur in the graphite structure are assumed to be stacked together with a constant distance, c, between the sheets as the only degree of regularity present. There is no relationship between the relative rotations or translations within the planes of adjacent sheets (Biscoe and Warren [1942]).

The Patterson function can be sketched immediately as in Fig. 7.6(a) where we show a planar section perpendicular to the plane of the sheets. In the plane through the origin of $P(\boldsymbol{r})$ parallel to the sheets, say the x-y plane, each sheet will give the two-dimensionally regular set of Patterson peaks corresponding to its hexagonal structure. But since the orientations in this plane are random, the Patterson for a single sheet will be rotated to give a set of sharp rings in the x-y plane. The vectors with z coordinates equal to nc, for $n \neq 0$, will all have equivalent weight since the probability of an interatomic vector occurring is independent of the x, y, coordinates. Hence the Patterson contains a set of structureless planes of density at equal intervals, c, parallel to, but not including, the x-y plane through the origin. The relative weighting of these planes will fall off with distance from the origin if the finite extent of the assemblies of sheets is taken into account.

In taking the Fourier transform to get $J(\boldsymbol{u})$ we divide $P(\boldsymbol{r})$ into two parts. A set of parallel structureless planes of spacing c, including one through the origin of $P(\boldsymbol{r})$, gives a set of sharp peaks of separation $1/c$ in the w direction in reciprocal space. Subtracting the plane through the origin of $P(\boldsymbol{r})$ gives a negative continuous line in the w direction through the origin of $J(\boldsymbol{u})$ but this is compensated by the general positive overall back-ground intensity arising from the origin peak of $P(\boldsymbol{r})$. The series of sharp rings in the x-y plane of $P(\boldsymbol{r})$ gives a set of concentric cylinders in reciprocal space with the w direction as axis. These show up as sets of parallel lines in the section of $J(\boldsymbol{u})$ drawn in Fig. 7.6(b).

Since the assemblies of parallel sheets may be assumed to be of limited extent, the sharp peaks and the cylinders of scattering power will be broadened by convolution with appropriate shape-transforms. If the distance between sheets is not quite regular, the sharp peaks on the central line of $J(\boldsymbol{u})$ will be progressively broadened in the way suggested by Fig. 7.5(b).

The appearance of a powder pattern obtained from a large number of randomly oriented assemblies of sheets may be derived by considering the $J(\boldsymbol{u})$ function to be rotated about the reciprocal lattice origin and then taking the intersection with the Ewald sphere. The result is suggested in Fig. 7.6(c). The sharp peaks will give sharp rings in the powder pattern. The cylinders of scattering power will give broadened rings with high intensity at the minimum scattering angle corresponding to the radius of the cylinder and intensity falling off to higher angles more and more slowly. These components, sketched separately in Fig. 7.6(c), will be added together to give the total intensity.

The complete randomness we have assumed in the relative orientations and translations of adjacent sheets represents an over-simplification. Any correlations between sheets will have the result of introducing some structure into the planes in $P(\boldsymbol{r})$ which do not pass through the origin with a consequent modulation of the scattering power in the cylinders of $J(\boldsymbol{u})$ and a modification of the simple form of the broad peaks in the powder pattern.

More complicated forms of disorder in orientation and relative translation in one, two or three dimensions arise frequently in poorly crystallized materials. We have attempted to demonstrate the value of the Patterson function approach in offering a rapid and usually adequate means for describing the state of order, for determining the resultant form of the reciprocal space distribution and so of the observable intensities. In Section IV we will apply these methods and also the more conventional ones to the detailed study of several cases of diffraction from imperfectly ordered systems which are of particular interest.

Problems

1. Consider a one-dimensional array of atoms in the form of regularly spaced AB molecules where the separation of the A and B atoms in a molecule is one third of the separation of the centers of the molecules. At intervals of approximately T seconds, each molecule flips end-for-end (flipping time very short compared with T). There is no correlation in the flips of adjacent molecules. Find the elastic and inelastic scattering from this array and the energy changes which would be observed if sufficient energy resolution were available.

2. It is postulated that extra atoms may be introduced into the lattice of a face-centered cubic metal by the formation of a "split-interstitial"; e.g. the atom with fractional coordinates 1/2, 1/2, 0 is replaced by two atoms having coordinates 1/2, 1/2, 1/4 and 1/2, 1/2, −1/4. In a crystal containing N atoms, a small number, n, of split interstitials is introduced, distributed at random in position and at random among the three symmetrically equivalent directions. How will this modify the sharp Bragg reflections and what diffuse scattering will be produced? (Ignore lattice relaxation effects.)

3. A complex oxide has a "tungsten-bronze" type of structure consisting for the most part of an almost close-packed array of oxygens with metal atoms in octahedral positions. Along the c-axis of the unit cell there is an open channel of about 4 Å diameter containing no oxygen atoms. In this channel, there are equivalent sites at, 0, 0, 1/4 and 0, 0, 3/4 in which a heavy metal atom may sit. In each such channel (one per unit cell) a heavy atom sits at one or other of these two sites but not at both. The choice of site varies at random from one channel to the next. How does this random distribution affect the sharp Bragg reflections and what diffuse scattering is produced? How will the diffraction effects be affected if there is some tendency for ordering, such that, if a heavy atom sits at 0, 0, 1/4 in one channel, this site will tend to be empty in neighboring channels?

4. Long straight identical molecules having a periodic structure along their lengths, are stacked together in a close-packed (hexagonal) array, but there is no correlations of the positions of the molecules in the direction of their lengths. Sketch the Patterson function for this array and deduce the form of the distribution of scattering power in reciprocal space. How are these functions in real and reciprocal space affected if there are irregular variations in the distances between molecules so that no average hexagonal stacking arrangement can be defined for more than a few repeat distances?

Section III

DYNAMICAL SCATTERING

CHAPTER 8

Diffraction by perfect crystals

8.1. Multiple coherent scattering

Scattering from any three-dimensional object must, in principle, involve multiple scattering processes. Radiation scattered from one part of the object will pass through other parts of the object and will be scattered again.

We have seen that the kinematical, or single-scattering, approximation is very useful and has reasonable validity for a wide range of scattering experiments. Now, it is time to investigate the more general case of dynamical scattering in which the coherent interaction of multiply scattered waves is taken into consideration.

For incoherent sources of finite dimensions and for an appreciable range of wavelengths, the observed intensities may be obtained by summing, incoherently, the intensities for individual point sources and single wavelengths. Hence we may start by considering perfectly coherent incident radiation for which the amplitudes of all multiply-scattered radiation are added.

If the atoms of the sample are spaced with complete regularity, the relative phases of the waves scattered by the different atoms will be definable so that wave amplitudes may be added in a meaningful way. Any disorder in the atom positions or any imperfections of the crystal which prevent the definition of a regular periodic lattice, will introduce phase changes in the scattering from the various atoms. If this happens with sufficient frequency there will be an averaging of the phase-sensitive components of the intensity and the more striking dynamical effects may be lost (see Chapter 16).

Hence as a first step we explore the nature of the dynamical scattering effects by considering the ideal case of a perfect crystal in coherent, monochromatic radiation. The strength of the diffracted waves produced in a crystal will depend on the scattering cross-sections or atomic scattering factor of the atoms and the direction of the incident radiation with respect to the crystal axes. The importance of the multiple scattering effects will depend on these factors and also on the dimensions of the sample. As may be expected, the variety of effects is great and there are considerable differences for the different radiations we have been discussing. However there is a common body of theory which allows us to appreciate the main types of phenomena to be observed and we present this in the simplest form which serves these ends.

An important difference between the X-ray and electron diffraction cases comes from the fact that for X-rays polarization effects can not be ignored. With the scattering through large angles the polarization factor for each scattering process may vary from zero to unity. Multiple scattering sequences which would be equivalent for the low-angle electron cases must be clearly differentiated. However, since adequate accounts of these complications are given in the literature they will be ignored here except that, occasionally, we will quote some results for X-rays to show how they differ from the results of the simpler scalar-wave diffraction theory.

A second important difference for the different radiations arises from the relative strengths of the interactions with matter. For X-rays and neutrons, by the time the amplitude of the scattered wave has built up to the magnitude for which multiple scattering is important and the kinematical approximation fails, the radiation has traversed sufficient thickness of crystal for sharply-defined Bragg reflections to be produced and the probability that more than one Bragg reflected beam will be produced at a time is very small. Then it is possible to make use of the assumption valid for most cases that only two beams need to be considered at any one time, the incident beam and a beam diffracted from one set of lattice planes.

For electrons, on the other hand, strong scattering may occur for passage of the radiation through only the first few atom-thicknesses of crystal, i.e. for a slice of crystal thin enough to be regarded as a two-dimensional phase-grating which will give rise to several tens or hundreds of diffracted beams simultaneously. In order to take account of the multiple coherent interactions of all these diffracted beams, an n-beam dynamical theory must be used. The fact that for particular orientations there is destructive interference which weakens all but two of the beams, means that for electron diffraction also the "two-beam approximation" has some relevance. We proceed to consider this approximation with the full understanding that, at least for the electron case, it represents an assumption which must be justified later by a more complete n-beam treatment.

8.2. Theoretical approaches

The theoretical approaches which have been used for the formulation of the dynamical theory of diffraction by crystals may be divided into two general classes; those based on the formulation of wave-mechanics as a differential equation, the wave equation in a crystal lattice, and those based on the integral equation formulation. Approaches by means of quantum field theory for electron diffraction have been made by Ohtsuki and Yanagawa [1965, 1966] and a modern approach to X-ray diffraction theory has been given, for example, by Kuriyama [1970, 1972], but these will not be discussed here.

The integral equation methods follow from the ideas mentioned in Chapter 1. They may be considered to represent mathematically the progression of radiation through the crystal. An incident plane wave is successively scattered in the crystal and the multiply scattered components are added up according to their relative amplitudes and phases to form the out-going waves. The use of the Born series, equations (1.17), (1.22) may be interpreted as considering scattering by successive volume elements. The incident wave (zero-order term) is scattered by each volume element of the crystal to give the singly-scattered wave amplitude (first-order term) which is scattered again by each volume element to give the doubly-scattered wave and so on. This approach was used by Fujiwara [1959] for electron diffraction. Although the convergence of the Born series is notoriously bad, Fujiwara was able to obtain series-solutions for the scattering from a crystal which allowed important general conclusions to be drawn including the nature of the modifications to the scattering theory required when relativistic effects for high energy incident electrons are included (Fujiwara [1961]).

For the particular case of medium or high energy electrons (or other small-angle scattering) it is possible to take advantage of the fact that the propagation of the wave is close to the forward direction only and consider the scattering by successive planes of infinitesimal thickness. This was the approach of Cowley and Moodie [1957] using the concept of transmission through an infinite number of two-dimensional objects as suggested in Part (5) of Chapter 3.

It has been said, with very little justification that this approach resembles the first treatment of the scattering of X-rays by crystals given by Darwin [1914] and the related method used for the calculation of electron microscope intensities given by Howie and Whelan [1961]. In these treatments individual planes of atoms are considered to diffract incident plane waves to give a set of diffracted beams i.e., the conditions for Fraunhofer diffraction, rather than Fresnel diffraction are assumed to be relevant in the inter-atomic distances. In the original Darwin treatment it was assumed that an incident plane wave would be reflected from a plane of atoms to give only a single diffracted beam. The justification of this in terms of expediency and reasonableness is clear, but since we know that a two-dimensional grating gives rise to many diffracted beams, a more complete justification in terms of n-beam diffraction theory would seem appropriate. A more comprehensive and updated account of the Darwin approach to X-ray diffraction has been given by Borie [1966] and Warren [1969] and the approach to electron diffraction and microscopy is described by Hirsch et al. [1965].

The differential equation approach was used in the initial formulation of X-ray diffraction theory by Ewald [1916, 1917] and von Laue [1935] and in the first formulation of electron diffraction theory by Bethe [1928]. Recent accounts of this approach have been given by Zachariasen [1945], James [1950], Authier [1970], Pinkster [1978], and Batterman and Cole [1964] for X-rays and by Morse [1930], Hirsch et al. [1965], Heidenreich [1964], and Kambe and Moliere [1970] for electrons. The Bethe theory has been expressed in matrix form (Sturkey [1957]; Niehrs [1959a]; Fujimoto [1959]) and from this comes the scattering matrix method (Niehrs [1959b]; Sturkey [1962]) which has much in common with the integral equation methods in that progression of an electron wave through successive slices of crystal can be represented by repeated application of a scattering matrix. A further approach which has something of this dual character is that of Tournarie [1960, 1961]. An illuminating discussion of the nature and interrelation of the various approaches has been given by Goodman and Moodie [1974].

For our present purposes it is convenient to begin our discussion of dynamical effects by reference to the Bethe theory. Although referring to the scalar electron waves it provides, by analogy, an indication of the X-ray diffraction effects when allowances for the polarization effects is made. Also it allows a straightforward means for deriving the results for the simple two-beam case which is adequate for describing most dynamical effects observed with X-rays and neutrons and forms a reasonable first approximation for many of the electron diffraction phenomena. Readers familiar with matrix theory may prefer to use the scattering matrix approach, from which the two beam approximation may be derived very simply as shown by Rez [1977] and outlined in Subsection 10.2.5.

8.3. Bethe theory

8.3.1. The dispersion equations

The Schrödinger equation (1.5) may be written in the form

$$\{\nabla^2 + K_0^2 + \vartheta(\boldsymbol{r})\}\,\psi(\boldsymbol{r}) = 0 \tag{1}$$

where we have put

$$K_0^2 = 2meE/\hbar^2; \qquad \vartheta(\boldsymbol{r}) = 2me\varphi(\boldsymbol{r})/\hbar^2 = 2K_0\sigma\varphi(\boldsymbol{r}) .$$

Then K_0 is the wave number, or the modulus of $\boldsymbol{K}_0$, the incident wave in vacuum, and $K_0 = 2\pi/\lambda$.

We have departed somewhat from our previous convention for notation in order to conform with the notation initiated by Bethe which has become standard for this type of dynamical theory.

If we now Fourier transform (1), the transform of the second differential, $\nabla^2\psi$, gives a term in $k^2\Psi(\boldsymbol{k})$ where $\boldsymbol{k}$ is the wave vector of the wave in the crystal, so that we obtain

$$(K_0^2 - k^2)\,\Psi(\boldsymbol{k}) + v(\boldsymbol{k}) * \Psi(\boldsymbol{k}) = 0 . \tag{2}$$

If the potential distribution term $\vartheta(\boldsymbol{r})$ is periodic, the Fourier transform will be given by

$$v(\boldsymbol{k}) = \sum_g v_g\, \delta(\boldsymbol{k} - 2\pi\boldsymbol{g}) , \tag{3}$$

where $\boldsymbol{g}$ is a reciprocal lattice vector.

According to Bloch's theorem the wave field in the crystal must have the periodicity of the lattice and $\Psi(\boldsymbol{k})$ must therefore be of the form

$$\Psi(\boldsymbol{k}) = \sum_h \Psi_h\, \delta(\boldsymbol{k} - \boldsymbol{k}_h) ,$$

where the wave vector $\boldsymbol{k}_h$, corresponding to the reciprocal lattice point $\boldsymbol{h}$, is defined as

$$\boldsymbol{k}_h = \boldsymbol{k}_0 + 2\pi\boldsymbol{h} , \tag{4}$$

and $\boldsymbol{k}_0$ is the wave vector corresponding to the reciprocal lattice origin. Then inserting (3) into (2) we obtain the relationship between the wave numbers $\boldsymbol{k}_h$ and wave amplitudes $\Psi_{\boldsymbol{h}}$, or $\Psi(\boldsymbol{h})$, for the set of diffracted waves in the crystal,

$$(K_0^2 + v_0 - k_{\boldsymbol{h}}^2)\,\Psi_{\boldsymbol{h}} + \sum_{\boldsymbol{g}}{}' \, v_{\boldsymbol{h}-\boldsymbol{g}}\,\Psi_{\boldsymbol{g}} = 0 , \tag{5}$$

where the prime on the summation indicates that we have omitted the term for $\boldsymbol{g} = \boldsymbol{h}$, namely $v_0\,\Psi_{\boldsymbol{h}}$, and included v_0 in the coefficients of $\Psi_{\boldsymbol{h}}$ in the first term. This is the fundamental reciprocal space equation for the waves in a crystal, known as the "dispersion equation" because it relates the wave numbers or momentums of the waves to their energies.

We may further simplify the notation by putting $K_0^2 + v_0 = \kappa^2$. Then $\boldsymbol{\kappa}$ is the wave vector for the incident wave $\boldsymbol{K}_0$ after it has passed from vacuum ($v_0 = 0$) into a medium of constant potential v_0 equal to the average potential in the crystal lattice.

The mean refractive index for an electron in the crystal is given by

$$n^2 = \kappa^2/K_0^2 = 1 + \varphi_0/E\,,$$

$$\text{or}\quad n \approx 1 + \varphi_0/2E\,. \tag{6}$$

If, subject to all our reservations and restrictions on the use of the concept of absorption for electrons in crystals, we introduce absorption in a phenomenological way by making $\varphi(\boldsymbol{r})$ complex, then the refractive index n and the wave number $K_{\boldsymbol{h}}$ will be complex. Similarly all the coefficients v_h and wave vectors $\boldsymbol{k}_h$ will be complex. We expand our definitions to include these complications.

The equation (5) may be written as a matrix equation, thus:

$$\begin{pmatrix} \kappa^2 - k_0^2 & v_{0h} & v_{0g} \\ \vdots & \ddots & \vdots \\ v_{h0} & \cdots\ \kappa^2 - k_h^2\ \cdots & v_{hg} \\ \vdots & \ddots & \vdots \\ v_{g0} & \cdots\ v_{gh} & \cdots\ \kappa^2 - k_g^2 \\ & & \ddots \end{pmatrix} \begin{pmatrix} \Psi_0 \\ \vdots \\ \Psi_h \\ \vdots \\ \Psi_g \end{pmatrix} = 0 \tag{7}$$

where, for convenience, we have written $v_{hg} = v_{\boldsymbol{h}-\boldsymbol{g}}$. In the absence of absorption the matrix is Hermitian since for a real potential $v_{hg} = v_{gh}^*$. If the crystal has a center of symmetry the matrix will be real symmetric since $v_{hg} = v_{gh}$. If absorption is present, a quantity $\mathrm{i}\mu_0$ must be added to the diagonal elements and $\mathrm{i}\mu_{hg}$ to the off-diagonal elements.

8.3.2. Solutions of the equations

The object is then to solve the set of non-linear equations (5) or the matrix equation (7) to obtain the wave vectors $\boldsymbol{k}_h$ and the Fourier coefficients of the wave functions, $\Psi_{\boldsymbol{h}}$, for the waves in the crystal, subject to the boundary conditions. Since there is no limitation to the number of reciprocal lattice points, there will be, in principle, an infinite number of solutions and so an infinite number of wave vectors $\boldsymbol{k}_h^i$ and amplitudes Ψ_h^i for each reciprocal lattice point. Alternatively we may say that for the solution number i, there will be a set of wave vectors, $\boldsymbol{k}_h^i$ and a set of amplitudes Ψ_h^i, one for each reciprocal lattice vector. This set is known as the "Bloch wave" number i since it represents one of the solutions for a wave in the crystal which, according to Bloch's theorem must be of the form

$$\psi^i(\boldsymbol{r}) = \sum_{\boldsymbol{h}} \Psi^i_{\boldsymbol{h}} \exp\{i\boldsymbol{k}^i_h \cdot \boldsymbol{r}\} , \tag{8}$$

where $\boldsymbol{k}_h$ is given by (4).

8.3.3. Boundary conditions

For a crystal limited by a planar surface, the boundary conditions imply the equality of the tangential components of the wave vectors on the two sides of the boundary. If a plane wave, K_0 is incident on the crystal the projection on the surface of κ, the incident wave in the crystal, must be equal to the projection of K_0. This is just Snell's law of refraction. Hence we may draw the diagram, Fig. 8.1, which is a reciprocal space representation with the crystal boundary (real space) sketched in in order to define the surface normals used to apply the boundary conditions. The vector $\boldsymbol{\kappa}$ is drawn from a point L the "Laue point", to the reciprocal lattice origin, O, such that its projection on the surface is the same as that of K_0. The i Bloch wave solution of the dispersion equations will give a set of vectors $\boldsymbol{k}^i_0 + 2\pi\boldsymbol{b}$ drawn to the reciprocal lattice points, $\boldsymbol{b}$, from a point L^i, distance ξ^i (the "anpassung") from L along the surface normal drawn through L. The diagram is drawn for the case of electron waves for which $\kappa, k^i_0 > K_0$. For X-rays K_0 is larger than κ.

If an Ewald sphere is drawn, centered on the point L and of radius $|\boldsymbol{\kappa}|$ it will miss the reciprocal lattice points by an amount $\zeta_{\boldsymbol{h}}$ (the "excitation error") measured along the Ewald sphere radius. Since in general the anpassung ξ^i will be very small compared with $|\boldsymbol{\kappa}|$, the wave vectors $\boldsymbol{k}^i_h$ will be very nearly parallel to the corresponding Ewald sphere radii and we may write

$$k^i_h = \kappa - \zeta_h - \xi^i \cos\theta_h \tag{9}$$

where θ_h is the angle with the surface normal.

From the equations (5) or (7) it is clear that the solutions will give the k^i_h as functions of both the incident magnitude κ and the Fourier coefficients of potential v_{hg}. Hence from (9) we see that the anpassung, ξ^i, will depend on both the incident beam magnitude and directions and the v_{hg} but the dependence on v_{hg} will decrease as the diagonal terms of (7) become larger. As the incident beam direction changes and the surface normal through L and L^i changes its position relative to O in (Fig. 8.1), the locus of the point L will be

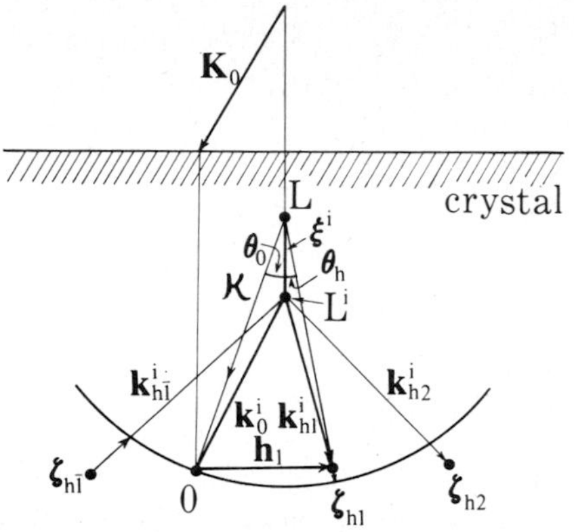

Fig. 8.1. Diagram illustrating the wave vectors for waves within a crystal when an incident wave, $\boldsymbol{K}_0$, strikes the surface. The excitation errors, $\zeta_{\boldsymbol{h}}$, for the lattice points $\boldsymbol{h}$, and the *anpassung* ξ^i, Laue point L^i and wave vectors $\boldsymbol{k}_h^i$ for one Bloch wave are shown.

a sphere centered on O of radius κ. Each point L^i will trace out a surface, known as the "i branch" of the dispersion surface.

Some idea of the form of this multi-branch dispersion surface may be obtained by considering the limiting case of (5) or (7) for which all v_h tend to zero. Then the solution to the dispersion equation is $\kappa^2 - k_h^2 = 0$ for all h, i.e., the dispersion surface is a set of spheres, $|\boldsymbol{k}_h| = |\boldsymbol{\kappa}|$, one centered on each reciprocal lattice point, as suggested by Fig. 8.2(a). As the off-diagonal elements, v_{hg}, of the matrix in (7) are increased from zero the points or lines of intersection of these spheres will be modified, giving rise to a succession of non-intersecting surfaces, or branches of the dispersion surface, as suggested by Fig. 8.2(b) for a small part of the surface for a very simple case. For each branch of the dispersion surface, a surface normal will have two intersections, so that if N reciprocal lattice points are included in the considerations there will be $2N$ intersections and hence $2N$ Bloch waves. Of these N will correspond to forward scattering and N to back-scattering. Complications are introduced, particularly for longer wavelengths and hence small Ewald sphere radii, when the dispersion surface may be cut by the surface normal only in imaginary points.

For the general n-beam case the form of the dispersion surface is difficult to describe and in fact the solution of the equations (5) or (7) may not be made in general but only under simplifying assumptions chosen to give reasonable approximations to particular experimental conditions.

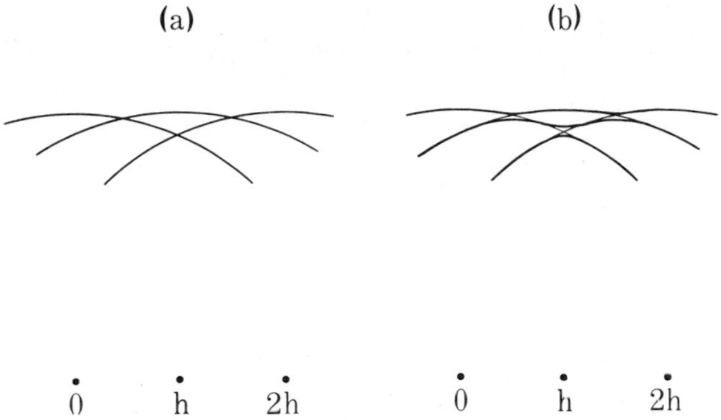

Fig. 8.2. The dispersion surface for waves in a crystal corresponding to the set of reciprocal lattice points 0, ***h***, 2***h***..., for (a) the kinematic case of very weak interaction and (b) for dynamical scattering with appreciable interactions.

8.4. Two-beam approximation

8.4.1. Bloch waves and dispersion surfaces

Since for most important cases in X-ray and neutron diffraction and for selected cases in electron diffraction the maximum number of strong diffracted beams presented is two, it is a useful approximation to assume that only two wave amplitudes Ψ_0 and Ψ_h are not zero. It may be emphasized that this is not an approximation to a general solution in the normal sense. It is the solution of a different and simpler problem: the assumption of a universe in which only two waves *can* exist. Then the matrix equation (7) simplifies immediately to give

$$\begin{pmatrix} \kappa^2 - k_0^2 & v_{0h} \\ v_{h0} & \kappa^2 - k_h^2 \end{pmatrix} \begin{pmatrix} \Psi_0 \\ \Psi_h \end{pmatrix} = 0 . \tag{10}$$

For a non-trivial solution the determinant of the matrix must be zero, giving, in general, four solutions for the wave vectors. However two of these solutions correspond to the back-scattering of electrons and may usually be ignored for high voltage transmission diffraction. An alternative derivation of the two beam approximation, starting from the scattering matrix formulation, is given in Subsection 10.2.5.

There will be two forward scattering solutions with two Bloch waves $i = 1, 2$.

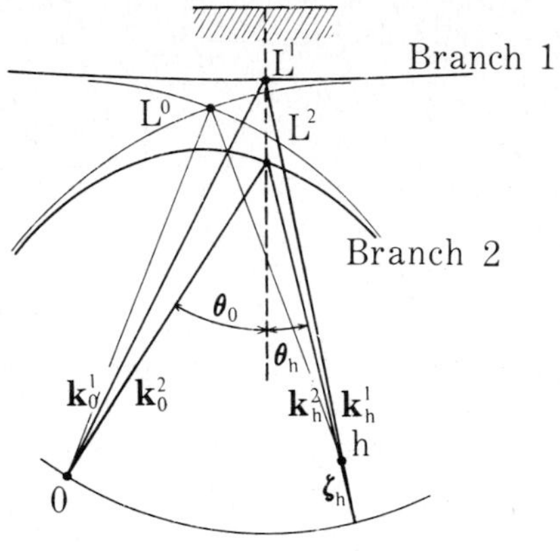

Fig. 8.3. The dispersion surface construction for the 2-beam case.

For $i = 1$ there will be wave amplitudes Ψ_0^1, Ψ_h^1 and wave vectors $\boldsymbol{k}_0^1$, $\boldsymbol{k}_h^1$ and similarly for $i = 2$. The dispersion surface will have two branches which approximate to spheres about the O and $\boldsymbol{h}$ reciprocal lattice points except near their line of intersection. This is all suggested in the diagram, Fig. 8.3, where we have numbered the branches of the dispersion surface in order of decreasing k_0^i (Humphreys and Fisher [1971]). The dispersion surface section shown is symmetrical about the perpendicular bisector of the vector $\boldsymbol{h}$. The spheres about O and $\boldsymbol{h}$ meet in the point L_0 which in three dimensions becomes a circle. Introducing the boundary conditions at an entrance surface defines the perpendicular through L which cuts the dispersion surface at the "tie-points" L^1 and L^2.

If the $\boldsymbol{h}$ vector is parallel to the crystal surface and the angle of incidence is adjusted to make L coincide with L_0 then we have the simplest symmetrical case in which incident and diffracted beams make equal angles with the surface, the excitation error ζ_h is zero, the Bragg condition for the reflection is exactly satisfied and $|\boldsymbol{k}_0^i| = |\boldsymbol{k}_h^i|$.

Then the condition that the determinant of the matrix in (10) should be zero gives, for the case of no absorption,

$$\kappa^2 - k_h^2 = (v_{0h}\, v_{h0})^{1/2} = |v_h| \,,$$

or, since κ and k_h differ by a relatively small amount

$$\kappa - k_h^i = \pm |v_h|/2\kappa \,. \tag{11}$$

Then the "anpassung" for this case becomes

$$\xi^i = \pm \frac{|v_h|}{2\kappa \cos\theta_h} \tag{12}$$

Thus the minimum separation of the two branches of the dispersion surface is proportional to $|v_h|$.

For this symmetrical diffraction condition, the two Bloch-wave solutions, corresponding to the two branches of the dispersion surface will be equally excited. However, with increasing deviation from the Bragg condition it is seen that for one or other of the Bloch waves the wave vector $\boldsymbol{k}_0^i$ becomes closer to $\boldsymbol{\kappa}$, the incident wave vector without diffraction, while for the other $\boldsymbol{k}_0^i$ deviates more and more from $\boldsymbol{\kappa}$. It is to be expected that the value of $\boldsymbol{k}_0^i$ involving less difference from $\boldsymbol{\kappa}$ will be favored especially as the strength of the diffraction effect decreases. Hence from Fig. 8.3, as the point L moves from left to right of the diagram, the greatest amplitude will be in Bloch wave 1 at first and then switch over to Bloch wave 2 as the diffraction condition for the reflection $\boldsymbol{h}$ is passed.

In these considerations and in Fig. 8.3, we have taken into account only one of the two intersections of the surface normal with the dispersion surface. There will be a further intersection diametrically opposite that shown. For many purposes this does not matter. The intersection shown is the one for which the incident beam in the crystal is approximately in the direction of the incident beam in vacuum and the diffracted beams are in the "forward" direction. However the other intersection can become important in the case of very long wavelength radiation or when the surface normal is rotated by about 90° with respect to our figure so that it becomes almost tangential to the dispersion surface as in the so-called "Bragg case" of diffraction from planes almost parallel to the surface.

8.4.2. Conduction electrons – energy representation

At this stage we may pause to note that the two-beam model we have used here is almost exactly the same as that used, with possibly less justification, for the consideration of the behavior of nearly-free conduction electrons in crystalline solids. In most text books on solid state physics the wave equation (1) is set up for an electron in a periodic lattice and the two-beam assumption is made immediately. The main difference from our treatment is that the aim is to establish the energy levels of the system rather than the direction and amplitudes of diffracted beams. Hence the equation (10) is written in a form such as

$$\begin{pmatrix} \epsilon_\kappa - \epsilon_{k_0} & U_{-h} \\ U_h & \epsilon_\kappa - \epsilon_{k_h} \end{pmatrix} \begin{pmatrix} \Psi_0 \\ \Psi_h \end{pmatrix} = 0 \tag{13}$$

where ϵ_κ is the kinetic energy, equal to $\hbar^2\kappa^2/2m$ and the solution gives the energies ϵ^i of the electrons. In particular it is found that when the Bragg condition is satisfied for a reflection $\boldsymbol{h}$ (when the electron wave vector extends to the Brillouin zone boundary, the plane bisecting the vector $\boldsymbol{h}$ at right angles) the electron energy has two values $\epsilon_0 \pm |U_h|$ where $U_h = cv_h$ volts and c is the appropriate constant. Then as the wave-vector of the conduction electron, $\boldsymbol{K}$, passes through a Brillouin zone boundary, there is a discontinuity in its energy of amount $2|U_h|$. These results provide the basis for the "energy-band" picture widely used in solid-state physics and electronics.

8.4.3. X-ray diffraction; polarization

At this stage, also, we take note of the parallel development for X-ray diffraction, based on wave equations for the electric or magnetic field vectors derived from Maxwell's equations. From the equation equivalent to (10), the condition of zero value for the determinant is (Batterman and Cole [1964])

$$\begin{vmatrix} K_0^2(1-\Gamma F_0) - \boldsymbol{k}_0\cdot\boldsymbol{k}_0 & -k^2P\Gamma F_{\bar{h}} \\ -k^2P\Gamma F_h & K_0^2(1-\Gamma F_0) - \boldsymbol{k}_h\cdot\boldsymbol{k}_h \end{vmatrix} = 0 \tag{14}$$

where the value K_0 for the vacuum wave vector $\boldsymbol{K}_0$ is modified inside the crystal by the factor $1 - \Gamma F_0$ where F_0 is the zero structure factor and $\Gamma = e^2\lambda^2/4\pi\epsilon_0 mc^2 \cdot \pi V$. The polarization factor P is equal to unity for the σ polarization state (field vector $\boldsymbol{E}$ perpendicular to the plane of incidence) and equal to $\cos 2\theta$ for the π polarization state ($\boldsymbol{E}$ lying in the plane of incidence). Writing $\boldsymbol{k}_0\cdot\boldsymbol{k}_0$ and $\boldsymbol{k}_h\cdot\boldsymbol{k}_h$ emphasizes that these quantities are complex.

8.5. The Laue (transmission) case

8.5.1. Electron diffraction for a thin crystal

We have so far considered only the system of waves set up when an incident beam enters through a planar surface into a semi-infinite periodic crystal field. Next we consider the special cases which are suggestive of important real experimental conditions. For the relatively simple two-beam model, there are two situations for which a result can be obtained readily. The first is the "Laue case" of transmission (without any back scattering) through a perfect, parallel-sided crystal plate of infinite extent in two dimensions. The second, the "Bragg case" is for reflection from the planar surface of a semi-infinite crystal. By use of reasonable approximations it is possible to use the results for these idealized cases to discuss a wide variety of experimental situations.

The case of transmission through a thin parallel-sided plate with no back-scattering, is treated by setting up the wave equation in the crystal in terms of the two Bloch waves and then applying the boundary conditions in appropriately simplified form at the two faces. Using equation (9), the equation (10) becomes

$$\begin{pmatrix} 2k\xi\cos\theta_0 & v_{0h} \\ v_{h0} & 2k\xi\cos\theta_h + 2k\zeta_h \end{pmatrix} \begin{pmatrix} \Psi_0 \\ \Psi_h \end{pmatrix} = 0\,, \tag{15}$$

where we have assumed all $\boldsymbol{k}$ vectors to be of approximately the same length and have retained only the essential differences between k values using (9). Solution of this equation gives the anpassung ξ as

$$\xi(2k\cos\theta_h) = -k\zeta_h \pm \left[k^2\zeta_h^2 + |v_{0h}|^2 \frac{\cos\theta_h}{\cos\theta_0}\right]^{1/2}, \tag{16}$$

and the reflection coefficient

$$C = \frac{\Psi_h}{\Psi_0} = \frac{\cos\theta_0}{v_{0h}\cos\theta_h}\left\{-k\zeta_h \pm \left[k^2\zeta_h^2 + |v_{0h}|^2 \frac{\cos\theta_h}{\cos\theta_0}\right]^{1/2}\right\}. \tag{17}$$

The total wave in the crystal is given by the sum of the two Bloch waves

$$\psi(\boldsymbol{r}) = \sum_{i=1,2} \alpha_i(\exp\{\mathrm{i}\boldsymbol{k}_0^i\cdot\boldsymbol{r}\} + C_i\exp\{\mathrm{i}\boldsymbol{k}_h^i\cdot\boldsymbol{r}\}), \tag{18}$$

where C_1, C_2 are given by equation (17) with the + and – signs. Then since at the entrance surface there is unit amplitude in the $\boldsymbol{k}_0$ direction and no amplitude in the diffracted direction, the boundary conditions imply that

$$\alpha_1 + \alpha_2 = 1$$

$$\alpha_1 C_1 + \alpha_2 C_2 = 0\,,$$

so that

$$\alpha_i = \frac{(-1)^i}{C_i}\left(\frac{C_1 C_2}{C_1 - C_2}\right). \tag{19}$$

At the exit surface, since there is assumed to be no wave reflected back into the crystal from any incident wave, each crystal wave is considered to pass directly into the vacuum unchanged except that the crystal wave vectors $\boldsymbol{k}_h^i$ become the vacuum wave vectors $\boldsymbol{K}_h$. This adds no new boundary conditions. Combining (16) with the wave amplitudes found by solution of (10) then gives the crystal wave amplitudes after passage through a thickness H of crystal and, adding the $i = 1$ and 2 contributions to the amplitudes for the O and $\boldsymbol{h}$ beams in vacuum then gives the desired result. This result is conveniently expressed in terms of two new parameters, the deviation parameter, w, giving the deviation from the exact Bragg reflection condition, and the "extinction distance" ξ_h (not to be confused with the anpassung ξ^i) which is inversely proportional to the structure amplitude $|v_h|$.

These quantities are defined by reference to Fig. 8.4, which represents the portion of the dispersion surface close to L_0. Since this region is normally very small in comparison to the radius of the two spheres which intersect there, the sections of the two spheres may be considered as straight lines and the section of the dispersion surface of interest is a hyperbola with these straight lines as assymptotes. The distance between branches of the dispersion surface in the direction of the surface normal is $2D$ given by

$$D = \pm(q^2 + t^2)^{1/2}\,, \tag{20}$$

where q by generalization of (12) is given by

$$q = \frac{(v_h v_{-h})^{1/2}}{2\kappa(\cos\theta_0 \cos\theta_h)^{1/2}} \tag{21}$$

and t is a measure of the deviation, $\Delta\theta$ from the Bragg angle, θ_B through the relationship

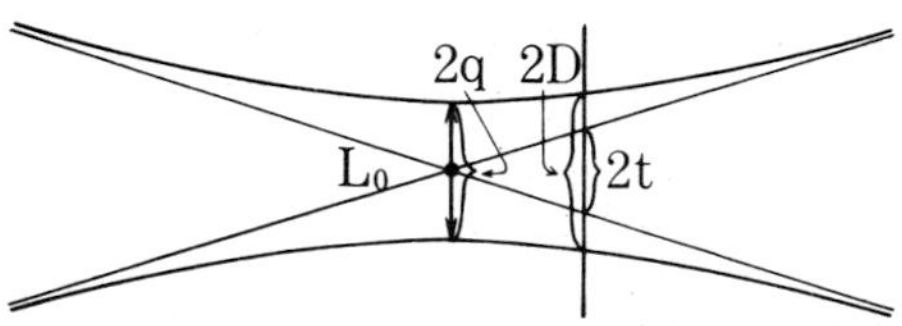

Fig. 8.4. Definition of the quantities q, D and t for two-beam diffraction.

$$\Delta\theta = \frac{2t\cos\theta_h}{\kappa\sin 2\theta_B}\,. \tag{22}$$

Then the parameters w and ξ_h are defined by

$$w = t/q\,;\qquad \xi_h = 1/2q\,. \tag{23}$$

8.5.2. Small angle approximation

If we may make the small-angle approximations common in transmission electron diffraction and put $\cos\theta_0 \approx \cos\theta_h \approx 1$ and $\sin 2\theta_B \approx 2\sin\theta_B$, we obtain

$$\xi_h = \frac{\kappa}{(v_h v_{-h})^{1/2}} = \frac{1}{2\sigma\Phi_h} \tag{24}$$

$$w = \frac{\Delta\theta\cdot\kappa}{d_h(v_h v_{-h})^{1/2}} = \frac{\Delta\theta\cdot\xi_h}{d_h}\,, \tag{25}$$

or, since $d_h = |\boldsymbol{b}|^{-1}$ and $\zeta_h = \Delta\theta\cdot|\boldsymbol{b}|$,

$$w = \xi_h\cdot\zeta_h\,. \tag{26}$$

Then the constants obtained from the boundary conditions are

$$\alpha_1 = \tfrac{1}{2}[1 + w/(1+w^2)^{1/2}]\,,$$

$$\alpha_2 = \tfrac{1}{2}[1 - w/(1+w^2)^{1/2}]\,,$$

$$C_1 = w - (1+w^2)^{1/2}\,,$$

$$C_2 = w + (1+w^2)^{1/2}\,, \tag{27}$$

and the intensities of the transmitted and diffracted beams are

$$I_0 = |\alpha_1 \exp\{\mathrm{i}\boldsymbol{k}_0^1\cdot\boldsymbol{r}\} + \alpha_2\exp\{\mathrm{i}\boldsymbol{k}_0^2\cdot\boldsymbol{r}\}|^2$$
$$= (1+w^2)^{-1}\left[w^2 + \cos^2\left\{\frac{\pi H(1+w^2)^{1/2}}{\xi_h}\right\}\right], \tag{28}$$

$$I_h = |\alpha_1 C_1 \exp\{\mathrm{i}\boldsymbol{k}_h^1\cdot\boldsymbol{r}\} + \alpha_2 C_2\exp\{\mathrm{i}\boldsymbol{k}_h^2\cdot\boldsymbol{r}\}|^2$$
$$= (1+w^2)^{-1}\sin^2\left\{\frac{\pi H(1+w^2)^{1/2}}{\xi_h}\right\}, \tag{29}$$

where in each case we have added the two components of the crystal wave which combine to form the vacuum wave. Thus the intensities of both the transmitted and diffracted beams oscillate with crystal thickness H. At the exact Bragg angle the periodicity is the extinction length, ξ_h.

8.5.3. Laue case with absorption

Here we have not excluded the absorption which would have the effect of making complex any quantity which involves v_h. It is probably better to introduce the absorption coefficients μ_0, μ_h explicitly and write these results in terms of real parameters w and ξ_h in which case (28) and (29) become

$$I_0 = \tfrac{1}{2}\exp\{-\mu_0 H\}\left[\left(1+\frac{w^2}{1+w^2}\right)\cosh\left(\frac{\mu_h H}{(1+w^2)^{1/2}}\right) - \frac{2w}{(1+w^2)^{1/2}}\sinh\frac{\mu_h H}{(1+w^2)^{1/2}} + \frac{1}{(1+w^2)}\cos\left\{\frac{2\pi H(1+w^2)^{1/2}}{\xi_h}\right\}\right] \tag{30}$$

and

$$I_h = \frac{\exp\{-\mu_0 H\}}{2(1+w^2)}\left[\cosh\left(\frac{\mu_h H}{(1+w^2)^{1/2}}\right) - \cos\left\{\frac{2\pi H(1+w^2)^{1/2}}{\xi_h}\right\}\right]. \tag{31}$$

Thus the intensities of both beams are decreased by the average absorption effect. Apart from this the effect of absorption is to prevent a completely sinusoidal oscillation of intensities even for $w = 0$. There is a background, non-oscillatory term which becomes increasingly important for large crystal thicknesses. The significance of this term will be discussed in the following chapter.

8.6. Bethe potentials

In this original paper, Bethe [1928] took into account the fact that, especially in the electron diffraction case, the condition that only two beams should exist in the crystal is never fully satisfied. Some weak beams are always present, corresponding to reciprocal lattice points for which the excitation error is reasonably large but not large enough to extinguish the contribution entirely. For a particular class of reflections, the "systematic" set, the excitation errors are constant and the "systematic interactions" of the corresponding weak waves with the strong diffracted waves are always the same for any direction of the incident beam satisfying the Bragg angle for the reflection ***h***. This sys-

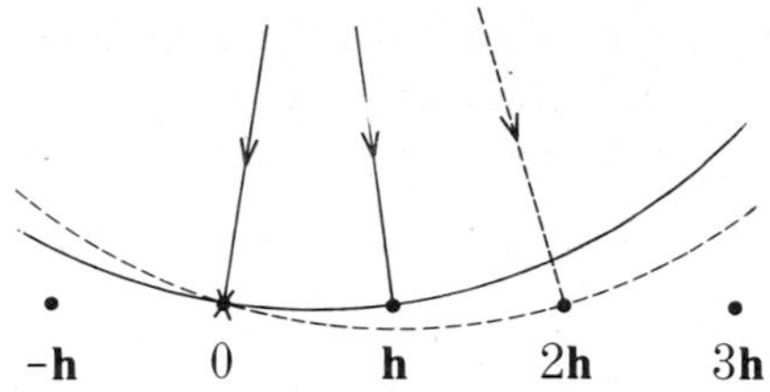

Fig. 8.5. The excitation of a systematic set of reflections. When the Bragg condition for one reflection is satisfied, the other reflections of the systematic set always have the same excitation errors.

tematic set consists of the set having integral or sub-integral multiples of the lattice vector $\boldsymbol{h}$ as suggested by Fig. 8.5.

The Bragg condition is satisfied for $\boldsymbol{h}$ for any orientation of the incident beam for which the Ewald sphere passes through $\boldsymbol{h}$ i.e., for any rotation of the plane of the figure about the O–$\boldsymbol{h}$ line. For any such orientation, the excitation errors for a reflection $n\boldsymbol{h}$ will always be the same.

As the Ewald sphere is rotated about the O–$\boldsymbol{h}$ line, it will pass through other "non-systematic" reflections at various irregular intervals. When the Ewald sphere is close to any non-systematic reflection then a definite n-beam situation will exist which must be handled accordingly. But if it is not too close to the reciprocal lattice point a weak beam will be produced which may be handled in much the same way as for the weak "systematic" set.

The amplitude of a weak beam $\mathbf{g}$ may be expressed approximately from (5) in terms of the amplitudes of the two strong beams only:

$$(K_0^2+v_0-k_g^2)\Psi_g + v_g\Psi_0 + v_{g-h}\Psi_h = 0$$

or

$$\Psi_g = -\frac{(v_g\Psi_0+v_{g-h}\Psi_h)}{\kappa^2-k_g^2}, \tag{32}$$

where we imply that the denominator must be large.

Then in reducing the set of equations (5) or (7) to give the two-beam case we include the contributions (32) to the summation in (5) and instead of (10) we obtain

$$\begin{pmatrix} \kappa_0^2-k_0^2 & U_{0h} \\ U_{h0} & \kappa_h^2-k_h^2 \end{pmatrix}\begin{pmatrix} \Psi_0 \\ \Psi_h \end{pmatrix} = 0, \tag{33}$$

where

$$\kappa_0^2 = K_0^2 + U_{00} ,$$

$$\kappa_h^2 = K_0^2 + U_{hh} ,$$

and

$$\begin{aligned} U_{00} &= v_0 - \sum_g{}'' \frac{v_g v_{-g}}{\kappa^2 - k_g^2} , \\ U_{hh} &= v_0 - \sum_g{}'' \frac{v_{g-h} v_{h-g}}{\kappa^2 - k_g^2} , \\ U_{0h} &= v_h - \sum_g{}'' \frac{v_g v_{h-g}}{\kappa^2 - k_g^2} , \\ U_{h0} &= v_{-h} - \sum_g{}'' \frac{v_{-g} v_{g-h}}{\kappa^2 - k_g^2} , \end{aligned} \tag{34}$$

and the double primes on the summations indicate that the values $g = 0, h$ are excluded.

Thus the effect of the weak beams is taken into account, to a first approximation by modifying the values for the potential Fourier coefficients v_0 and v_h in the two-beam solution. The modified potentials are often referred to as "Bethe potentials".

It is interesting to note that while, as we shall see later, the Bethe potentials give a good account of n-beam dynamical effects for some particular cases of experimental interest, the higher-order approximations generated by a repeated application of the process of (32) give no further improvement but in some cases, at least, give a much worse agreement with complete n-beam dynamical treatments.

It was pointed out by Miyake [1959] that in the limiting case of zero thickness of a crystal the use of the Bethe potentials gives the wrong result since the diffracted intensity given by (29) should be proportional to v_h^2 as for kinematical scattering and not $|U_{0h}|^2$. This limiting case has been discussed in detail by Gjønnes [1962] who showed how a consistent treatment of weak beams may be derived.

8.7. The Bragg case

When an incident beam is diffracted from planes parallel or nearly parallel to a flat surface of a crystal so large that it may be considered semi-infinite, the diffracted beams observed are those which reenter the vacuum on the same side of the crystal as the incident beam. Again we can obtain a relatively simple solution if we assume a two-beam case with only one strong reflected beam, but the results are quite different in form from the transmission case.

To treat this case we must return to the equation (15) and solve it for the given boundary conditions without making the simplifying approximations leading to equations (24) to (27), since the angles θ_0 and θ_h of the beams with the surface normal cannot be assumed small and equal. For high energy electrons these angles will approach $\pi/2$ and the cosines will usually be of opposite sign.

For reflection from a crystal surface, $\cos\theta_h$ will be of opposite sign to $\cos\theta_0$ so that the square root term in (16) and (17) will be imaginary for a range of ζ_h values between the limits

$$\zeta_h = \pm \frac{v_{0h}}{k} \left| \frac{\cos\theta_h}{\cos\theta_0} \right|^{1/2} . \tag{35}$$

Within this range the wave vectors in the crystal will have imaginary values, corresponding to strongly damped, non-propagating waves. In the absence of absorption there is total reflection within this range with the reflectivity falling off rapidly on either side as suggested in Fig. 8.6. In the presence of absorption the reflectivity is less than 100 percent and is asymmetric, as indicated.

In terms of the dispersion surface representation, the crystal surface in the reflection case is almost perpendicular to the h vector. The surface normal

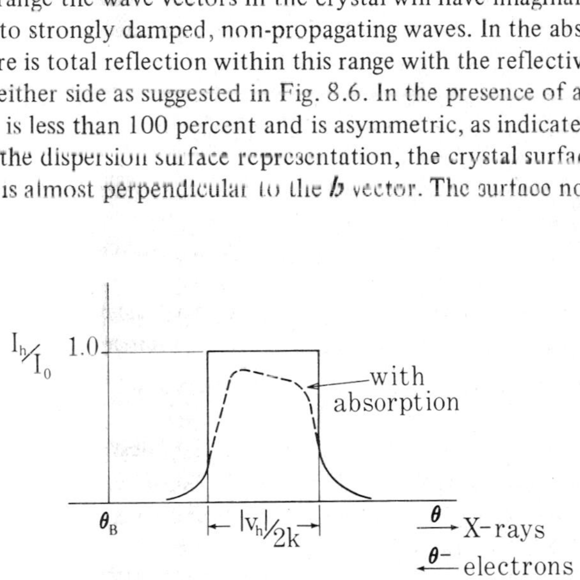

Fig. 8.6. The intensity of reflection as a function of the deviation from the Bragg angle for no absorption and with absorption (dotted line) for the Bragg case of reflection from a large perfect crystal.

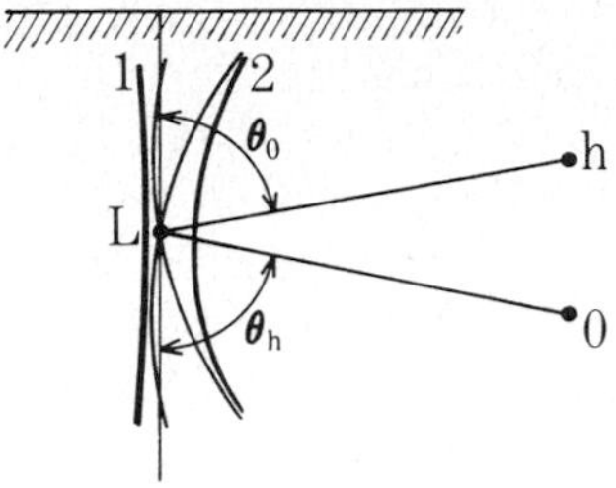

Fig. 8.7. The dispersion surface construction for the two beam Bragg case, when the surface normal may pass through the gap between the two branches.

through the Laue point L may cut the Branch 1 of the dispersion surface in two points, or it may cut Branch 2 in two points, or it may pass through the gap between the branches, thus giving imaginary components of the wave vectors corresponding to the exponentially damped waves in the crystal (Fig. 8.7). The angular range for total reflection is then given, for the symmetrical case $|\cos\theta_0| = |\cos\theta_h|$, by the width of the gap (35).

It is straightforward to include the Bethe potentials in the two-beam treatment as for the transmission electron diffraction case (see Pinsker [1964]). However it may well be that, as in the case of transmission through very thin crystals, the validity of the Bethe potentials is questionable when only a few layers of atoms are involved in the diffraction process, as is often the case for reflection electron diffraction.

For diffraction of high voltage (20 to 100 kV) electrons from surfaces the scattering angles are of the order of a few degrees. Consequently the angle made by the incident beam with the surface $(\pi/2-\theta_0)$ is of the order of one degree. The penetration of the beam into the crystal is severely limited by absorption or by the diffraction when a strong reflection is excited and may be only a few Å. Another important factor to be considered in this case is that, although the refractive index of the crystal may be only slightly greater than unity for electrons, the refraction effects will be considerable for such small angles of incidence. Waves diffracted from planes parallel to the surface having spacings as small as 2 or 3 Å may suffer total internal reflection and be unable to leave the crystal. Diffracted waves reaching the crystal at slightly larger angles will be refracted so that they are displaced by large distances in the diffraction patterns.

Because the diffraction angles are so small and the refraction effects relatively large, the reflection method of diffraction for high-energy electrons, RHEED, is extremely sensitive to slight deviations from exact planarity of the surface as well as to the composition of the topmost layers of atoms of the crystal, (Menadue [1972]; Colella and Menadue [1972]) so that extreme care must be exercised in obtaining a flat clean surface if any attempt is to be made to relate the theoretical predictions to experimental observations.

A further complication is that even elementary considerations show that a simple two-beam approximation cannot be adequate for any practical experimental situation and hence a complete n-beam dynamical treatment is necessary (Colella [1972]). Similarly the use of low energy electrons (10 to 500 eV) with near-normal incidence on the surface (LEED) offers an essentially n-beam diffraction situation, with additional complications, and so will not be considered here. An excellent summary of the theory used for LEED is given in the book by Pendry [1974].

As in other cases, however, the two beam approximation is adequate for most considerations of Bragg-case reflection of X-rays and neutrons from surfaces of large perfect crystals. The original result for a non-absorbing crystal predicting the region of total reflection, was obtained by Darwin [1914]. More complete recent discussions have been given by James [1950] and Batterman and Cole [1964]. The agreement between the two-beam theory and experimental measurements is excellent. For neutrons, see Goldberger and Seitz [1947].

CHAPTER 9

Dynamical diffraction effects

9.1. Thickness fringes and rocking curves – electron diffraction

9.1.1. Intensity formulas

We have seen in Chapter 8 that when the usual small-angle approximations of high energy electron diffraction are applied in the two-beam approximation the intensities of the transmitted and diffracted beams from a flat parallel sided crystal plate may be written in simple form, especially for the case of no absorption and centro-symmetric crystals:

$$I_0 = (1 + w^2)^{-1} \left[w^2 + \cos^2 \left\{ \frac{\pi H (1 + w^2)^{1/2}}{\xi_h} \right\} \right], \tag{1}$$

$$I_h = 1 - I_0 = (1 + w^2)^{-1} \left[\sin^2 \left\{ \frac{\pi H (1 + w^2)^{1/2}}{\xi_h} \right\} \right], \tag{2}$$

where, under these special conditions, $\xi_h = \kappa/|v_h| = (2\sigma\Phi_h)^{-1}$ and $w = \xi_h \cdot \zeta_h = \Delta\theta \cdot \xi_h/d_h$, which is a measure of the deviation from the Bragg angle. If the incident beam is exactly at the Bragg angle, $w = 0$ and

$$I_h = 1 - I_0 = \sin^2 \{\pi H |v_h|/\kappa\}\,. \tag{3}$$

This is the well-known pendellosung or pendulum solution of Ewald. The energy is passed back and forward between the incident and diffracted beams in much the same way as energy is exchanged between coupled pendulums. The coupling is provided in this case by the scattering from one beam to the other with a strength proportional to $|v_h|$.

9.1.2. Real space picture

An alternative picture of the diffraction process is provided by considering the wave-fields in real space in the crystal lattice. An incident plane wave ψ_0

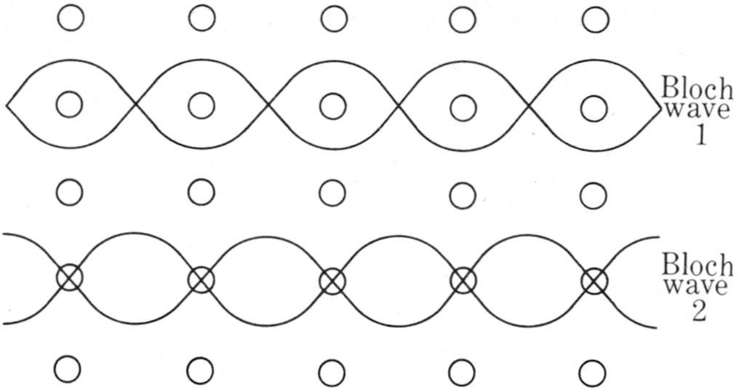

Fig. 9.1. Illustrating the relation of the two Bloch waves to the lattice planes for a two-beam approximation.

and a diffracted wave ψ_h proceeding in directions inclined at $+\theta_B$ and $-\theta_B$ to the reflecting planes will interfere to form a standing wave field having the periodicity of the lattice planes. But we have seen that under these conditions there are two solutions to the wave equation and two wave-fields or Bloch waves. These two Bloch waves correspond to the two standing solutions which have the full symmetry of the diffraction conditions in the lattice, namely one having nodes half-way between atomic planes and one with nodes on the atomic planes as suggested in Fig. 9.1. At the Bragg angle these two solutions are equally probable and have equal amplitudes. However, since the waves are in a potential field modulated by the lattice wave of amplitude φ_h, they will experience not the average potential φ_0 but the modified potential $\varphi_0 \pm \varphi_h$ where the + sign refers to Bloch wave 1.

Thus the values of the refractive index and the wave number of the two Bloch waves will differ (c.f. Fig. 8.3). After these two Bloch waves have passed through a thickness H of crystal they will differ in phase by an amount proportional to $H(\varphi_h/E_0)$. Hence, when the two contributions to the diffracted wave in the crystal are added to form the diffracted wave in vacuum they may either reinforce or cancel each other, depending on the thickness. As the crystal thickness is increased the diffracted intensity will then vary sinusoidally with a periodicity $H = \kappa/v_h = \xi_h$. The $\pi/2$ phase difference between transmitted and diffracted waves in each case ensures that when there is reinforcement for the diffracted beam there is cancellation for the transmitted beam, and the sum of the intensities of the two beams is constant.

Thus the pendulum solution may be looked upon as the result of double-refraction in the crystal, somewhat similar to, although different in principle from, the optical case of transmission of a polarized wave through an anisotropic crystal.

9.1.3. Rocking curves

The effect of a deviation from the Bragg angle ($w \neq 0$) is to reduce the average diffracted beam intensity by a factor $(1 + w^2)^{-1}$ and to reduce the periodicity of the oscillations with thickness by multiplying ξ_h by $(1 + w^2)^{-1/2}$.

For a fixed thickness the variation of diffracted intensity with deviation from the Bragg angle is best seen by putting $w = \xi_h \zeta_h$ as in (8.26) and writing (2) as

$$I_h = \left(\frac{\xi_h^{-2}}{\xi_h^{-2} + \zeta_h^2} \right) \sin^2 [\pi H(\xi_h^{-2} + \zeta_h^2)^{1/2}] \tag{4}$$

Then if ζ_h^2 is much greater than $\xi_h^{-2} = |\upsilon_h|^2/\kappa^2$, this is approximately equal to

$$I'_h = \frac{|\upsilon_h|^2}{\kappa^2} \pi^2 H^2 \frac{\sin^2(\pi H \zeta_h)}{(\pi H \zeta_h)^2} \tag{5}$$

which is exactly the expression for the kinematical diffraction from a parallel-sided plate of thickness H as a function of the excitation error ζ_h.

Thus the two-beam solution tends to the kinematical intensity formula if the strength of the interaction with the lattice, represented by $|\upsilon_h|$ goes to zero or if, for finite υ_h, the excitation error is sufficiently large.

If the angle of incidence of the electron beam is varied and the diffracted intensity is recorded to give the "rocking curve" for the reflection, the deviation from the simple $(\sin^2 x)/x^2$ curve of the kinematical theory will be most apparent near the exact Bragg condition. For $w = 0$, or $\zeta_h = 0$, the intensity will oscillate sinusoidally as given by (3) as the thickness increases instead of increasing continuously with H^2 as given by (5).

9.1.4. Extinction contours

It is not convenient in practice to record a rocking curve by rotating a thin perfect crystal in an electron beam. Instead either of two methods may be used. Firstly the bright or dark-field images of a uniformly bent parallel-sided crystal plate may be observed in an electron microscope. For a plane parallel illuminating beam the angle of incidence of the beam with the lattice planes

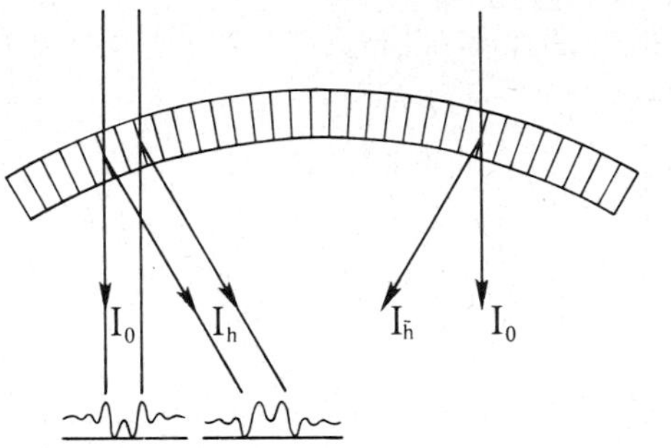

Fig. 9.2. The bright-field and dark-field images of a bent crystal show the variation of diffracted intensity as a function of angle of incidence.

then varies with distance across the crystal as suggested by Fig. 9.2. The intensity of the transmitted and diffracted beams at any point of the exit surface will depend on the excitation error for the lattice planes at that point and for uniform curvature of the crystal the excitation error will vary almost linearly across the crystal. Then if the diffracted beams are imaged, the intensity in the resulting image will show the variation given by (4) as a function of distance across the crystal which takes the place of ζ_h. Imaging of the direct transmitted beams, I_0, will give the bright-field image which in the present case with absorption neglected will show the intensity variation $1 - I_h(\zeta_h)$.

Thus the image of a bent crystal will be traversed by "extinction contours" which are dark in bright field and bright in dark field and have the form of strong and subsidiary weak fringes marking the parts of the crystal which are in the correct orientation to give a diffracted beam.

An image of a crystal, bent in two dimensions, is shown in Fig. 9.3. The extinction contours tend to occur in sets of parallel lines corresponding to the sets of positive and negative higher orders of a strong reflection, i.e. the "systematic" sets such as ... $-3h$, $-2h$, $-h$, ... h, $2h$, $3h$ ... . The fact that the sets of parallel lines do not in general show the form of the simple contours suggested by the two beam theory and illustrated in Fig. 9.2 is partly a consequence of the omission of the effect of absorption from the two beam theory given above, and is due, in part, to the occurrence of n-beam dynamical diffraction effects which are specifically excluded from the two-beam treatment. Further evidence of n-beam dynamical scattering effects, in this case involving "non-systematic" interactions, is given at points where non-parallel extinction contours intersect. Here the intensities of the contours are by no means additive but may fluctuate wildly.

Fig. 9.3. Electron micrograph of a thin crystal of gold, showing extinction contours related to the bending of the crystal in two dimensions. The center of the "star" corresponds to the [100] direction.

9.1.5. Convergent beam diffraction

A second means for observing the rocking curves for diffraction from a thin parallel sided crystal is by use of the convergent beam diffraction technique originally introduced by Kossel and Mollenstedt [1939]. Here an electron beam of finite aperture is focussed on the specimen. If the beam is defined by a circular aperture, each spot of the diffraction pattern is spread into a circular disc as suggested by Fig. 9.4(a). For each angle of incidence represented in the

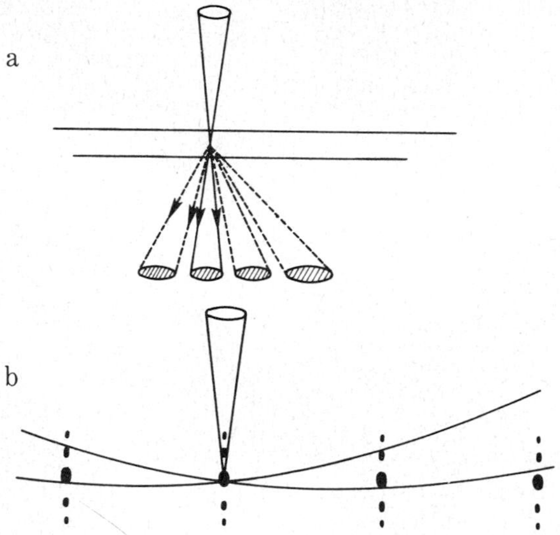

Fig. 9.4. The formation of a convergent-beam diffraction pattern illustrated (a) in the real-space configuration and (b) in reciprocal space where the Ewald sphere for each incident direction produces its own section of the scattering power distribution.

convergent beam the appropriate Ewald sphere may be drawn and the diffraction intensities found, as suggested in the (kinematical) reciprocal space diagram, Fig. 9.4(b). The variation of intensity across the central spot and diffracted spot discs in the diffraction pattern will then give the variation of intensities with angle of incidence. Fig. 9.5 gives a typical convergent beam diffraction pattern from a thin perfect crystal of MgO.

For a two-beam case, the incident and diffracted beam discs will be crossed by parallel fringes of intensity given by equation (4) for I_h and $I_0 = 1 - I_h$. On this basis MacGillavry [1940] interpreted the convergent beam patterns obtained by Kossel and Mollenstedt from thin mica crystals and from the form of the intensity variations was able to deduce values of $|v_h|$ for the various reflections by use of equation (4) or the equivalent. The values she derived were, fortuitously, in excellent agreement with the values calculated from the structure of mica as then known.

The convergent beam diffraction technique has more recently been refined by Hoerni [1950] and further by Goodman and Lehmpfuhl [1964] and Cockayne et al. [1967]. As will be recounted later (Chapter 14) these last two groups of authors have developed the technique as a means for making accu-

Fig. 9.5. Convergent beam diffraction pattern for a thin perfect crystal of MgO showing the variation of intensity with angle of incidence for the $2h, 0, 0$ reflections. Some perturbation of the fringes by non-systematic interactions is visible.

rate measurements of intensities, demonstrating that the two-beam results form only a first approximation to the n-beam diffraction situation and making use of the n-beam dynamical diffraction effects to derive highly accurate values for the Fourier coefficients v_h or Φ_h.

9.1.6. *Diffraction and imaging of crystal wedges*

Diffraction from a wedge-shaped crystal can be treated, with reasonable approximations, by an extension of the treatment of transmission through a parallel-sided crystal. This has been done in detail by Kato [1952]. The boundary conditions for an exit face not parallel to the entrance face are imposed to give the wave entering the vacuum. From elementary principles we can say that since the crystal has two values for the effective refractive index corresponding to the potential coefficients $v_0 \pm v_h$, both incident and diffracted waves will be refracted by the "prism" of crystal to give two waves coming out in slightly

Fig. 9.6. Near-sinusoidal thickness fringes in the dark-field image of a silicon crystal wedge. The image is obtained using the [111] reflection when the Bragg condition is satisfied for the [222].

different directions. Then each spot in the diffraction pattern will be split into a pair of two close spots. At the exit face of the crystal the two wave fields will interfere to give a sinusiodal variation of intensity with thickness when viewed with either the transmitted or diffracted beams. Thus electron microscope images in either bright or dark field will show a pattern of sinusoidal fringes across the image of the wedge, as in Fig. 9.6.

An alternative explanation of this effect which requires further justification but gives essentially the correct result, is based on the assumption that for each thickness of the wedge the intensities of the incident and diffracted beams will be exactly the same as for a parallel-sided crystal of the same thickness. Then the variation of intensity across the bright or dark-field image of the wedge will be given by the equation (1) and (2) and the distance in the image is related to the thickness H through the wedge angle and the orientation of the crystal with respect to the beam.

These thickness fringes were first observed in the images of MgO smoke crystals which form almost perfect cubes and so present up to six wedge-shaped regions to the incident beam (Heidenreich [1942]), (Kinder [1943]). The corresponding splitting of the diffraction spots due to the double-refraction effect was observed in diffraction patterns of MgO smoke by Cowley and Rees [1946, 1947] and Honjo [1947]. The six wedge-shapes regions of an MgO cube give rise to a star-shaped group of six radial pairs of spots surrounding the flat-crystal diffraction spot position. In recent years a more detailed study of the intensity distributions in images of wedge crystals has shown that strong deviations from the simple sinusoidal intensity variations of the two-beam frequently occur. Accurate measurements of the intensity distributions have shown excellent agreement with n-beam dynamical calculations and have been used as a means for deriving values for the Fourier coefficients of potential v_h or Φ_h (for a review see Cowley [1969]).

Similarly the more thorough study of the refraction fine structure of diffraction spots by Moliere and Wagenfeld [1958] and Lehmpfuhl and Reissland [1968] have shown that a multiplicity of spots, rather than only two may be produced by wedge. The latter authors recorded the splitting as a function of crystal orientation, or ζ_h, and so could map out the lengths of the various wave vectors in the crystal and then plot directly the form of the dispersion surface, showing a number of branches.

9.1.7. Absorption effects for wedges

With the introduction of absorption into the two beam solution the intensities of the incident and diffracted beams transmitted through a thin crystal are modified as indicated in equations (8.30) and (8.31). There is an over-all loss of intensity of both beams due to the mean absorption coefficient μ_0. Then a non-oscillating term is added which gives a background to the sinusoidal oscillations. For simplicity we consider the special case that the Bragg condition is exactly satisfied, i.e. $w = 0$. Then

$$I_0 = \tfrac{1}{2} \exp\{-\mu_0 H\} \left[\cosh \mu_h H + \cos\left(\frac{2\pi H}{\xi_h}\right)\right], \tag{6}$$

$$I_h = \tfrac{1}{2} \exp\{-\mu_0 H\} \left[\cosh \mu_h H - \cos\left(\frac{2\pi H}{\xi_h}\right)\right]. \tag{7}$$

These intensities are sketched as a function of thickness in Fig. 9.7 which thus represents the form to be expected for the thickness fringes in the images of a wedge crystal and is seen to give something like the intensity distribution

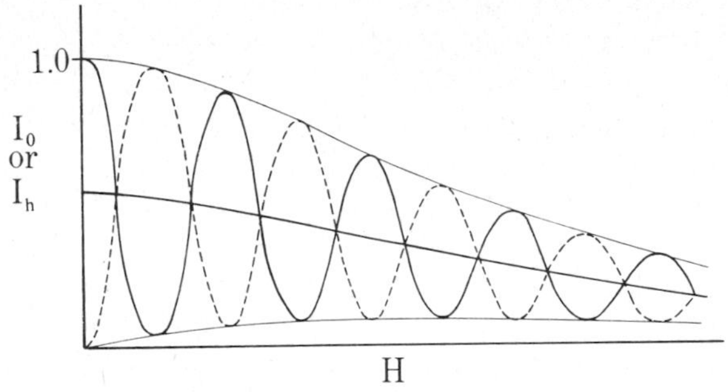

Fig. 9.7. Diagram suggesting the form of the intensity variation of incident and diffracted beams for a two-beam case with absorption and a wedge crystal of thickness H.

of the fringes in Fig. 9.6. The first term in each of (6) and (7) gives the non-oscillating "center curve" of the profile as

$$\tfrac{1}{4}[\exp\{-(\mu_0+\mu_h)H\}+\exp\{-(\mu_0-\mu_h)H\}]\ .$$

Of these two terms the first will die out more rapidly if μ_h has the same sign as μ_0. Then for H large the center curve is $\exp\{-(\mu_0-\mu_h)H\}$. This decreases more slowly with H than the amplitude of the oscillating part which is proportional to $\exp\{-\mu_0 H\}$.

In principle then it is possible to deduce the values of μ_0 and μ_h from the form of the thickness fringes. Attemps to do this by Uyeda et al. [1965] gave some results but showed deviations from the two-beam formulas.

In terms of the picture of Bloch waves transmitted through the crystal, Fig. 9.1, it can be seen that, if the absorption process takes place when electrons pass close to the atoms, the two Bloch waves will be absorbed differently. Since the probability of finding an electron at a certain position is proportional to the square of the wave function modulus, the electrons of Bloch wave 1 will be more likely to be in the vicinity of atoms and so will be more strongly absorbed, with an absorption coefficient $\mu_0+\mu_h$, while those of Bloch wave 2 will spend most of their time between the planes of atoms and so be less absorbed, with an absorption coefficient $\mu_0-\mu_h$. Then when they leave the crystal the contributions from the two Bloch waves to the diffracted beam will not have equal amplitude and so will not be able to produce interference fringes of maximum contrast.

We saw before that the two Bloch waves have different refractive indices in the crystal and so give vacuum waves in slightly different directions from a wedge shaped crystal. With absorption, the diffraction spot will be split into two components of unequal intensity. The wave refracted most, from Bloch wave 1, will also be attenuated most. This was first observed and analysed by Honjo and Mihama [1954].

Absorption also has the effect of making the rocking curve for a crystal in the transmitted beam, i.e. the extinction contour for the bright field image of a bent crystal, no longer symmetric. This is readily seen from the presence of the asymmetric sinh term in equation (8.27). It follows also from the dispersion curve picture of Fig. 8.3. There we stated that at the Bragg angle both branches of the dispersion surface are equally excited, but for an incident beam such that L is to the left of L_0, branch 1 predominates and on the other side of the Bragg angle branch 2 will predominate. Since absorption makes the two branches non-equivalent, the intensities will be different on the two sides of the Bragg angle.

The asymmetry will be reversed when we go from the $\boldsymbol{h}$ to the $\bar{\boldsymbol{h}}$ reflection. Hence we see in the pattern of extinction contours, Fig. 9.3, that the region between two strong parallel, $\boldsymbol{h}$ and $\bar{\boldsymbol{h}}$, fringes tends to be darker than the region outside of them.

9.2. Dynamical effects of X-ray and neutron diffraction

9.2.1. Techniques for X-ray diffraction

Apart from the polarization factor which appears in (8.14) and the inapplicability of the small angle approximation, it would appear that the dynamical effects for X-ray and neutron diffraction should be exactly the same as for electron diffraction and, in fact, thickness fringes have been observed for both X-rays and neutrons. However differences in the properties of the radiations and the conditions for the experimentally feasible observations introduce complications which are not involved in the electron diffraction situations.

For electrons we consider a small crystal bathed in incident radiation which can be so well collimated that it can be approximated by a plane wave. The thickness fringes which have periodicities of the order of hundreds or sometimes thousands or Å are viewed with an electron microscope. The diffraction angles are so small that the difference in path of incident and diffracted beams through the crystals can often be neglected. For X-rays the same possibilities for collimation of the beam and formation of the image by microscope methods are not available. The thickness fringe periodicities are of the order of hundreds of microns and so can be recorded directly on photographic film and then enlarged optically, but the specimens are of the order of millimeters in dimension

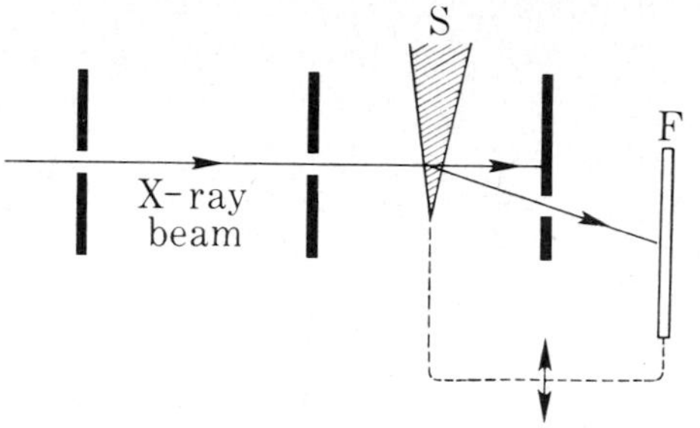

Fig. 9.8. The method used to obtain X-ray "projection topographs". The sample, S, in this case a wedge crystal, and the film F are rigidly coupled and move back- and forward together.

and it is not practicable to produce well-collimated beams of reasonable intensity over such areas. Two special experimental arrangements have been devised. The equipment introduced by Lang [1959] for obtaining X-ray "projection topographs" of crystals for the study of dislocations and other crystal defects may be employed. The principle of this is illustrated in Fig. 9.8. A set of fine slits define the X-ray beam incident on the sample S and the beam is diffracted at a given angle. The sample S and film F are coupled and moved back and forward uniformly together. The film then records the variation of diffracted intensity as a function of position on the specimen and for a wedge-shaped crystal this shows thickness fringes (Kato and Lang [1959]). In the second dimension, along the length of the slits, although the beam diverges from the source this does not change the angle of incidence on the diffracting planes

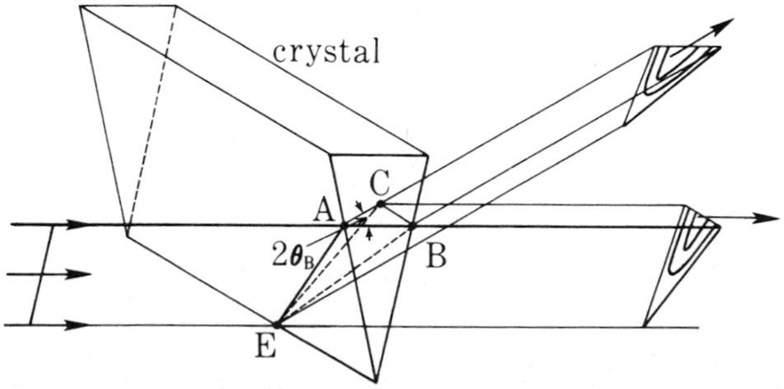

Fig. 9.9. The production of a "section topograph" when a thin, flat collimated X-ray beam intersects a wedge-shaped crystal.

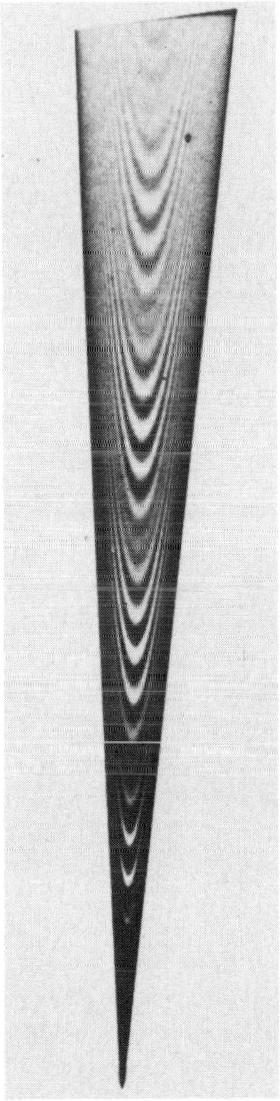

Fig. 9.10. Reproduction of a "section topograph" obtained from a silicon wedge using a 400 reflection by the method of Fig. 9.9. (From Kato [1969]: after T. Kajimura).

appreciably. Hence in this direction the diffraction conditions are uniform. The resolution in the "topograph" depends on the source size in relation to the distances from source to specimen to film.

The second type of experiment introduced by Kato and Lang [1959] is that of the "section topograph" in which a thin flat, well-monochromated X-ray beam cuts across a wedge-shaped crystal at a considerable angle to the edge, as suggested in Fig. 9.9. The incident beam strikes the crystal along the line AE. The direct transmitted beam leaves the crystal along the line EB, but after being diffracted through an angle of $2\theta_B$ and back into the incident beam direction, the beam leaving the crystal in that direction comes from the whole of the triangle EBC. Similarly the diffracted beam comes from the whole of the triangle EBC. Hence a photographic film placed after the crystal shows two triangles of intensity. For the electron diffraction case one would expect these triangles to be crossed by equal thickness fringes parallel to the crystal edge. Instead they show the hook-shaped fringes which can be seen in the reproduction of a section topograph from a silicon wedge shown in Fig. 9.10.

9.2.2. Energy flow

The above results serve to emphasize the difference in the experimental situation for X-rays and electrons. Firstly, because the diffraction angles are large and the width of the beam is small compared with the crystal dimensions, the path of the beam through the crystal can be traced in detail. As can be seen in Fig. 9.10, for thin crystals the beam intensity is spread fairly evenly over the possible range of directions, $\boldsymbol{k}_0$ to $\boldsymbol{k}_h$, in the crystal. For thicker crystals the energy is concentrated in a particular direction. This is the direction of energy flow, given by the Poynting vector. It was shown by von Laue [1952] that the Poynting vector is

$$\boldsymbol{S} = \boldsymbol{k}_0 |\boldsymbol{E}_0|^2 + \boldsymbol{k}_h |\boldsymbol{E}_h|^2 , \tag{8}$$

where $\boldsymbol{E}_0$ and $\boldsymbol{E}_h$ are the amplitudes of the electrical field vector in the crystal, equivalent to the scalar amplitudes ψ_0 and ψ_h.

Thus at the exact Bragg angle, when the amplitudes of incident and transmitted beams are equal, the direction of energy flow is half-way between $\boldsymbol{k}_0$ and $\boldsymbol{k}_h$ i.e. it is parallel to the diffracting planes. Then for a thick crystal the path of the X-ray beam may be sketched as in Fig. 9.11. With increasing deviation from the Bragg angle the direction of energy flow reverts to that of the incident beam.

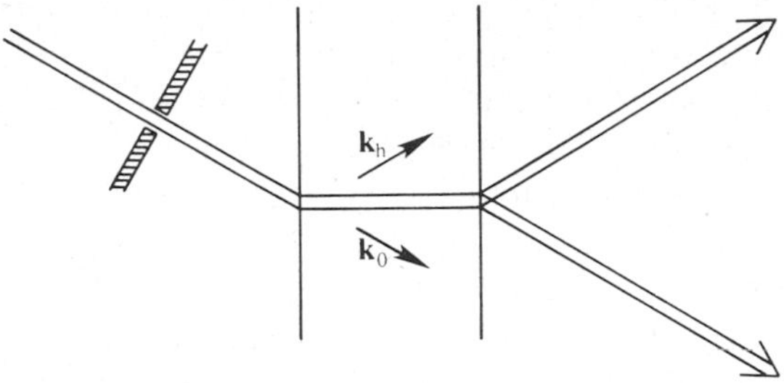

Fig. 9.11. Energy flow and diffracted beam configurations for a narrow X-ray beam incident on a thick perfect crystal.

9.2.3. Dispersion surface picture

A further difference from the electron case arises from the relative strengths of the interaction with the crystal. In terms of the dispersion surface representation we have seen that the deviation of the dispersion surface from the two intersecting spheres of Fig. 8.3 depends on this strength of the interaction since, for example, the anpassung at the Bragg angle is, from (8.12), equal to $|v_h| \cos \theta_h / 2\kappa$. For the Bragg case of reflection from the surface of a large crystal, for example, the angular width of rocking curve for the reflection is equal to twice this quantity. For X-rays this is of the order of 10^{-5} radians or less whereas for electrons it is of the order of 10^{-2} radians.

On the other hand the incident electron beam may be collimated to have an angular spread of 10^{-5} radians or less but for X-rays the divergence of the radiation from each point of the source gives a variation of incident angle across the illuminated region of the specimen (about 20 μm wide) of about 10^{-4} radians. Thus for electrons the plane wave approximation is good, but for X-rays one must consider a coherent spherical wave from each point of the source, with a variation of incident angle considerably greater than the angular width of the Bragg reflection. Then in the dispersion surface picture it is not possible to consider a single incident direction defining two tie-points on the two branches of the surface as in Fig. 8.3. Instead it must be considered that the whole region of the dispersion surface around L_0 is excited simultaneously and coherently. This situation was realized by Kato and Lang [1959] and Kato [1961] showed how to integrate over the spherical wave front and obtain expressions giving a reasonable account of the features of the section topographs. The intensity of the thickness fringes obtained in the projection topographs is then derived by integration of the section topograph along equal-thickness lines.

A further complication of the X-ray diffraction situation is indicated by the periodic fading of the fringes in the section topograph, Fig. 9.10. This has been shown to be a result of polarization effects. The two components of the displacement vector of the electro-magnetic field, polarized parallel and perpendicular to the diffraction plane, have different wave vectors and amplitudes in the crystal as a result of the factor P in the dispersion equations (see equation (8.14)). Hence two independent sets of fringes having slightly different periodicities are produced. The periodic modulation of the fringe contrast then results from the beating together of these two periodicities.

The spherical wave theory of Kato permits the interpretation of the X-ray thickness fringes with sufficient accuracy to allow them to be used for the determination of structure factors with high accuracy. Only in exceptional circumstances is there any complication arising from the limitations of the two-beam approximation since n-beam diffraction situations can be readily avoided. The main limitation of the method arises from the need to use crystals which are very nearly perfect, with no dislocations and no distortion. Such crystals can be obtained for silicon and germanium but for very few other substances. A summary of the methods and results has been given by Kato [1969] and Pinsker [1978].

9.2.4. Neutron diffraction

While neutron diffraction experiments may be devised to parallel those with X-ray diffraction, the most successful experiments with neutrons have been those of Shull [1968] which were designed to take advantage of the particular properties of the neutron source. The beam from a nuclear reactor is "white radiation" with a smooth distribution of intensity with wavelength. In the absence of the strong monochromatic "characteristic" lines of the X-ray spectrum it is wasteful of the rather scant supply of neutrons to make use of a strongly monochromated beam. Instead Shull used the wavelength as the variable for this experiments. With an experimental arrangement in which the beam paths were restricted to those of Fig. 9.11, the wavelength diffracted was varied by varying the angle of incidence. Then because the extinction length (8.21) is a function of wavelength, the transmitted intensity fluctuated as the number of extinction lengths in the crystal changed with crystal orientation. The results showed very good agreement with theory.

9.3. Borrmann effect

The anomalous transmission of X-rays through crystals, first observed by Borrmann [1941, 1950] is, at first sight, a surprising phenomenon: the transmission of a thick perfect crystal may increase by a very large factor when the crystal is rotated into the correct orientation for a strong Bragg reflection. This is in contrast with the thin crystal case for which the intensity of the transmitted beam is decreased when energy is taken from it to form a diffracted beam.

The phenomenon is a consequence of the effects of absorption on dynamical diffraction and can be understood to a first approximation by an extension of our discussion of absorption effects in 9.1.7. The two Bloch waves, as suggested by Fig. 9.1, have different absorption coefficients since for Bloch wave 2 the electrons pass between the rows of atoms while for Bloch wave 1 they are travelling mostly in the vicinity of the atoms and so have a higher probability of absorption. From equations (6) and (7) the oscillatory part of the transmitted beams in both the incident and diffracted directions is decreased by the exponential factor $\exp\{-\mu_0 H\}$ while the non-oscillatory part of both in composed of the two parts, having effective absorption coefficients $\mu_0 \pm \mu_h$. With increasing thickness the part with largest absorption coefficient will die away first, then the oscillatory part, so that for sufficiently large thicknesses only the part with absorption coefficient $\mu_0 - \mu_h$ will remain. The intensity in the incident and diffracted beam directions will then be the same. Provided that μ_h is an appreciable fraction of μ_0, the intensity of each may easily exceed that of a beam in a non-diffracting orientation for which the absorption coefficient is μ_0.

For X-rays the absorption process is highly localized since it comes mostly from the excitation of electrons from the inner shells of atoms. Thus the Fourier transform of the absorption function will die away very slowly with distance from the origin of reciprocal space, and the value corresponding to the diffracted beam direction μ_h may be very little less than that for the forward direction μ_0.

The complications of the X-ray case by the vector nature of the amplitudes and the polarization effects modify the simple scalar result. The absorption coefficients for the two wave fields are given by (Batterman and Cole [1964])

$$\mu = \mu_0 \left(1 \pm P \frac{2|\boldsymbol{E}_0|\,|\boldsymbol{E}_h|}{E_0^2 + E_h^2} \frac{\mu_h}{\mu_0} \right) . \qquad (9)$$

The two values of P for the two polarizations imply that there will be four portions of the beam having different absorption coefficients. The part with the smallest absorption coefficient may have an absorption coefficient which is very small compared with μ_0 and may give a beam traversing a large thickness of crystal and being almost perfectly polarized.

For example, for transmission of Cu K_α radiation through a crystal of Ge of thickness 1 mm, the absorption without any diffraction gives an attenuation of $\exp\{-38\}$. At the Bragg condition for the (220) reflection the ratio μ_h/μ_0 is 0.95 and the attenuations on the four parts of the beam are

σ polarization, branch 2: $\exp\{-1.9\}$,

π polarization, branch 2: $\exp\{-12.5\}$,

π polarization, branch 1: $\exp\{-63.5\}$,

σ polarization, branch 1: $\exp\{-74\}$.

Hence virtually all the transmitted energy comes from the one branch with σ polarization. The enhancement of transmission relative to the non-diffracting crystal is by a factor of $(1/4)\exp\{+36.1\}$.

Because of the smallness of the angular range over which X-ray reflections take place for perfect crystals, the Borrmann effect transmission gives X-ray beams which are very well collimated as well as being almost perfectly polarized. This has provided the incentive and basis for a considerable expansion of the possibilities for experiments involving precision measurements on near-perfect crystals and a means for a more complete study of the diffraction, absorption and scattering processes of X-rays.

For electron diffraction the absorption comes mostly from plasmon excitation, which contributes to μ_0 only, and thermal diffuse scattering which gives a μ_h value falling off moderating rapidly with scattering angle (Hall and Hirsch [1968]). The ratio of μ_h to μ_0 is usually small and the Borrmann effect not very striking.

One interesting consequence of the variation of absorption processes under Borrmann diffractions conditions is that changes in absorption will in turn affect the intensity of any secondary radiation produced as a result of the absorption. Thus when electrons are strongly diffracted in a thick crystal there is a variation of the X-ray emission with angle of incidence (Duncumb [1962]; Miyake et al. [1968]). Likewise Knowles [1956] showed that for neutron diffraction the establishment of standing wave fields which reduced the density the neutrons in the vicinity of the nuclei of the atoms have the effect of decreasing the inelastic interaction of the neutrons with the nuclei which give rise to a γ-emission. For X-rays the corresponding observation made by Batterman [1962, 1964] is that the intensity of secondary fluorescence emission from the atoms can be used as an indication of the electric field strength of the incident X-radiation on the atomic planes. A number of further observations of related nature involving Kikuchi- or Kossel-type patterns, especially with electrons, have been reported recently and will be discussed in Chapter 14.

Problems

1. A cube-shaped crystal of MgO is oriented to give a 220 reflection with the incident beam very nearly parallel to the body diagonal of the cube. Find the form and dimensions of the group of fine-structure spots formed around the normal diffraction spot position assuming a two-beam dynamical approximation and taking $\varphi_0 = 13.5$ volts, $\varphi_{220} = 5.0$ volts. How will the form of the group of spots vary with deviation from the Bragg condition? (See Cowley, Goodman and Rees [1957].)

2. Find the intensity distribution of the convergent beam diffraction pattern from a flat plate-shaped crystal of MgO set to give the 220 reflection when the thickness of the crystal is equal to (a) one half of the extinction distance and (b) twice the extinction distance. Assume the two beam approximation with no absorption.

3. Find how the intensity distributions in problem 2 will be modified in the presence of absorption if $\mu_{220}/\mu_0 = 0.2$. Find the sum of the transmitted and diffracted beam intensities and hence the energy lost by absorption as a function of incident angle. Hence, on the assumption that a constant fraction of the absorption gives rise to the emission of soft X-rays, find the variation of X-ray emission with orientation of the incident electron beam.

4. A beam of electrons of energy 50 keV strikes a flat surface of a perfect single crystal of MgO at a grazing incidence of a few degrees. Assuming a mean inner potential $\varphi_0 = 13.5$ volts find an expression for the change of angle of the incident beam when it enters the crystal. If the surface is parallel to the 220 plane, find the orientation of the incident and diffracted beams, outside of the crystal, for the exact Bragg condition for 220 and 440 reflections. Assuming $\varphi_{220} = 5.0$ volts and $\varphi_{440} = 1.6$ volts find the angular width of the rocking curve for these reflections, assuming the two-beam approximation.

CHAPTER 10

Extension to many beams

10.1. Dynamical *n*-beam diffraction

We have seen that for X-ray diffraction from a perfect crystal the angular width of a reflection is of the order of 10^{-5} radians. For a crystal of simple structure the angular separation of reflections is of the order of 10^{-1} radians. The occurrence of more than one reflection at a time is a very special case which, experimentally, must be sought with particular care. When three-beam or many-beam effects do occur they give some striking modifications of intensity, as seen in the Borrmann effect studies made with wide angle convergent beams (e.g. Borrmann and Hartwig [1965]). A theoretical study of the three-beam case has been made by Ewald and Heno [1968] (also Heno and Ewald [1968]).

For electron diffraction, as we have seen, the simultaneous appearance of more than one diffracted beam is the rule rather than the exception. If special care is taken to choose an appropriate orientation it is possible that as much as 99 per cent of the energy of the Bragg beams may be contained in two strong beams for a thick crystal. On the other hand for thin crystals in principal orientations, such as that giving the diffraction pattern, Fig. 6.3 or Fig. 13.4(c), the number of beams occuring simultaneously may amount to several hundred. Because of the strong scattering of electrons by all but the lightest of atoms, the amplitudes of the scattered beams from even a thin crystal may be sufficiently strong to ensure that there is considerable multiple coherent scattering and hence we must deal with the dynamical theory for a very large number of diffracted beams. With the advent of high-voltage electron microscopes, with accelerating voltages up to 1 MeV or more, the occurrence of patterns having a very large number of diffracted beams has become even more common (Uyeda [1968]). The wave-length decreases even more rapidly with accelerating voltage than would be inferred from the non-relativistic relationship, since the relativistic form is

$$\lambda = \left\{ \frac{2m_0 eE}{h^2} \left(1 + \frac{eE}{2m_0c^2}\right)\right\}^{-1/2} . \qquad (1)$$

On the other hand the interaction constant σ is given by

$$\sigma = \frac{\pi}{E\lambda} \frac{2}{1 + (1 - \beta^2)^{1/2}} = \frac{2\pi m_0 e\lambda}{h^2} \left(1 + \frac{h^2}{m_0^2 c^2 \lambda^2}\right)^{1/2} \tag{2}$$

where $\beta = v/c$. As the voltage increases the value of σ decreases more and more slowly, tending to a limiting constant value.

With the decrease of wavelength the Ewald sphere becomes more nearly planar so that it may intersect a greater number of the extended reciprocal lattice points. At the same time the strength of interaction with atoms and hence the strength of dynamical interactions ceases to decrease. As a result, the number of interacting beams increases rapidly with accelerating voltage.

Hence the need for n-beam dynamical treatment arises most urgently for electrons and because of the variety of observed effects and theoretical complications, a number of different approaches have been applied. Most of these take advantage of the fact that for high voltage electrons the simplifying approximation of small angle scattering is possible. Hence in this chapter we discuss n-beam diffraction theory entirely from the point of view of high-voltage electrons.

One of the most powerful and effective approaches to the n-beam diffraction problem is the so-called physical optics approach, based on the type of approach to diffraction contained in our first few chapters. Since this has been expounded in print to only a limited extent we reserve a special chapter, the next, to its description. In the present chapter we treat other approaches, some more widely known and used, which follow more immediately from the treatment of the last two chapters.

The application of the Born series method by Fujiwara [1959] involves the evaluation of the series of equations (1.17) and (1.22), where the scattering potential is that of a flat plate-like crystal, with

$$\varphi(\boldsymbol{r}) = c \sum_h v_h \exp\{2\pi i \boldsymbol{h} \cdot \boldsymbol{r}\}$$

within the limits $0 \leqslant z \leqslant H$.

This leads to the same type of general expressions for the n-beam diffraction amplitudes as will be derived by other methods. Because of the notoriously poor convergence of the Born series, this approach has not led directly to practical means for calculating intensities. Hence, while referring the reader to Fujiwara's interesting papers we do not offer any more detailed discussion.

When more than two beams occur in a crystal, the intensities in diffraction patterns and in images will depend on both the magnitudes and relative phases of the structure amplitides. This follows because in coherent interference effects it is the amplitudes, not the intensities of the waves which are added. If the dynamical scattering process could be inverted, the structure amplitudes and hence the atom positions could be deduced directly and unambiguously from the diffraction data.

It seems that in principle it should be possible to invert the dynamical diffraction process. By using the variables of incident beam direction, specimen thickness and wavelength of the radiation it should be possible to obtain sufficient experimental data to over-determine the problem. So far no general solution has been found.

A major step in this direction, however, is the demonstration that for a centrosymmetric crystal, a complete inversion of the diffracted intensity distribution is possible for the three-bean case (Moodie [1979]). If convergent beam diffraction patterns are used, so that incident beam orientation is a variable, it may be shown that within the three beam pattern it is possible to distinguish what are essentially two-beam intensity distributions along particular lines and from the positions of these unique lines it is possible to derive the magnitudes of v_h, v_g, v_{h-g} and the sign of the product $v_h v_g v_{h-g}$.

Extensions of this type of solution to include more beams or structures not having a center of symmetry are difficult and may well involve the introduction of additional experimental variables.

10.2. Extension of Bethe theory – transmission

10.2.1. Matrix formulation

From the starting point of the Schrödinger equation for an electron in a periodic potential, equation (8.1), we derived the general matrix equation (8.7) relating the amplitudes Ψ_h and wave vectors $\boldsymbol{k}_h$ of the wave-field set up when an incident beam enters a crystal.

If we include N reciprocal lattice points there will be $2N$ solutions given by the condition that the determinant of the matrix is zero. The second order terms on the diagonal of the $N \times N$ matrix create the difficulty. In general it can be shown that the $2N$ solutions are given by use of a $2N \times 2N$ matrix, having only first order terms, in an equation of the form

$$\det\left|\begin{pmatrix} \boldsymbol{B} - x\boldsymbol{I} & -\boldsymbol{Q} \\ \boldsymbol{I} & \boldsymbol{O} - x\boldsymbol{I} \end{pmatrix}\right| = 0$$

where $\boldsymbol{B}$, $\boldsymbol{Q}$, $\boldsymbol{I}$ and $\boldsymbol{O}$ are $N \times N$ matrices; $\boldsymbol{I}$ and $\boldsymbol{O}$ being the identity and null matrices (Collela [1972-; Moon [1972]).

The $2N$ eigenvalues correspond to waves travelling both forward and backward in relation to the incident beam. If as in the case of high-energy electron scattering we can make the assumption that back-scattering is negligible, N of the solutions can be ignored. The problem is simplified by reducing the diagonal terms to the first order by taking only the one value of $\kappa^2 - k_h^2$ which corresponds to a wave almost parallel to the incident beam, and making a small angle approximation.

The cosines of the diffracting angles are put equal to unity. Since in general the Fourier coefficients of the crystal potential, Φ_h, are very much smaller than the accelerating potential of the incident beam, the anpassung values ξ^i will be very much less than κ. Hence, neglecting second order terms in ξ^i we may write, from (8.9),

$$k_h = \kappa - \zeta_h - \xi \,, \qquad (3)$$

and

$$\kappa^2 - k_h^2 = 2\kappa\xi + 2\kappa\zeta_h - \zeta_h^2 \equiv - x + p_h \,, \qquad (4)$$

where we have put

$$x = -2\kappa\xi \qquad (5)$$

$$p_h = 2\kappa\zeta_h - \zeta_h^2 \,, \qquad (6)$$

so that the various values of x correspond to the values of ξ^i for the various Bloch waves while the values for p_h depend on the excitation errors for the various reciprocal lattice points and are the same for all Bloch waves.

The matrix equation (8.7) can then be re-written in the simplified, linear form

$$\boldsymbol{M\Psi} = x\boldsymbol{\Psi} \qquad (7)$$

where

$$\boldsymbol{M} = \begin{pmatrix} \cdots p_0 & \cdots v_{0h} & \cdots v_{0g} & \cdots \\ \vdots & \vdots & \vdots & \\ v_{h0} & \cdots p_h & \cdots v_{hg} & \cdots \\ \vdots & \vdots & \vdots & \\ v_{g0} & \cdots v_{gh} & \cdots p_g & \cdots \\ \vdots & \vdots & \vdots & \end{pmatrix} \tag{8}$$

Then the eigenvalues x^i give the values of ξ^i for the different Bloch waves. The eigen vectors $\boldsymbol{\Psi}^i$ have components $\Psi_0^i \ldots \Psi_h^i \ldots$, which are the amplitudes of the individual plane waves which make up the Bloch waves.

In the absence of absorption, the matrix $\boldsymbol{M}$ is Hermitian. For a centro symmetric structure it is real, symmetric.

10.2.2. Small angle approximation

Frequently, because of the short wavelength of the electrons and consequent flatness of the Ewald sphere, the only reflections which are important are those corresponding to reciprocal lattice points in one plane almost perpendicular to the incident beam. For these conditions, as pointed out by Fujimoto [1959], the solution of the matrix equation is independent of the magnitude of κ but depends only on the projection of κ on the operative plane of the reciprocal lattice.

This is seen from the geometry of Fig. 10.1 for which the angles θ_0 and θ_h have been greatly exaggerated. To a reasonable approximation we may write

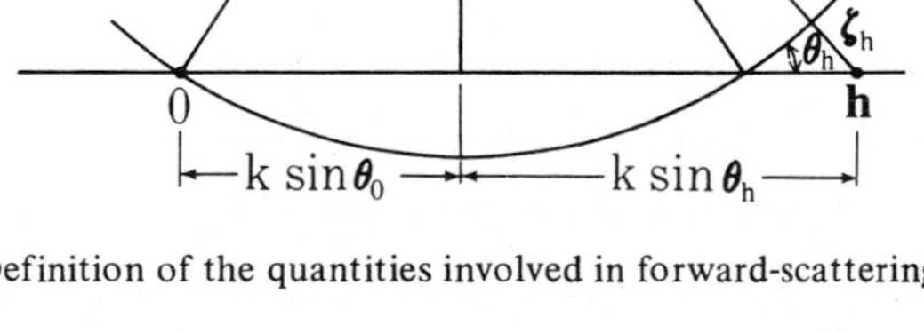

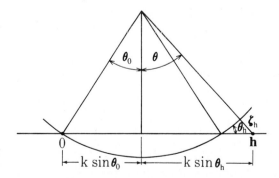

Fig. 10.1. Definition of the quantities involved in forward-scattering geometry.

$$\zeta_h = (\kappa \sin\theta_h - \kappa \sin\theta_0) \sin\theta_h . \tag{9}$$

Then p_h ($= 2\kappa\zeta_h - \zeta_h^2$) can be expressed entirely in terms of the quantities $\kappa \sin\theta_h$ and $\kappa \sin\theta_0$, which are projections on to the plane of the $\boldsymbol{h}$ reciprocal lattice points, and $\sin^2\theta_h$ which is negligible. Hence the matrix $\boldsymbol{M}$ is independent of the magnitude of κ and the amplitudes of the diffracted beams should be the same for different wavelengths and energies of the incident beam in this approximation, given equivalent orientations. This conclusion does not appear to have been adequately tested experimentally.

10.2.3. Bloch waves and boundary conditions

We have seen that there are, in general, an infinite number of solutions of the equation (7). The solution number i gives the amplitudes Ψ_h^i of the plane waves which make up the Bloch wave number i. The intensities of the diffracted beams for any real experimental situation are determined by the distribution of energy between the various Bloch waves and this is determined by the boundary conditions. Thus to get the amplitude of each wave in the crystal and so the amplitudes of the vacuum waves emerging from the crystal we add the Bloch waves with a weighting factor α_i so that the amplitude of the wave corresponding to the $\boldsymbol{h}$ reciprocal lattice point in the Bloch wave i is written

$$u_h^i = \alpha_i \Psi_h^i . \tag{10}$$

The boundary conditions at the entrance surface ensure that the sum of the amplitudes of all beams is zero except in the incident beam direction, i.e.

$$\sum_i u_h^i = \sum_i \alpha_i \Psi_h^i = \delta_{h0} \tag{11}$$

where δ_{h0} is the Kronecker delta (= 1 for $h = 0$, and 0 for other h). Multiplying (11) by Ψ_h^{j*} and summing over h gives

$$\sum_h \sum_i \alpha_i \Psi_h^i \Psi_h^{j*} = \sum_h \delta_{h0} \Psi_h^{j*} . \tag{12}$$

Provided that the Bloch waves are orthogonalized and normalised, $\Sigma \Psi_h^i \Psi_h^{j*} = 0$ unless $i = j$, and $\Sigma_h \Psi_h^j \Psi_h^{j*} = 1$. Then (12) becomes

$$\alpha_j \sum_h \Psi_h^j \Psi_h^{j*} = \Psi_0^{j*}$$

or

$$\alpha_j = \Psi_0^{j*} \tag{13}$$

and

$$u_h^i = \Psi_0^{i*} \Psi_h^i \,. \tag{14}$$

Thus the multiplier for the amplitudes of each Bloch wave is the complex conjugate of the zero Fourier coefficient of that Bloch wave.

For a parallel sided plate-shaped crystal, all waves in the crystal corresponding to the one reciprocal lattice point $\boldsymbol{h}$ will be refracted into the same direction at the exit face to form the vacuum diffracted wave $\boldsymbol{K}_h$. In the crystal the amplitude of this wave will be

$$\Psi_h = \sum_i u_h^i \exp\{i\boldsymbol{k}_h^i \cdot \boldsymbol{r}\} \,. \tag{15}$$

Refraction at the exit face gives a change in length of the projection of the wave vector on the surface normal, as suggested in Fig. 10.2. In the small angle approximation, the difference in projection of $\boldsymbol{\kappa}$ and $\boldsymbol{K}_h$ is $\xi^i = x^i/2\kappa$, and the difference in projection of $\boldsymbol{K}_0$ and $\boldsymbol{\kappa}$ is $v_0/2\kappa$. Then if the surface normal is unit vector $\boldsymbol{z}$ and $\boldsymbol{r}\cdot\boldsymbol{z} = z$, we have

$$\begin{aligned} \boldsymbol{k}_h^i \cdot \boldsymbol{z} &= \boldsymbol{\kappa}\cdot\boldsymbol{z} - \xi^i \\ \boldsymbol{\kappa}\cdot\boldsymbol{z} &= \boldsymbol{K}_h \cdot \boldsymbol{z} - v_0/2\kappa \,. \end{aligned} \tag{16}$$

Then, since the x and y components of $\boldsymbol{k}_h^i$ and $\boldsymbol{K}_h$ are equal, we can insert in (15),

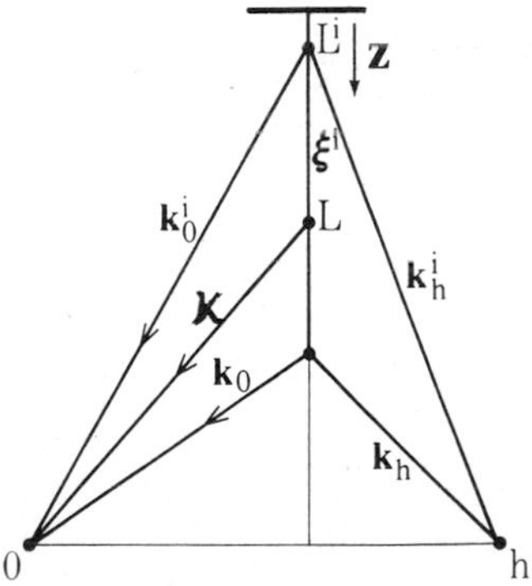

Fig. 10.2. The relationship of wave vectors in a crystal and in space for a two-beam case.

$$\exp\{\mathrm{i}\boldsymbol{k}_h^i\cdot\boldsymbol{r}\} = \exp\{\mathrm{i}\boldsymbol{K}_h\cdot\boldsymbol{r}\}\exp\{\mathrm{i}v_0 z/2\kappa\}\exp\{\mathrm{i}\xi^i z\}\,. \tag{17}$$

Hence the vacuum wave function for the h reflection wave is

$$\psi_h^v(\boldsymbol{r}) = \left[\sum_i \Psi_0^{i*}\Psi_h^i \exp\left\{\frac{\mathrm{i}zx^i}{2\kappa}\right\}\right]\exp\{\mathrm{i}\boldsymbol{K}_h\cdot\boldsymbol{r}\}\,. \tag{18}$$

The exponential term $\exp\{\mathrm{i}v_0 z/2\kappa\}$ is neglected because it is a common phase term for all beams. The intensity of the diffracted beam is then given by the square of the modulus of the summation in the square bracket. In order to calculate the amplitudes and intensities of the diffracted beams it is therefore necessary to find the eigen-values x^i and eigenvectors $\boldsymbol{\Psi}^i$ for the matrix equations (7) and (8).

As in Chapter 8 we may include absorption explicitly and deal with real quantities p_h and v_{hg} by adding the Fourier coefficients of the absorption function μ_0 and μ_h so that (8) becomes

$$\boldsymbol{M} = \begin{pmatrix} \cdots p_0 \cdots v_{0h} \cdots v_{0g} \cdots \\ \cdots v_{h0} \cdots p_h \cdots v_{hg} \cdots \\ \cdots v_{g0} \cdots v_{gh} \cdots p_g \cdots \end{pmatrix} + \mathrm{i}\begin{pmatrix} \cdots \mu_0 \cdots \mu_{0h} \cdots \mu_{0g} \cdots \\ \cdots \mu_{h0} \cdots \mu_0 \cdots \mu_{hg} \cdots \\ \cdots \mu_{g0} \cdots \mu_{gh} \cdots \mu_0 \cdots \end{pmatrix} \tag{19}$$

Then both of these matrices are Hermitian, but the sum of the two is general, complex.

10.2.4. The scattering matrix

Proceeding further, we now express the result in a rather different form, that of a scattering matrix. For a particular eigen-value, x^i, we have from (7)

$$\boldsymbol{M}\boldsymbol{\Psi}^i = x^i\boldsymbol{\Psi}^i\,. \tag{20}$$

Since x^i is a simple number, applying the matrix to the right-hand side of (20) gives

$$\boldsymbol{M}^2\boldsymbol{\Psi}^i = (x^i)^2\boldsymbol{\Psi}^i\,,$$

and, in general

$$\boldsymbol{M}^n\boldsymbol{\Psi}^i = (x^i)^n\boldsymbol{\Psi}^i\,. \tag{21}$$

Then for the $\boldsymbol{h}$ component of the i Bloch wave

$$(x^i)^n \Psi_h^i = \sum_g \Psi_g^i (M^n)_{hg} \,. \tag{22}$$

where $(M^n)_{hg}$ is the component in the h column and g row of the matrix M^n.

In order to apply the boundary condition, we multiply both sides of (22) by Ψ_0^{i*} and sum over i to give

$$\sum_i (x^i)^n \Psi_h^i \Psi_0^{i*} = \sum_i \sum_g \Psi_g^i \Psi_0^{i*} (M^n)_{hg}$$

$$= \sum_g \left(\sum_i \Psi_g^i \Psi_0^{i*} \right) (M^n)_{hg} = (M^n)_{h0} \,, \tag{23}$$

since the boundary condition (11) with $\alpha_i = \Psi_0^{i*}$ gives

$$\sum_i \Psi_0^{i*} \Psi_h^i = \delta_{h0} \,.$$

Then, expanding the exponential in (18) we have

$$\Psi_h^v(\boldsymbol{r}) = \sum_i \Psi_0^{i*} \Psi_h^i \exp\left\{\frac{\mathrm{i}zx^i}{2\kappa}\right\} \exp\{\mathrm{i}\boldsymbol{K}_h \cdot \boldsymbol{r}\}$$

$$= \sum_i \Psi_0^{i*} \Psi_h^i \sum_n \frac{1}{n!} \left(\frac{\mathrm{i}zx^i}{2\kappa}\right)^n \exp\{\mathrm{i}\boldsymbol{K}_h \cdot \boldsymbol{r}\}$$

$$= \sum_n \frac{1}{n!} \left(\frac{\mathrm{i}z}{2\kappa}\right)^n (M^n)_{h0} \exp\{\mathrm{i}\boldsymbol{K}_h \cdot \boldsymbol{r}\} \,. \tag{24}$$

By convention we write for matrices, as for ordinary functions

$$\exp\{\mathrm{i}\boldsymbol{M}\} = \boldsymbol{1} + \mathrm{i}\boldsymbol{M} - \frac{1}{2!}\boldsymbol{M}^2 + \dots \tag{25}$$

and the matrix components are

$$[\exp\{\mathrm{i}\boldsymbol{M}\}]_{hg} = (\boldsymbol{1})_{hg} + \mathrm{i}(\boldsymbol{M})_{hg} - \frac{1}{2!}(\boldsymbol{M}^2)_{hg} + \dots \,. \tag{26}$$

Hence, from (24), the amplitude of the $\boldsymbol{h}$ wave in vacuum, leaving the crystal of thickness z is

$$\left[\exp\left\{\frac{\mathrm{i}z}{2\kappa}\boldsymbol{M}\right\}\right]_{h0} ,$$

and in general we may write

$$\begin{pmatrix} \Psi_0 \\ \Psi_1 \\ \cdot \\ \cdot \\ \cdot \\ \Psi_h \\ \cdot \\ \cdot \\ \cdot \end{pmatrix} = \exp\left\{\frac{\mathrm{i}z\boldsymbol{M}}{2\kappa}\right\} \begin{pmatrix} \Psi_0^{(0)} \\ 0 \\ 0 \\ \cdot \\ \cdot \\ \cdot \\ 0 \\ \cdot \\ \cdot \\ \cdot \end{pmatrix}, \tag{27}$$

where $\Psi_0^{(0)}$ is the incident wave amplitude and operation on the vector $\boldsymbol{\Psi}^{(0)}$ with the exponential of the matrix gives the vector $\boldsymbol{\Psi}^{(1)}$ which has components Ψ_h, the amplitudes of the waves leaving the crystal. Thus propagation of a wave field through a crystal of thickness z is represented by the operation of the scattering matrix

$$\boldsymbol{S} = \exp\left\{\frac{\mathrm{i}z\boldsymbol{M}}{2\kappa}\right\}. \tag{28}$$

In the application of this concept to the calculation of dynamical intensities by Sturkey [1962], the scattering matrix is first calculated for a thin slice of crystal of thickness D. Then transmission through successive identical slices of crystal will be simulated by the repeated application of the scattering matrix so that we may write for n slices

$$\boldsymbol{\Psi}^{(n)} = \boldsymbol{S}^n \boldsymbol{\Psi}^{(0)}$$

and

$$\boldsymbol{S}^n = \exp\left\{\frac{\mathrm{i}Dn\boldsymbol{M}}{2\kappa}\right\}. \tag{29}$$

For non-identical slices of crystal the same approach is valid if for each slice the operation on the wave field vector is made by means of the appropriate matrix. In this way it is possible to apply the formulation to problems of faults and imperfections or variations of structure of a crystal. A scattering matrix must be generated for each type of slice involved.

10.2.5. Derivation of the two-beam approximation

It has been shown by Rez [1977] and Goodman [private communication] that the two-beam approximation for transmission through a crystal plate follows very readily from the scattering matrix formulation. The wave function after traversing a distance z is given by equation (28) as

$$\begin{pmatrix} \boldsymbol{\Psi}_0 \\ \boldsymbol{\Psi}_h \end{pmatrix} = \exp\left\{\frac{izM}{2\kappa}\right\}\begin{pmatrix} 1 \\ 0 \end{pmatrix}. \tag{30}$$

If v_h is real, i.e. for the centrosymmetric case without absorption, the matrix M is written, from (8),

$$M = \begin{pmatrix} -\frac{1}{2}p_h & v_h \\ v_h & \frac{1}{2}p_h \end{pmatrix} = -\tfrac{1}{2}p_h\begin{pmatrix} 1 & 0 \\ 0 & -1 \end{pmatrix} + v_h\begin{pmatrix} 0 & 1 \\ 1 & 0 \end{pmatrix} \tag{31}$$

where we have subtracted out $\frac{1}{2}p_h\,\mathbf{1}$ which adds the same phase factor to each beam.

The matrices in the final part of (31) are Pauli spin matrices $\boldsymbol{\sigma}_1$ and $\boldsymbol{\sigma}_2$ for which $\boldsymbol{\sigma}_1^2 = \boldsymbol{\sigma}_2^2 = \mathbf{1}$ and $\boldsymbol{\sigma}_1 \cdot \boldsymbol{\sigma}_2 = 0$ so that the terms in the expansion of the exponential, (25), take the form

$$M = A + B = a\boldsymbol{\sigma}_1 + b\boldsymbol{\sigma}_2$$

$$M^2 = A^2 + B^2 = (a^2 + b^2)\mathbf{1} = r^2\,\mathbf{1}$$

$$M^3 = (A^2 + B^2)(A + B) = M \cdot (a^2 + b^2)$$

$$M^4 = (A^2 + B^2)^2 = r^4 \cdot \mathbf{1}\,,$$

Here we have put $r^2 = a^2 + b^2 - p_h^2/4 + v_h^2$. Summing the odd and even terms of the series separately gives

$$\exp\left\{\frac{izM}{2\kappa}\right\} = \cos\left(\frac{zr}{2\kappa}\right) \cdot \mathbf{1} + ir^{-1}\sin\left(\frac{zr}{2\kappa}\right) \cdot M\,.$$

By putting $p_h = 2\kappa\zeta_h$ (assuming $\zeta_h \ll \kappa$) and $r^2 = \kappa^2\zeta_h^2 + v_h^2$, the wave amplitudes at the depth z are then given from (30) as

$$\psi_0 = \cos\left(\frac{zr}{2\kappa}\right) - i\frac{p_h}{2r}\sin\left(\frac{zr}{2\kappa}\right)$$

$$= \cos\left\{\frac{z}{2}(\zeta_h^2 + v_h^2/\kappa^2)^{1/2}\right\} - i\,\frac{\zeta_h}{(\zeta_h^2 + v_h^2/\kappa^2)^{1/2}}\,\tfrac{1}{2}\sin\left\{\frac{z}{2}(\zeta_h^2 + v_h^2/\kappa^2)^{1/2}\right\}$$

$$\psi_h = v_h/r\,\sin\left(\frac{zr}{2\kappa}\right)$$

$$= \frac{v_h/\kappa}{(\zeta_h^2 + v_h^2/\kappa^2)}\sin\left\{\frac{z}{2}(\zeta_h^2 + v_h^2/\kappa^2)^{1/2}\right\}.$$

Putting the crystal thickness $H = z/2\pi$, to be consistent with the usage $\kappa = 2\pi/\lambda$, then makes this result identical with (8.28) and (8.29).

The case of v_h complex $(= v_h + iv_h^i)$ can be treated by a simple extension, using the three Pauli spin matrices

$$M = -p_h \begin{pmatrix} 1 & 0 \\ 0 & -1 \end{pmatrix} + v_h \begin{pmatrix} 0 & 1 \\ 1 & 0 \end{pmatrix} + v_h^i \begin{pmatrix} 0 & i \\ -i & 0 \end{pmatrix}$$

$$= a\boldsymbol{\sigma}_1 + b\boldsymbol{\sigma}_2 + c\boldsymbol{\sigma}_3 \, .$$

10.3. The Darwin-type approach

The Darwin treatment of the X-ray diffraction by reflection from a face of a large perfect crystal (Darwin [1914]) involved the establishment of a transmission and reflection coefficient for each plane of atoms and then the summation of transmitted and diffracted amplitudes at each plane. The application to transmission electron diffraction of this type of theory was made by Howie and Whelan [1961] primarily for the purpose of finding the contrast in electron microscope images of defects. The amplitudes of the diffracted beams are considered to be continuous functions of the distance in the beam direction and are related by a set of differential equations. It is essentially a perfect-crystal theory for each slice of the crystal although variations in orientation of the diffracting planes from one slice to another may be included.

For the simple two beam case the relationship between transmitted and scattered amplitudes, Ψ_0 and Ψ_h, corresponding to incident and diffracted beam directions, for propagation in the z direction is given by

$$\frac{\mathrm{d}\Psi_0}{\mathrm{d}z} = \mathrm{i}\sigma\Phi_h\Psi_h \, ,$$

$$\frac{\mathrm{d}\Psi_h}{\mathrm{d}z} = \mathrm{i}\sigma\Phi_h\Psi_0 + 2\pi\mathrm{i}\zeta_h\Psi_h \, . \tag{32}$$

In words, these equations state that the variation of the transmitted beam amplitude with thickness is given by scattering from the $\boldsymbol{h}$ beam with a phase change of $\pi/2$ and a scattering strength proportional to the Fourier coefficient of potential Φ_h. The diffracted beam amplitude is modified by scattering from the transmitted beam but also has a phase change because the wave vector $\boldsymbol{k}_h$ drawn to the reciprocal lattice point is not the same as that to the reciprocal lattice origin, $\boldsymbol{k}_0$, but differs from it by the excitation error, ζ_h.

Absorption may be taken into account by making the potential coefficients Φ_h complex so that in the equations, (32) the quantity $i\Phi_h$ is replaced by $i\Phi_h - \Phi'_h$ and there is an overall decrease in both beams due to the average absorption coefficient proportional to Φ'_0. If one plane of the crystal is translated by a vector $\boldsymbol{R}$ relative to the host lattice then this will change the phase of the diffracted beam by an amount proportional to $\boldsymbol{b} \cdot \boldsymbol{R}$. Thus if the displacement is a function of z, namely $\boldsymbol{R}(z)$, the equations (32) with absorption, will become

$$\frac{d\Psi_0}{dz} = -\sigma\Phi'_0\Psi_0 + \sigma(i\Phi_h - \Phi'_h)\Psi_h ,$$

$$\frac{d\Psi_h}{dz} = \sigma(i\Phi_h - \Phi'_h)\Psi_0 + \{-\sigma\Phi'_0 + 2\pi i(\zeta_h + \beta_h)\}\Psi_h , \tag{33}$$

where

$$\beta_h = \frac{d}{dz}(\boldsymbol{b} \cdot \boldsymbol{R}(z)) . \tag{34}$$

The progressive changes of the amplitudes Ψ_0 and Ψ_h through the crystal may then be followed by integration of (33), i.e. by finding the successive modifications of phase and amplitude of Ψ_0 and Ψ_h together for infinitesimal increments of z.

Like the original Darwin approach, this two-beam version suffers from the difficulty that a slice of infinitesimal thickness will give a very large number of simultaneous beam and a two beam situation is only approached after transmission through a crystal thickness comparable to the extinction distance $2\sigma\Phi_h$ or more. Hence it is more appropriate in principle and, in the case of electrons, for practical considerations to make use of the n-beam form of (33) which may be expressed in matrix form as

$$\frac{d}{dz}\boldsymbol{\Psi} = 2\pi i(\boldsymbol{A} + \boldsymbol{\beta})\boldsymbol{\Psi} \tag{35}$$

where $\boldsymbol{\Psi}$ is a column vector whose elements Ψ_h are the amplitudes of the diffracted waves, $\boldsymbol{\beta}$ is a diagonal matrix whose elements are given by (34) and $\boldsymbol{A}$ is a matrix whose elements are

$$A_{hh} = \zeta_h + i\sigma\Phi'_0/4\pi ,$$

$$A_{hg} = \sigma(\Phi_{h-g} + i\Phi'_{h-g})/4\pi . \tag{36}$$

Then, from (35) the change in any amplitude depends on all other amplitudes, the strength of the interactions depending on the potential coefficients Φ_{hg} and the excitation errors of all beams except the incident beam. The amplitudes of all the beams are integrated through the crystal simultaneously.

10.4. Special cases – beam reduction

In general the only way to gain an appreciation of the results of the dynamic interaction of a large number of beams in a crystal is by performing many detailed n-beam calculations for various crystals having a variety of thickness and orientations and attempting to analyse the results. There are special cases, however, for which the n-beam diffraction result may be understood by comparison with a simpler, related result for relatively few beams. These are cases of high symmetry in the diffraction pattern for which members of sets of beams are equivalent in that they have equal excitation errors and interact through equivalent structure factor values, Φ_{hg}. The way in which such sets of equivalent beams may be merged to give, for each set, a single representative beam, was demonstrated by Gjonnes by use of the integral equation representation (Gjonnes [1966]) and this approach was used by Fisher [1968]. A different approach through the matrix formulation of equation (10.8) was given by Fukuhara [1966].

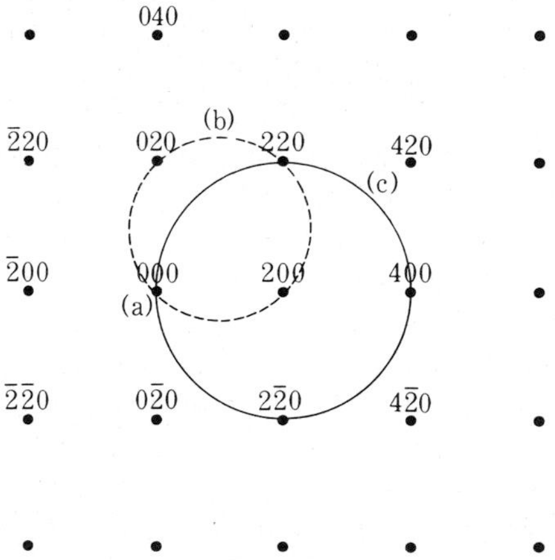

Fig. 10.3. Illustrating the interaction of the Ewald sphere with the $hk0$ plane of reflections for a F.C.C. crystal in three special cases where beam-reduction techniques can be used.

For purposes of illustration we refer to the particular case of an electron beam almost parallel to the cube axis of a face-centered cubic structure. The plane of $hk0$ reciprocal lattice spots near to the Ewald sphere is sketched in Fig. 10.3.

For the situation (a), when the incident beam is exactly parallel to the c-axis, the four beams 200, 020, $\bar{2}00$ and $0\bar{2}0$ will be exactly equivalent having the same excitation error, interacting between themselves through the Fourier coefficients of the types Φ_{220} and Φ_{400}, interacting with the zero beam through coefficients Φ_{200}, interacting with the set of 200 beams through Φ_{200} and Φ_{420} and so on. Thus the four 200-type beams may be combined to give an equivalent single beam with definable interactions with the 000 beam, the representative beam of the 220 set, and so on.

If a 5-beam calculation is to be made, including only the 000 and the 200 set, this is immediately reduced to a 2-beam calculation. Similarly the problem is greatly reduced if more beams are included. Fisher [1968] calculated the intensities for this orientation (with no absorption) for a 49-beam case including reflections out to the 660. By symmetry the number of effective beams to be considered is reduced to 10. The result of the calculation shown in Fig. 10.4 indicates the interesting situation that there is something resembling a two beam solution in that the zero beam oscillates out of phase with all the diffracted beams, varying together. This result may be influenced by the special circumstance that the excitation errors for the $hk0$ reflections will be proportional to the square of d_{hk0}^{-1} and so to $h^2 + k^2$. Hence the phase changes introduced, in (36) for example, by the excitation errors will be integral multiples of 2π when that for the 200 reflections is 2π. An investigation of such special cases by Fejes, Iijima and Cowley [1973] suggests, however, that the situation is more complicated and that the periodicity of the intensity variation with thickness depends on the presence of a particular relationship between scattering amplitudes and excitation errors.

A second special case indicated by (b) in Fig. 10.3 is for a tilt of the incident beam so that the Ewald sphere cuts exactly the 000, 200, 220 and 020 reciprocal lattice points. Then the four corresponding beams will be equivalent in that they have the same excitation error and interact among themselves and with successive concentric sets of beams, through the same interaction potentials. By grouping these sets of reflections together the effective number of beams is again reduced. In this case the 000 beam will be distinguished from the rest of the set when the boundary conditions for the entrance surface are applied but the beam-reduction is effective in simplifying the problem up to that stage.

Similarly the case (c) of Fig. 10.3 with the Ewald sphere passing through

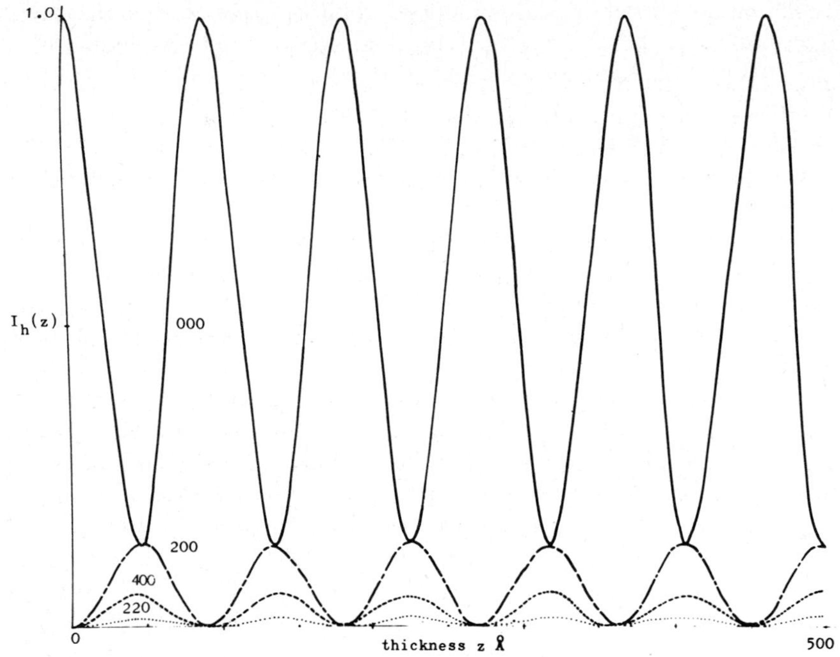

Fig. 10.4. Calculation of the intensities of $hk0$ reflections as a function of thickness for the incident beam exactly parallel to c-axis of a F.C.C. copper-gold alloy (disordered Cu_3Au) with no absorption (Fisher [1968]). 100 keV.

the points 000, 220, 400 and $2\bar{2}0$ is reducible by symmetry for the infinite crystal formulation. Fukuhara [1966] gives a number of other special cases appropriate for square and hexagonal symmetry of the diffraction pattern.

10.5. Computing methods

In order to obtain a numerical solution to the matrix equation (7) subject to the boundary conditions, any of the standard programs available for modern digital computers may be used to diagonalize the matrix $\boldsymbol{M}$. Programs based on the Jacobi and QR methods have been used, among others. The problem is greatly complicated in the presence of absorption, when the matrix elements are complex, unless simplifying assumptions based on the relative smallness of the imaginary components are made. However, even for the full problem of a non-centrosymmetric crystal with absorption the amount of computation is not beyond the capabilities of a moderate-size computer for which a 50-beam problem, involving the diagonalization of a 50 × 50 matrix, would take only a few minutes (Fisher [1971]).

From the diagonalized matrix the eigen-values x^i and the eigenvector components Ψ_h^i may be deduced. Substituting in (18) then allows the diffracted amplitudes for all beams, $\boldsymbol{h}$, to be derived very rapidly for any number of crystal thicknesses, z.

Since the computing time required increases very rapidly with the number of diffracted beams involved, it is of interest to determine the minimum number of beams which should be included in order to obtain a required degree of accuracy. This number will vary widely in practice, depending on the size of the unit cell projected in the beam direction (i.e. on the density of points in the plane of reciprocal space involved), on the wavelength of the electrons and the atomic numbers of the atoms present.

To give an indication on this point. Fisher [1968] made the calculation for the case (a) of Fig. 10.3 with the incident beam exactly parallel to the c-axis of a face centered cubic disordered copper-gold alloy. For this material the

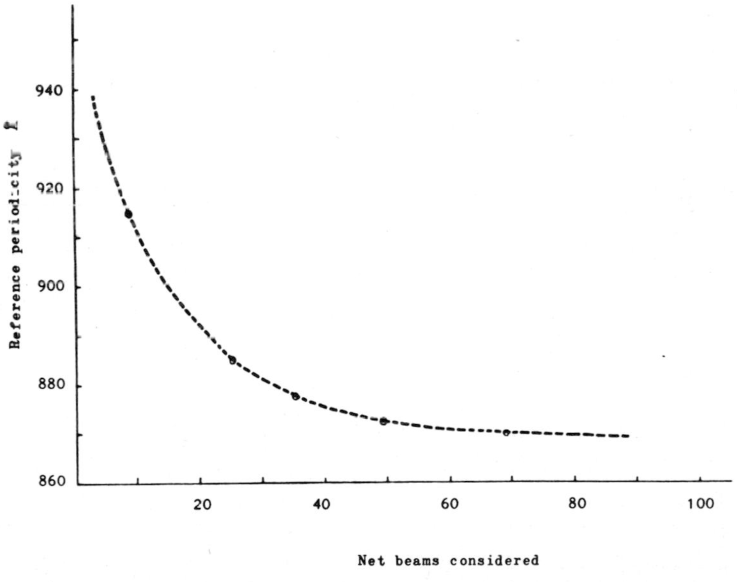

Fig. 10.5. Variation of the periodicity of the intensity oscillations of the beams for CdS, ⟨120⟩ orientation, 100 keV, as a function of the number of beams included in the calculation (Fisher [1968]).

reciprocal lattice points in the plane of interest are relatively widely spaced, being 0.5 $Å^{-1}$ apart, but the average atomic number is moderately high. The general features of the intensity distributions as shown in Fig. 10.4 were established for less than 10 beams but at this stage the relative intensities were badly wrong.

The periodicity of the oscillations of the incident beam intensity, the "extinction distance", was used as an indication of the accuracy of the computation and is plotted against the number of beams considered in Fig. 10.5 for CdS in ⟨120⟩ orientation. For the case of Fig. 10.4, from a value of about 128 Å for the simplest, 5-beam case, the periodicity decreases to a limiting value of about 90 Å. For an accuracy of 1 or 2 per cent in predicting this periodicity it is necessary to include 40 or 50 beams.

These particular cases of high symmetry do not necessarily give a good indication for the general case, but experience with calculations for other orientations, such as that of Fig. 10.6, corresponding to case (c) of Fig. 10.3, shows that the requirements for accuracy of intensities are fairly well indicated by Fig. 10.5.

It is to be expected that for comparable accuracy the number of beams to be included should vary roughly as the square of the average unit cell dimension and in proportion to the atomic number. However such indications are necessarily very approximate. The only way to test whether, in any particular case, the number of beams taken is sufficient is to increase the number of beams by a considerable factor and ensure that such an increase makes no appreciable difference to the intensities of the beams of interest.

Some fore-knowledge of the form of the diffraction pattern is, of course, desirable as a guide to the selection of the diffracted beams to include in the calculations. For example, for a crystal tilted away from a principle orientation the "laue circle" of spots given by the intersection of the Ewald sphere with the principle reciprocal lattice plane will be strong and should be included. Also care should be taken for complicated structures to include the often widely-spaced, strong reflections coming from a prominent sub-lattice of the structure.

As mentioned previously, Sturkey [1962] made use of the scattering matrix formulation, equation (27), to calculate diffracted beam intensities. The scattering matrix for a small crystal thickness H, raised to the nth power, gives the scattering from a crystal of thickness nH. In raising a matrix to higher powers errors accumulate rapidly and the requirements for accuracy in the initial thin-crystal matrix are increased. Considerable care is needed to ensure that the number of beams included and the accuracy of the calculations for the initial thin-crystal calculation are sufficient.

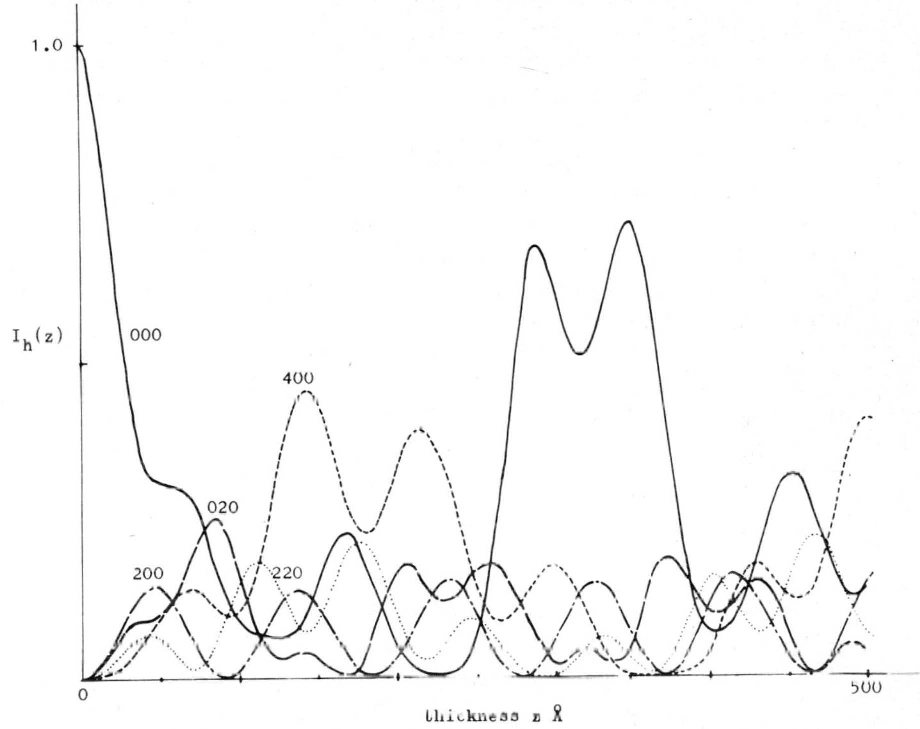

Fig. 10.6. Calculated intensities of *hk*0 reflections as a function of thickness for the incident beam tilted slightly away from the *c* axis of the F.C.C. copper gold alloy of Figs. 10.3 and 10.4 to give the Ewald sphere intersection indicated by (c) in Fig. 10.3.

Calculation based on the Howie–Whelan equations (35) involve the progressive integration of the amplitudes through the crystal. This may be done by the use of either digital or analogue computers. The use of an analogue computer as described by Johnson [1968] has considerable advantages in speed and flexibility although the number of beams which can be considered is limited by the amount of hardware available.

In this type of computing, the progress of the beams of the *z*-direction is represented by the time variable. For each beam included one circuit is used, consisting of two integrators to compute the real and imaginary parts of the amplitude Ψ_h, the elements of the column vector $\mathbf{\Psi}$, inverters to generate negatives and potentiometers to represent the excitation errors. Interactions between beams are simulated by coupling of the circuits through potentiometers representing the real and imaginary parts of the structure amplitudes Φ_{h-g} and Φ'_{h-g}. The computer is run so that one second of real time may correpond, for example, to transmission through 100 Å of crystal. The intensity variations of the beam are displayed on cathode ray oscilloscope tubes.

In order to include distortions of the crystal with displacments of the crystal planes by vectors $\boldsymbol{R}(z)$, the phase factor involved in equation (35) may be introduced by means of suitable function generators. In this way many cases of interest for the electron microscopy of defects may be simulated. The speed with which calculations of reasonable accuracy may be made and displayed suggested that the analogue method could be a very useful adjunct to the more precise digital computer calculations.

10.6. Column approximation

The use of equations such as (35) to derive the form of electron microscope images or diffraction patterns from defects in crystals depends on the use of the "column approximation". This may be described by means of the diagram, Fig. 10.7. The amplitudes of the waves emerging from a crystal at a point P are assumed to be dependent only on the structure of a column of diameter D centered on P and extending through the crystal in the direction of energy propagation. Then if the contents of the column can be approximated by a function dependent on the z-coordinate only, the intensity at P will be the same as for a crystal of infinite lateral extent of structure having the same z-dependence, for which the intensity may be calculated by equations such as (35). The requirement is that the crystal structure should not vary appreciably across the width of the column. Then the intensity distribution across the exit

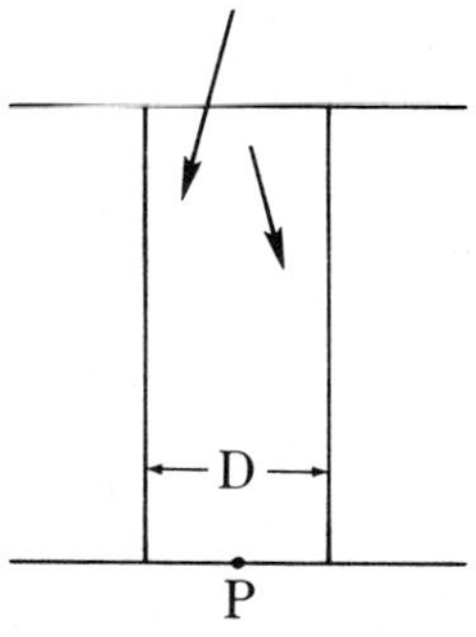

Fig. 10.7. Diagram illustrating the assumptions of the "column approximation".

face as seen in an electron microscope will give a representation of the changes of diffraction conditions with a resolution roughly equal to the column diameter.

The minimum diameter of the column may be estimated in a number of ways which give comparable results. For example in a near-two-beam case the energy is transferred from incident to diffracted beam or vice versa, in a distance $\xi_h/2$. Thus the lateral movement of the beam before its direction is changed is of the order of $\xi_h\theta_h$. For $\xi_h = 200$ Å and $\theta_h = 10^{-2}$ the product is 2 Å, but we should allow a factor of 2 or 3 greater than this to give the half width of the column as approximately 5 Å.

Alternatively, if there is only weak scattering, the main contribution to the column width comes from the spread of the electron wave by Fresnel diffraction. As a measure of this spread we may take the condition

$$\exp\left\{\frac{ikx^2}{2R}\right\} = \exp\{i\pi\}$$

or

$$x = (R\lambda)^{1/2} .$$

Thus for $\lambda = 0.04$ Å and a crystal thickness of 500 Å we have that the spread of the wave due to Fresnel diffraction is indicated as $x \approx 4.5$ Å.

More complete and sophisticated discussions of the effective width of the column by Jouffrey and Taupin [1967] and Howie and Basinski [1968] give results of the same order of magnitude. Hence for electron microscopy with a resolution not better than 10 Å and for perturbations of the crystal lattice which do not have strong variations over distances of 10 Å the column approximation would not give rise to any serious error. The experimental observations under these conditions have been almost universally of a rough non-quantitative nature and within these limitations appear to give reasonable agreement with column approximation calculations.

A further application of the column approximation is to the intensities for a crystal for which the lateral variation is that of crystal thickness rather than crystal distortion. This is the basis for our simplified discussion of the origin of thickness fringes in electron micrographs in Chapter 9.

Problems

1. Show that the differential equations (10.32) are consistent with the two-beam form of the Bethe equations (8.10) without absorption and the solutions

for a flat plate crystal (8.25) and (8.26). Similarly show that the n-beam form (10.33) is consistent with the dispersion equation (8.7).

2. Write down the form of the scattering matrix (10.29) appropriate for a very thin crystal in a first approximation. How is (10.30) then related to a kinematical approximation and to a phase-grating approximation? (See Sturkey [1962].)

CHAPTER 11

The "physical optics" approach

11.1. Propagation of electrons in crystals

11.1.1. Transmission through thin slices

In Chapter 3 we considered transmission of an incident wave through two dimensional phase- and amplitude-objects and the propagation between such objects, described in terms of Fresnel diffraction concepts. The effect on the incident wave of a two-dimensional object was represented by multiplication by a transmission function, $q_n(x,y)$ and transmission through space by convolution with a propagation function $p_n(x,y)$. The extension to the consideration of three-dimensional objects was indicated in equation (3.31) for real space and equation (3.32) for reciprocal space.

The relevance of this approach for a consideration of the scattering of electrons from crystals was indicated by the accumulation of evidence that for very thin single crystals there may be hundreds of simultaneous diffracted beams and the diffraction conditions are more closely related to those for a two dimensional grating than to those for an infinite periodic lattice which is the starting point of Bethe theory.

In the formulation of n-beam diffraction theory by Cowley and Moodie [1957] transmission of electrons through a sample is represented by transmission through a set of N two dimensional phase- and amplitude-objects separated by distances Δz. The total phase change and amplitude change of the electron wave in a slice of the specimen of thickness Δz is considered to take place on one plane. Then the propagation of the wave from one such plane to the next is by Fresnel diffraction in vacuum. It has been shown (Moodie [1972]) that in the limiting case that the thickness of the slice Δz goes to zero and the number of slices N goes to infinity in such a way that $N\Delta z = H$, where H is the specimen thickness, this form of description becomes a rigorous representation of the scattering process, completely consistent with the more conventional quantum mechanical descriptions. Several authors have recently provided derivations of the multislice formulation from established quantum mechanical bases such as that of Fineman (Jap and Gaeser [1978], Ishizuka and Uyeda [1977]).

The phase change of the electron wave transmitted through a thin slice depends on the value of the refractive index,

$$n = \left\{1 + \frac{\varphi(\boldsymbol{r})}{E_0}\left(1 + \frac{eE_0}{2m_0c^2}\right)\right\}^{1/2} \approx 1 + \frac{\varphi(\boldsymbol{r})}{2E}, \tag{1}$$

so that the phase difference with respect to a wave in vacuum for a slice thickness Δz at $z = z_n$ is

$$\left[\left(1 + \frac{\varphi}{2E}\right) - 1\right] \Delta z \frac{2\pi}{\lambda} = \frac{\pi}{\lambda E} \Delta z \, \varphi(x, y, z_n) . \tag{2}$$

We put $\pi/\lambda E$ equal to the interaction constant, σ, which varies with voltage as indicated by (10.2). We may also introduce an absorption function $\mu(\boldsymbol{r})$ depending on the circumstances of the experiment to be described.

Then the transmission function for the plane $z = z_n$ representing the slice may be written

$$q_n(x, y, z_n) = \exp\{-\mathrm{i}\sigma\varphi(x, y, z_n)\,\Delta z - \mu(x, y, z_n)\,\Delta z\} . \tag{3}$$

The negative sign before the potential term is consistent with the use of the form $\exp\{\mathrm{i}(\omega t - \boldsymbol{k}\cdot\boldsymbol{r})\}$ for a plane wave. The opposite sign was used by Cowley and Moodie [1957] who therefore unwittingly produced a theory of positron diffraction.

In the limit that the slice thickness goes to zero, we may write

$$q_n(x, y, z_n) = 1 - \mathrm{i}\sigma\varphi(x, y, z_n)\,\Delta z - \mu(x, y, z_n)\,\Delta z , \tag{4}$$

and Fourier transforming in two dimensions gives

$$Q_n(u, v) = \delta(u, v) - \mathrm{i}\sigma\Phi_n(u, v)\,\Delta z - M_n(u, v)\,\Delta z , \tag{5}$$

where

$$\Phi_n(u, v) = \int \Phi(\boldsymbol{u}) \exp\{-2\pi \mathrm{i} w z_n\}\,\mathrm{d}w , \tag{6}$$

and

$$M_n(u, v) = \int M(\boldsymbol{u}) \exp\{-2\pi \mathrm{i} w z_n\}\,\mathrm{d}w , \tag{7}$$

where $\Phi(\boldsymbol{u})$ and $M(\boldsymbol{u})$ are the Fourier transforms of $\varphi(\boldsymbol{r})$ and $\mu(\boldsymbol{r})$, respectively. For convenience we may use a "kinematical" structure amplitude

$$F(\boldsymbol{u}) = \sigma\Phi(\boldsymbol{u}) - \mathrm{i}M(\boldsymbol{u}) . \tag{8}$$

so that

$$Q_n(u,v) = \delta(u,v) - \mathrm{i}F_n(u,v)\,\Delta z . \tag{9}$$

The Fresnel diffraction of the wave on transmission through the distance Δz is given by convolution with the propagation function which, in the small angle approximation, is given by

$$p(x,y) = \frac{\mathrm{i}}{R\lambda} \exp\{-\mathrm{i}k(x^2+y^2)/2\Delta z\} , \tag{10}$$

or, in reciprocal space we multiply by

$$P(u,v) = \exp\{\pi\mathrm{i}\lambda\Delta z(u^2+v^2)\} . \tag{11}$$

11.1.2. Three-dimensional objects

Transmission through a three-dimensional object is given by the limiting case, for N going to infinity, of equation (3.31), written in one dimension only;

$$\psi(x) = q_N(x)\underset{N}{[}\ldots\underset{3}{[}q_2(x)\underset{2}{[}q_1(x)\underset{1}{[}q_0(x) * p_0(x)\underset{1}{]} * p_1(x)\underset{2}{]} * p_2(x)\underset{3}{]} * \ldots\underset{N}{]} * p_N(x) , \tag{12}$$

where $q_N(x)$ multiplied by the N bracket represents the wave leaving the object and this wave is propagated by convolution with $p_N(x)$ to the plane of observation. If the radiation incident on the first slice of the object, which has transmission function $q_1(x)$, is a plane wave inclined at an angle α to the normal to the plane of the slices, the convolution $q_0 * p_0$ is replaced by

$$\exp\{2\pi\mathrm{i}(\sin\alpha)\,x/\lambda\} \tag{13}$$

which multiplies $q_1(x)$.

It is often convenient, especially when dealing with diffraction by crystals, to work with the reciprocal space expression given, in one dimension, by the limiting case of equation (3.32), namely,

$$\Psi(u) = \underset{N}{[}Q_N(u) * \ldots\underset{2}{[}Q_2(u) * \underset{1}{[}Q_1(u) * Q_0(u)P_0(u)\underset{1}{]}P_1(u)\underset{2}{]}P_2(u)\ldots\underset{N}{]}P_N(u). \tag{14}$$

For a plane wave incident, $Q_0(u)P_0(u)$ is replaced by $\delta\{u + (\sin\alpha)/\lambda\} \equiv \delta(u-u_0)$.

Both of the forms, (12) and (14), are appropriate to finding the wave function at or near the exit face of an object, and so are useful for finding the amplitude and intensity distributions in ideal electron microscope images. For this purpose the diffracted amplitudes from (14) are Fourier transformed.

The amplitudes of the diffracted beams from an object are given directly from (14) with the final phase factor $P_N(u)$ usually neglected, or may be obtained by Fourier transform of $\psi(x)$ from (12). Other forms for these expressions, involving some transmission functions in real space and some in reciprocal space may be generated for use in special cases by application of the relationships (3.33) to (3.37) (see Cowley and Moodie [1958]).

11.1.3. Diffraction by a crystal

In the case of diffraction by a crystal, the slice transmission functions $q_n(x,y)$ will be periodic so that we may write

$$q_n(x,y) = \sum_h \sum_k Q_n(h,k) \exp\left\{-2\pi i\left(\frac{hx}{a} + \frac{ky}{b}\right)\right\} \tag{15}$$

or

$$Q_n(u,v) = \sum_h \sum_k Q_n(h,k)\,\delta\left(u - \frac{h}{a}, v - \frac{k}{b}\right). \tag{16}$$

From (5) and (9) we see that $Q_n(u,v)$ will have a dominating delta function at the origin, plus, for all h, k (including $h, k = 0$),

$$Q_n(h,k) = -i\Delta z \cdot F_n(h,k)\,, \tag{17}$$

and

$$F_n(h,k) = \sigma\Phi_n(h,k) - iM_n(h,k)\,, \tag{18}$$

where

$$\Phi_n(h,k) = \sum_l \Phi_{hkl} \exp\left\{-2\pi i l \frac{z_n}{c}\right\}, \tag{19}$$

$$M_n(h,k) = \sum_l M_{hkl} \exp\left\{-2\pi i l \frac{z_n}{c}\right\}. \tag{20}$$

Measuring z from the entrance face of the crystal we have

$$z_n = (n-1)\,\Delta z \quad \text{and} \quad H = (N-1)\,\Delta z\,. \tag{21}$$

If now we return to the one-dimensional form of (14) for convenience, the bracket numbered 1, for an incident plane wave becomes $Q_1(u-u_0)$. The bracket 2 becomes

$$\underset{2\ 2}{[...]} = (-\mathrm{i}\Delta z)^2 [F_2(u) * F_1(u-u_0) \exp\{\pi\mathrm{i}\lambda\Delta z \cdot u^2\}]$$

$$= (-\mathrm{i}\Delta z)^2 \sum_{h_1} \sum_{h_2} F_1(h_1) F_2(h_2)$$

$$\times \exp\left\{\pi\mathrm{i}\lambda\Delta z \left(u - \frac{h_2}{a}\right)^2\right\} \delta\left(u - u_0 - \frac{h_1+h_2}{a}\right). \tag{22}$$

Here, for convenience, we have included the $h = 0$ term in the summation over h with a weight $\mathrm{i}/\Delta z$ so that it will have weight unity when multiplied by $(-\mathrm{i}\Delta z)$. We see from (22) that for a beam diffracted twice corresponding to a reciprocal lattice point $h = h_1 + h_2$, the phase factor depends on the square of $\{u - (h_2/u)\}$, i.e. on the square of

$$u_0 + \frac{h_1+h_2}{a} - \frac{h_2}{a} = \left(u_0 + \frac{h_1}{a}\right)$$

and so is not dependent on the diffraction indices in the same way as the total change in u as given by the δ-function. This, for the more general case of many slices, is the origin of much of the complication of the n-beam dynamical theory.

Similarly

$$\underset{3\ 3}{[...]} = (-\mathrm{i}\Delta z)^3 \exp\{\pi\mathrm{i}\lambda z_2 u^2\}$$

$$\times \sum_{h_1} \sum_{h_2} \sum_{h_3} F_1(h_1) F_2(h_2) F_3(h_3)\, \delta\left(u - u_0 - \frac{h_1+h_2+h_3}{a}\right)$$

$$\times \exp\left\{\pi\mathrm{i}\lambda\Delta z \left[\left(u_0 + \frac{h_1}{a}\right)^2 + \left(u_0 + \frac{h_1+h_2}{a}\right)^2\right]\right\}. \tag{23}$$

Extending this to the case of N slices and making use of (21) we may write the final amplitude as

$$\Psi(u) = (-\mathrm{i}\Delta z)^N \exp\{\pi\mathrm{i}\lambda H u^2\}$$

$$\times \sum_{h_1} \sum_{h_2} \dots \sum_{h_N} F_1(h_1) F_2(h_2) \dots F_N(h_N)\, \delta\left(u - \frac{h_1+h_2+\dots+h_N}{a} - u_0\right)$$

$$\times \exp\left\{2\pi\mathrm{i}\lambda \sum_{n=1}^{N} z_n \left[-\frac{u_0 h_N}{a} + \frac{h_N^2}{a^2} + \sum_{m=n+1}^{N} \frac{h_n h_m}{2a^2}\right]\right\}. \tag{24}$$

The delta function specifies the direction of the h diffracted beam where $h = \sum_{n=1}^{N} h_n$. What we wish to find is the amplitude of the h beam for all scattering processes, i.e. for all combinations of $h_1, h_2 ... h_N$ which lead to h.

The structure factors $F_n(h_n)$ or, more generally, $F_n(h_n, k_n)$, can all be expressed in terms of the one set of structure factors $F(h, k, l)$ for the crystal by use of the equations (18) to (20) which give

$$F_n(h,k) = \sum_l F(h,k,l) \exp\left\{2\pi i \frac{l z_n}{c}\right\} . \qquad (25)$$

Hence each summation in (24) is replaced by summations over h_n, k_n and l_n and the exponential of (25) is added to complicate the exponentials of (24).

11.1.4. General expression in terms of excitation errors

In order to simplify the result we introduce the quantity

$$\zeta_n = -\frac{\lambda}{2}\left\{\left(\frac{h^{(n)}}{a}\right)^2 + \left(\frac{k^{(n)}}{b}\right)^2\right\} - \frac{l^{(n)}}{c} + \frac{\lambda}{2}\left(\frac{u_0 h^{(n)}}{a} + \frac{v_0 k^{(n)}}{b}\right). \qquad (26)$$

This, in the small angle approximation, is exactly the excitation error for a reciprocal lattice point with coordinates

$$h^{(n)} = \sum_{r=1}^{n} h_r , \quad k^{(n)} = \sum_{r=1}^{n} k_r , \quad l^{(n)} = \sum_{r=1}^{n} l_r ,$$

since the Ewald sphere is approximated by the paraboloid

$$w = -\tfrac{1}{2}\lambda(u^2+v^2) + \tfrac{1}{2}\lambda(u_0 u + v_0 v) . \qquad (27)$$

The sign of ζ_n is chosen to conform with the convention that the excitation error should be positive for reciprocal lattice points inside the Ewald sphere, with the incident beam in the positive z direction. It is interesting to note that the Ewald sphere curvature, implied by the form (26) for the excitation error, is derived as a direct consequence of the form used for the propagation function and so represents the reciprocal space analog of Fresnel diffraction.

With the substitution (26) the general two-dimensional form of (24) becomes

$$\Psi(h,k) = (-\mathrm{i}\Delta z)^N \exp\{\pi \mathrm{i}\lambda H(u^2+v^2)\}$$

$$\times \sum_{h_1}\sum_{k_1}\sum_{l_1}\sum_{h_2}\sum_{k_2}\sum_{l_2} \dots \sum_{h_{N-1}}\sum_{k_{N-1}}\sum_{l_{N-1}}\sum_{l} F(h_1,k_1,l_1)F(h_2,k_2,l_2)$$

$$\dots F\left(h-\sum_{n=1}^{N-1} h_n, k-\sum_{n=1}^{N-1} k_n, l-\sum_{n=1}^{N-1} l_n\right) \exp\left\{-2\pi \mathrm{i}\left(H\zeta - \Delta z \sum_{n=1}^{N-1} \zeta_n\right)\right\}, \tag{28}$$

where ζ is the excitation error for the h, k, l reflection when $h = \sum_{n=1}^{N-1} h_n$, and so on. We note that while the structure amplitudes $F(h_n, k_n, l_n)$ refer to the nth scattering process only, the excitation errors ζ_n are cumulative and depend on the history of scatterings by all slices from 1 to n. This is the factor which makes it very difficult to simplify (28) further.

11.2. Multiple-scattering series

11.2.1. Zero-order scattering

If we now take the limit of N tending to infinity and Δz tending to zero, the number of summations in (28) becomes infinite. To proceed further we must systematize the choice of terms of the series. One way to do that is to note that, for each set of these summations over h_n, k_n, l_n, the zero term is very much larger than the rest since, from (17) and (18),

$$(-\mathrm{i}\Delta z) F(0,0,0) = 1 - \mathrm{i}\sigma\Delta z \Phi(0,0,0) - \Delta z M(0,0,0),$$

$$(-\mathrm{i}\Delta z) F(h,k,l) = -\mathrm{i}\sigma\Delta z \Phi(h,k,l) - \Delta z M(h,k,l). \tag{29}$$

Hence the largest single contribution to (28) comes where the $h, k, l = 0$ value is chosen from each set of summations. The next largest contributions correspond to the case when the zero term is chosen from all summations except one. These contributions give the singly-scattered radiation. By allowing successively two, three and more slices to give non-zero reflections we then generate the doubly, triply and multiply-scattered contributions and so obtain the equivalent of a Born series.

The zero order term of this series will be

$$\Psi_0(h,k) = \lim_{N\to\infty} \{(-\mathrm{i}\Delta z)^N F^N(0,0,0)\}$$

$$= \lim_{N\to\infty} [1 - \mathrm{i}\sigma\Delta z\,\Phi(0,0,0) - \Delta z M(0,0,0)]^N$$

$$= \exp\{-\mathrm{i}\sigma H\Phi(0,0,0) - HM(0,0,0)\}\,. \tag{30}$$

This corresponds to the transmission of the incident beam with a phase change appropriate to the mean inner potential and attenuation by the mean absorption coefficient.

11.2.2. Single scattering – kinematical approximation

The first-order term takes account of single scattering only. We may consider that the incident beam is transmitted without scattering through all slices but the mth. Then the summations of (28) are reduced by putting all $h_n, k_n, l_n = 0$ except that $h_m, k_m, l_m = h, k, l$. Then, from (26)

$$\zeta_n = \begin{cases} 0 & \text{for} \quad n < m\,, \\ \zeta & \text{for} \quad n \geqslant m\,. \end{cases} \tag{31}$$

Then summing for all values of m, (28) gives us

$$\Psi_1(h,k) = (-\mathrm{i}\Delta z)\,F^{N-1}(0,0,0) \sum_m \sum_l F(h,k,l) \exp\{-2\pi \mathrm{i} z_m\,\zeta\}\,. \tag{32}$$

In the limit that N tends to infinity, the summation over m is replaced by an integral over z from 0 to H giving

$$\Psi_1(h,k) = \exp\{-\mathrm{i}\sigma H\Phi(0,0,0) - HM(0,0,0)\}$$

$$\times\,(-\mathrm{i}) \sum_l F(h,k,l) \exp\{-\pi \mathrm{i} H\zeta\}\,[(\sin \pi H\zeta)/\pi\zeta]\,. \tag{33}$$

Assuming the excitation error to be large for all l except $l = 0$ and that $\Phi(h,k,0)$ and $M(h,k,0)$ are real, the intensity of the single-scattered diffracted beam is

$$|\Psi_1(h,k)|^2$$

$$= \exp\{-2HM(0,0,0)\}\,[\sigma^2\Phi^2(h,k,0) + M^2(h,k,0)]\,\frac{\sin^2(\pi H\zeta)}{(\pi\zeta)^2}\,, \tag{34}$$

and this is exactly the expression for kinematical scattering from a parallel-side plate-shaped crystal of thickness, H.

11.2.3. Multiple scattering

In much the same way it is possible to find the contribution to the amplitude when scattering, other than in the central beam, may take place in two, three or any number of slices. For example, the triple scattering term is found by considering that all $h_n, k_n, l_n = 0$ except that $h_m, k_m, l_m = h_1, k_1, l_1$, $h_p, k_p, l_p = h_2, k_2, l_2$ and $h_r, k_r, l_r = h-h_1-h_2, k-k_1-k_2, l-l_1-l_2$. Then

$$\zeta_n = \begin{cases} 0 & \text{for} \quad 0 < n < m \\ \zeta_1 & \text{for} \quad m \leqslant n < p \\ \zeta_2 & \text{for} \quad p \leqslant n < r \\ \zeta & \text{for} \quad r \leqslant n \leqslant N\,. \end{cases} \tag{35}$$

The term in the final exponent in (28) becomes

$$H\zeta \quad \Delta z \sum_{n=1}^{N-1} \zeta_n = z_1 \zeta_1 + z_2(\zeta_2 - \zeta_1) + z_3(\zeta - \zeta_2)\,. \tag{36}$$

In the limiting case, the amplitude expression is integrated over z_3 from z_2 to H, over z_2 from z_1 to H and over z_1 from 0 to H, giving with some rearrangement of terms,

$$\Psi_3(h,k) = \exp\{-\mathrm{i}\sigma H\Phi(0,0,0) - HM(0,0,0)\}$$

$$\times \sum_{h_1}\sum_{k_1}\sum_{l_1}\sum_{h_2}\sum_{k_2}\sum_{l_2}\sum_{l} F(h_1,k_1,l_1)F(h_2,k_2,l_2)F(h-h_1-h_2,k-k_1-k_2,l-l_1-l_2)$$

$$\times \frac{\exp\{-2\pi\mathrm{i}\zeta H\}}{(2\pi)^3}\left\{\frac{\exp\{2\pi\mathrm{i}\zeta_1 H\}-1}{\zeta_1(\zeta_1-\zeta_2)(\zeta_1-\zeta)} + \frac{\exp\{2\pi\mathrm{i}\zeta_2 H\}-1}{\zeta_2(\zeta_2-\zeta_1)(\zeta_2-\zeta)} + \frac{\exp\{2\pi\mathrm{i}\zeta H\}-1}{\zeta(\zeta-\zeta_1)(\zeta-\zeta_2)}\right\}. \tag{37}$$

Generalizing the result for n-times scattering gives

$$\Psi_n(h,k) = \exp\{-\mathrm{i}\sigma H\Phi(0,0,0) - HM(0,0,0)\}$$

$$\times \sum_l \sum_{h_1} \sum_{k_1} \sum_{l_1} \ldots \sum_{h_{n-1}} \sum_{k_{n-1}} \sum_{l_{n-1}} F(h_1,k_1,l_1)F(h_2,k_2,l_2)$$

$$\ldots F\left(h - \sum_{r=1}^{n-1} h_r, k - \sum_{r=1}^{n-1} k_r, l - \sum_{r=1}^{n-1} l_r\right) \frac{\exp\{-2\pi\mathrm{i}\zeta H\}}{(2\pi)^n}$$

$$\times \left[\frac{\exp\{2\pi\mathrm{i}\zeta_1 H\} - 1}{\zeta_1(\zeta_1-\zeta_2)\ldots(\zeta_1-\zeta_{n-1})(\zeta_1-\zeta)}\right.$$

$$+ \sum_{m=2}^{n-1} \frac{\exp\{2\pi\mathrm{i}\zeta_m H\} - 1}{\zeta_m(\zeta_m-\zeta_1)\ldots(\zeta_m-\zeta_{m-1})(\zeta_m-\zeta_{m+1})\ldots(\zeta_m-\zeta)}$$

$$\left. + \frac{\exp\{2\pi\mathrm{i}\zeta H\} - 1}{\zeta(\zeta-\zeta_1)(\zeta-\zeta_2)\ldots(\zeta-\zeta_{n-1})}\right]. \tag{38}$$

The total amplitude of the h, k, diffracted beam is then

$$\Psi(h,k) = \sum_{n=0}^{\infty} \Psi_n(h,k). \tag{39}$$

which is the Born Series type of expression for single, double and multiple scattering from planes in the crystal. The same result was obtained by different methods by Fujimoto [1959] and by Fujiwara [1959] with a little more generality in that the z components of the wave vectors for the individual reflections were included in the denominators to take account of the slight differences in direction of the corresponding waves.

As in the case of the more conventional Born Series, for scattering by volume elements, the convergence is slow when the scattering power or crystal thickness is too great for the kinematical single-scattering approximation to be used. Hence we seek other forms, more appropriate for situations in which the kinematical approximation can not be used.

11.3. General double-summation solution

11.3.1. General series solution

It was shown by Moodie (Cowley and Moodie [1962]) that the expression

(39) could be rewritten in the form

$$\Psi(\boldsymbol{h}) = \sum_{n=1}^{\infty} \boldsymbol{E}_n(\boldsymbol{h}) Z_n(\boldsymbol{h}) , \tag{40}$$

where we have used $\boldsymbol{h}$ to signify the pairs of indices h, k. The operator $\boldsymbol{E}_n(\boldsymbol{h})$ represents all the operations of summation over indices and multiplication by F quantities in the expression (38). This operates on the functions $Z_n(\boldsymbol{h})$ which are purely geometric, depending on the wavelength, unit cell dimensions and crystal size but independent of scattering amplitudes. Expanding the exponentials of (38) and manipulating the terms gives

$$Z_n(\boldsymbol{h}) = \sum_{r=0}^{\infty} \frac{(-2\pi i H)^{n+r}}{(n+r)!} h_r(\zeta, \zeta_1, \zeta_2 \ldots \zeta_{n-1}) , \tag{41}$$

where the functions $h_r(\ldots)$ are the complete homogeneous symmetric polynomials of degree r i.e.

$$\begin{aligned} h_1(\zeta, \zeta_1, \zeta_2 \ldots \zeta_{n-1}) &= \zeta + \zeta_1 + \zeta_2 + \ldots + \zeta_{n-1} , \\ h_2(\zeta, \zeta_1, \zeta_2 \ldots \zeta_{n-1}) &= \zeta^2 + \zeta_1^2 + \zeta_2^2 + \ldots + \zeta\zeta_1 + \zeta\zeta_2 + \ldots + \zeta_1\zeta_2 + \ldots \end{aligned} \tag{42}$$

Each polynomial contains, once only, each possible combination of variables, irrespective of ordering, which is of the prescribed power. The excitation errors, ζ_m, in these expressions are defined as in (26), and so depend on all indices for all scattering processes up to $n = m$.

Writing out (41) in detail then gives

$$Z_1(h) = (-2\pi i H) + \frac{(-2\pi i H)^2}{2!} \zeta + \frac{(-2\pi i H)^3}{3!} \zeta^2 + \ldots$$

$$Z_2(h) = \frac{(-2\pi i H)^2}{2!} + \frac{(-2\pi i H)^3}{3!} (\zeta + \zeta_1) + \frac{(-2\pi i H)^4}{4!} (\zeta^2 + \zeta_1^2 + \zeta_1 \zeta) + \ldots$$

$$\begin{aligned} Z_3(h) &= \frac{(-2\pi i H)^3}{3!} + \frac{(-2\pi i H)^4}{4!} (\zeta + \zeta_1 + \zeta_2) \\ &+ \frac{(-2\pi i H)^5}{5!} (\zeta^2 + \zeta_1^2 + \zeta_2^2 + \zeta\zeta_1 + \zeta\zeta_2 + \zeta_1\zeta_2) + \ldots \quad \text{etc.} \end{aligned} \tag{43}$$

Then adding the appropriate operators $\boldsymbol{E}_n(h)$ we obtain a two-dimensional array of terms with n increasing vertically and r horizontally. This two-dimensional array can then be summed in a number of ways. For example, we may sum the rows horizontally. This gives for the first row

$$\Psi_1(h) = \sum_l F(h, k, l) \left[\sum_r \frac{(-2\pi i H\zeta)^r}{r!} - 1 \right] \frac{1}{\pi\zeta}$$

$$= (-i) \sum_l F(h, k, l) \exp\{-\pi i H\zeta\} \left(\frac{\sin \pi H\zeta}{\pi\zeta} \right),$$

which is identical with (33) apart from the zero term. Similarly summing each row horizontally gives us the multiple-scattering series $\Psi_n(h)$ from which we started.

Fujimoto [1959] derived an expansion in terms of increasing powers of the crystal thickness H. This can be reproduced by selecting terms from (43) and summing along diagonal lines.

11.3.2. Phase grating approximation

If the terms of (43) are selected along the vertical lines, with the appropriate $\boldsymbol{E}$ operators, we obtain series for various r values. The first is

$$T_1(h) = \sum_{n=1}^{\infty} \frac{(-2\pi i H)^n}{(2\pi)^n n!}$$

$$\times \sum_{h_1} \sum_{k_1} \sum_{l_1} \sum_{h_2} \sum_{k_2} \sum_{l_2} \cdots \sum_{h_{n-1}} \sum_{k_{n-1}} \sum_{l_{n-1}} F(h_1, k_1, l_1) F(h_2, k_2, l_2)$$

$$\ldots F\left(h - \sum_{p=1}^{n-1} h_p, k - \sum_{p=1}^{n-1} k_p, l - \sum_{p=1}^{n-1} l_p \right). \qquad (44)$$

The multiple summation represents the amplitude of the (h,k) term of the $(n-1)$ fold self-convolution of

$$F(\boldsymbol{u}) = \sum_h \sum_k \sum_l F(h, k, l)\, \delta\left(u - \frac{h}{a}, v - \frac{k}{b}, w - \frac{l}{c} \right).$$

Hence the set of diffracted beam amplitudes $T_1(h, k, 0)$ represented by the function

$$T_1(u, v) = \sum_h \sum_k \sum_l T_1(h, k, l)\, \delta\left(u - \frac{h}{a}, v - \frac{k}{b} \right)$$

will be the Fourier transform of

$$\tau_1(x,y) = \sum_{n=1}^{\infty} \frac{(-2\pi \mathrm{i} H)^n}{(2\pi)^n n!} \{\sigma\varphi(x,y) - \mu(x,y)\}^n$$

$$= \exp\{-\mathrm{i}\sigma H\varphi(x,y) - H\mu(x,y)\}\,, \tag{45}$$

where $\varphi(x,y)$ and $\mu(x,y)$ are the values per unit thickness of the projections in the z direction of the potential and absorption functions $\varphi(\boldsymbol{r})$ and $\mu(\boldsymbol{r})$.

Thus the expression (44) given by summing the first vertical column from (43) gives the amplitudes of the diffracted beams when all excitation errors are assumed to be zero, as if the Ewald sphere were planar or the wavelength were zero. Thus it gives a high-energy approximation.

It is also the "phase-grating" approximation, given by assuming that the total scattering and absorption power of the crystal is concentrated on a single plane. More commonly the functions $\varphi(x,y)$ and $\mu(x,y)$ are taken to be the projections in the z direction of the potential distribution $\varphi(x, y, z)$ and the absorption function $\mu(x, y, z)$. The crystal is then replaced by a two-dimensional distribution having transmission function

$$q(x,y) = \exp\{-\mathrm{i}\sigma\varphi(x,y) - \mu(x,y)\}\,. \tag{46}$$

This phase-grating approximation can be considered as the first term of a series, which has been referred to as the "phase-grating series", for which the successive terms involve the successive vertical lines of terms in (43) and may be expected to introduce the three-dimensional diffraction effects with increasing accuracy. However for terms of this series beyond the first no convenient form of simplification has been found. In fact it is not easy to see that the higher terms can be very useful since they contain successively higher positive powers of the excitation errors which apparently emphasize the less important reflections; those for which the excitation errors are large.

Hence the general expression (40) forms a basis for a variety of series expansions each of which provides a relatively simple approximation to the general result valid in particular circumstances. The kinematical approximation is valid in the limits of the small structure factors for a given thickness. The phase-grating approximation is appropriate in the zero wavelength limit or in the limit of zero thickness.

The other simple approximation of proven usefulness, the two-beam approximation, is generated from the general expression by assuming that only two beams can exist; or that $\zeta_0 = 0$, ζ_h is small and ζ for any other reflection is very large. On this basis Cowley and Moodie [1957] reduced the expressions such as (38) to a set of terms which summed to give exactly the same amplitudes for the incident and diffracted beams as are given by Bethe theory.

11.4. Computing methods

11.4.1. "Slice method" calculations

On the basis of this "physical optics" approach to the diffraction problem, Goodman and Moodie [1965, 1974] (see also Lynch [1971]) developed a computing technique which enables the intensities of diffraction patterns or electron microscope images to be calculated with arbitrary accuracy by integrating the effects on a wave transmitted through a crystal. It is, of course, not feasible to follow the analytical process of considering the limiting case of an infinite number of slices of zero thickness. Instead a finite number of slices is used and a rigid testing routine is established to ensure that the slice thickness is small enough to give no appreciable errors.

For a slice of finite thickness the approximation of equation (4), the single scattering approximation, can not be used since we have seen previously that multiple scattering may be important even for a single heavy atom. Then the transmission function for a slice must be written in terms of the phase-grating approximation as

$$q_n(x,y) = \exp\{-\mathrm{i}\sigma\varphi_n(x,y) - \mu_n(x,y)\}\,, \tag{47}$$

$$\begin{aligned} Q_n(u,v) &= \mathcal{F}[q_n(x,y)] \\ &= \mathcal{F}[\exp\{-\mu_n(x,y)\}\cos\sigma\varphi_n(x,y)] \\ &\quad - \mathrm{i}\mathcal{F}[\exp\{-\mu_n(x,y)\}\sin\sigma\varphi_n(x,y)]\,. \end{aligned} \tag{48}$$

The propagation function used for transfer of the wave from one plane to another, $p_n(x,y)$, can be written in the small angle approximation (10) for most purposes without appreciable error. Usually for periodic objects the reciprocal space form is preferable.

The total effect of a crystal on an incident wave is then given by the repetition $(N-1)$ times of the unit process which converts a wave leaving the $(n-1)$th slice to the wave leaving the n slice, namely

$$\Psi_n(u,v) = \{\Psi_{n-1}(u,v)P_{n-1}(u,v)\} * Q_n(u,v)\,, \tag{49}$$

or, for a periodic object,

$$\Psi_n(h,k) = \sum_{h'} \sum_{k'} \Psi_{n-1}(h',k') P_{n-1}(h',k') Q_n(h-h',k-k') \,. \tag{50}$$

The process of multiplication and addition of a set of (complex) numbers represented by (50) is rapidly performed on a digital computer. Then starting from a known incident wave $\Psi_0(u,v)$, repetition of this process gives the diffracted amplitudes after transmission through any number of crystal slices. The real space amplitude $\psi_N(x,y)$ at the exit surface is obtained by summing the Fourier series with $\Psi_N(h,k)$ as coefficients and from this the in-focus microscope image is obtained. The out-of-focus images are given by multiplying the $\Psi_N(h,k)$ by the appropriate propagation function transform values $P_R(h,k)$.

It is convenient for the purposes of calculation if the $Q_n(h,k)$ and $P_n(h,k)$ values are the same for each slice. This is possible for a perfect crystal if the unit cell dimensions in the beam direction are sufficiently small, as is usually the case for relatively simple crystal structures. However it is not necessary for this to be the case. If the periodicity in the beam direction is large, the unit cell contents may be divided into a sequence of slices which is then repeated regularly in the calculations. Short-range or long-range changes in the periodicity or structure of the crystal which do not involve changes in the lateral periodicities may be simulated by varying the $Q_n(h,k)$ values, or the inter-slice distances occurring in $P_n(h,k)$, for individual slices or for any number of slices. Thus the effects of a wide range of perturbations of the crystal structure may be included without any essential modification of the computer program. In conjunction with the use of a column approximation, the program may be used to calculate the effects on diffraction patterns or microscope images of localized perturbations of the crystal lattice (for example, dislocation images).

Within each column the lateral displacements of the atoms are assumed constant and represented by multiplying $Q_n(h,k)$ by a phase factor. Displacements in the beam direction could be included by varying the $P_n(h,k)$ values. Perturbations of a crystal lattice for which a column approximation is not adequate may be treated by special methods to be described later.

11.4.2. Steps in a computation

In order to clarify the nature of the calculations, we now set out the steps involved for the simplest case of the incident beam parallel to the c-axis of a centro-symmetric orthorhombic crystal, having no absorption. The c-axis dimension is taken as the slice thickness. The program is set up as follows:

A. *Calculation of phase-grating amplitudes for slices.*

(a) The structure amplitudes $\Phi(h, k, 0)$ are calculated from the atomic scattering factors and atom positions.

(b) The unit cell projection $\varphi(x,y)$ is obtained by summing the Fourier series with $\Phi(h, k, 0)$ as coefficients.

(c) The real and imaginary parts of $\exp\{-i\sigma\varphi(x,y)\}$ are calculated.

(d) Fourier transform of $\cos\{\sigma\varphi(x,y)\}$ and $\sin\{\sigma\varphi(x,y)\}$ gives the real and imaginary parts of $Q(h,k)$.

B. *The propagation function.*

(a) The excitation errors, $\zeta_{\boldsymbol{h}}$, are found for the required range of indices h, k. For this the paraboloidal approximation to the Ewald sphere is usually sufficient.

(b) The phase factors $P(h,k) = \exp\{2\pi i c\zeta(h,k)\}$ are calculated.

C. *Iterative Calculation.*

(a) For the first slice the amplitudes for the wave leaving the slice are $Q(h,k)$.

(b) For the second slice, $\Psi_2(h,k)$ values are calculated by performing the summation (50).

(c) The summation (50) is repeated for each successive slice.

D. *Output.*

(a) The diffracted beam intensities as a function of crystal thickness are given by calculating $\Psi_n \Psi_n^*$ after each slice.

(b) In-focus images are found by summing the Fourier series to give $\psi_n(x,y)$, then calculating $\psi_n \psi_n^*$.

(c) For out-of-focus and aberrated images: the effect of defocus is given by multiplying the amplitudes $\Psi_n(h,k)$ by $P_R(h,k) = \exp\{2\pi i R\zeta(h,k)\}$ before summing the Fourier series of (b). The effect of lens aberrations is simulated by adding a phase shift depending on higher orders of h/a and k/b.

The addition of an absorption term makes the amplitudes $\Phi(h, k, 0)$ complex, but since the remainder of the computation is made with complex numbers in any case, this introduces very little further complication. If the incident beam is not exactly parallel to a crystal axis a tilt with components α_x, α_y may be included by adding a translation by amounts $c \tan\alpha_x$ and $c \tan\alpha_y$ to the propagation function which then becomes

$$P_\alpha(h,k) = \exp\left\{\pi i\lambda c\left(\frac{h^2}{a^2} + \frac{k^2}{b^2}\right)\right\} \exp\left\{-2\pi i c\left(\frac{h}{a}\tan\alpha_x + \frac{k}{b}\tan\alpha_y\right)\right\}. \quad (51)$$

Alternatively it may be convenient to assume that the incident beam direction is unchanged but that successive slices of crystal are translated to give the equivalent change in diffraction conditions, so that

$$Q_n(h,k) = Q_1(h,k) \exp\left\{2\pi i\left[\frac{hc}{a}(n-1)\tan\alpha_x + \frac{kc}{b}(n-1)\tan\alpha_y\right]\right\}. \tag{52}$$

More exactly, the repeat distances a and b should be modified by factors $\cos\alpha_x$ and $\cos\alpha_y$ and the distance c should be slightly increased. These tilted beam approximations are suitable only for tilts of a few degrees. For large tilts different projections of the structure should be considered.

11.4.3. Possible errors

One of the most important approximations in this computing method is that of taking a finite slice thickness. To test whether any assumed slice thickness is small enough, the calculation is repeated for smaller slice thicknesses and the results compared. In this way it has been found that for electrons of energy of the order of 100 keV and for moderately simple structures with moderately heavy atoms, no appreciable error results for slice thicknesses up to about 4 Å. For slice thicknesses of 8 to 10 Å errors of several per cent are introduced, especially for weak beams. Slice thicknesses of 12 or 15 Å give serious errors so that only the strongest beams show even qualitatively correct behavior.

A further important possible source of error is the failure to take a sufficient number of diffracted beams into account. For matrix method calculations we saw that the exclusion of significant beams distorts the intensity variations of those included since a false model is imposed on the crystal structure. For the slice-method calculations however, omission of some significant beams means that the interaction of these beams with other beams will be excluded. Energy is scattered into these beams but can not be scattered back, and so is lost to the system.

11.4.4. Consistency tests

The test imposed to determine whether sufficient beams are included is then to sum the intensities of all diffracted beams considered at each stage of the calculations. In the absence of absorption this sum should be constant and equal to the incident beam intensity with an accuracy of better than one part in 10^5 or more. As in the matrix calculations, it is found that for a two-dimensional diffraction pattern from a simple face-centered cubic metal, for example, at least 50 and preferably more beams should be included, and the number should be increased for substances having larger unit cells or for higher electron accelerating voltages.

The requirement of conservation of energy is part of a more general consistency requirement which should be met for each phase grating slice. In the absence of absorption

$$|q(x,y)|^2 = |\exp\{i\sigma\varphi(x,y)\}|^2 = 1\ . \tag{53}$$

In reciprocal space the corresponding relationship is

$$\sum_{h'}\sum_{k'} Q(h',k')\,Q(h+h',k+k') = \begin{cases} 1 & \text{if} \quad h,k=0 \\ 0 & \text{otherwise}\ . \end{cases} \tag{54}$$

Sets of $Q(h,k)$ values may be considered satisfactory if this relationship is satisfied with an accuracy of, typically, one part in a million. This is necessary in order that the errors will not accumulate to a significant extent in calculations involving the repeated use of these values.

11.4.5. One-dimensional calculations

Frequently, in order to reduce the amount of calculation required, those special orientations are considered for which only a single line of reflections in the diffraction pattern is strong i.e. the "systematic set" of beams corresponding to a line of reciprocal lattice points through the origin. Then we deal with the one-dimensional function,

$$q(x) = \exp\{-i\sigma\varphi(x) - \mu(x)\}\ , \tag{55}$$

where $\varphi(x)$ and $\mu(x)$ represent projections in the y and z directions. Then the summations are made over the single index h; the number of beams considered is reduced to from 7 to 15 for simple structures and the amount of calculation is very much less.

Justification for use of this one-dimensional form of calculation is not obvious, but extensive tests of its accuracy as compared with complete two-dimensional calculations for crystals in the same orientation have been carried out. For example McMahon [1969] compared one-dimensional systematic calculations for the $h00$ reflections of MgO with two-dimensional calculations for orientations chosen to minimize the number and strength of the non-systematic reflections. For the most favorable case, that illustrated in Fig. 15.7, the calculated convergent beam diffraction spot intensity profiles for the systematic and two-dimensional calculations agreed very well, but only if the structure amplitudes for the systematic case were taken one per cent too high.

For other slightly less favorable orientations use of the one-dimensional calculations gave larger deviations corresponding to 1.5 to 2 per cent error in structure amplitudes and for an arbitrary orientation, giving mainly the systematic set of reflections, the errors could be considerably higher. However accuracies of 1 or 2 per cent are required only for experiments such as these, designed to give highly accurate values of structure amplitudes, and for many purposes it may be assumed that one-dimensional calculations give a reasonably good indication of the diffraction intensities unless there are obvious perturbations possible from non-systematic interactions.

A comparison of this slice-method of computing and the matrix method mentioned in the previous chapter shows that for many purposes there is little advantage of one method over the other in computing time. In a series of tests carried out by P.M.J. Fisher, P.S. Turner and P.M. Warburton in Melbourne in 1968 calculations by the two methods were compared using in each case 49 beams from a copper crystal in [100] orientation and calculating intensities for thicknesses from 0 to 2000 Å at 4 Å intervals. The two methods gave intensities agreeing to within 10^{-5} of the total intensity and in each case the computing time was about 7 minutes on a medium-sized computer (an IBM 7044). For other types of calculations or on other types of computer either one or the other method may have advantages. The slice method may be better for thin crystals and for higher accelerating voltages and will have very definite advantages for the non-periodic objects to be considered next.

11.5. Intensities from non-periodic objects

Calculations of dynamical scattering from non-periodic objects, such as defects in crystals or small particles or molecules, have nearly all been made by use of the column approximation as discussed in Chapter 10. For each column of the sample, the calculation is made using one of the methods described in the last chapter or the slice method, which allow the modifications of the structure or displacements of unit cells to be included.

The principle limitation of this technique is that it is concerned only with the modifications of the amplitudes of the set of diffracted beams, $\boldsymbol{h}$, which correspond to the reciprocal lattice points for the perfect structure. As we have seen in Chapter 7, diffraction effects from crystal imperfections, as from general non-periodic objects, are not confined to this discrete set of beams. Much of the information on the nature of the defect or on the atomic configurations in a non-crystalline assembly of atoms, is contained in the continuous distribution of back-ground scattering in the diffraction pattern. This kinematical result will be equally relevant for the phase-grating scattering from each of the thin slices considered in the dynamical scattering formulation. Hence the diffuse scattering must be taken into account for any realistic consideration of the diffraction effects or images for any but very special kinds of deviations from perfect crystal periodicity.

Since it is not feasible to deal with continuous scattering functions in computer calculations, the distributions $Q_n(u,v)$ corresponding to non-periodic transmission functions $q_n(x,y)$ must be sampled at closely spaced points and so are replaced by the sets of values

$$\sum_H \sum_K Q_n(H,K)\,\delta\left(u - \frac{H}{A}, v - \frac{K}{B}\right), \tag{56}$$

where the intervals $1/A$ and $1/B$ are made small enough to allow a reasonable representation of the fluctuations of interest in $Q_n(u,v)$. The equivalent operation in real space is to consider that the region of interest in $q_n(x,y)$ is reproduced at regular intervals A, B thus

$$q_n(x,y) * \sum_m \sum_p \delta(x - mA, y - pB)\,. \tag{57}$$

Thus the crystal defect, or other arrangement of atoms of interest, is repeated regularly to form a superlattice having the diffraction pattern (56). Then the intensities of the diffraction pattern or image for this superlattice may be calculated using the perfect-crystal slice-method computer programs. The variation of scattering in the z direction is included by varying the contents of the successive slices.

Such a calculation may involve a very large number of diffracted beams corresponding to the very large unit cell size for the superlattice which must normally be assumed. The calculation may often be brought within reasonable bounds by use of special assumptions such as the consideration of a one-dimensional case. For example, it may be considered that the "core" of a dislocation, the region in which atom displacements are large, is of the order of 20 Å in diameter. If it is desired to calculate the image of such a dislocation with a resolution of 3 Å the diffraction pattern must be sampled out to $u = 0.33$ Å^{-1}. The 20 Å dimension of the core will give detailed structure in the diffraction pattern on a scale of 0.05 Å^{-1}. Then the diffuse scattering may be sampled at intervals of 0.01 Å^{-1}. The dynamical diffraction calculation is then made with $2 \times 33.3 = 67$ beams in one dimension. That means that, in real space, we assume parallel dislocations to exist at intervals of 100 Å. In the imaging process each dislocation will be imaged separately. There will be no overlapping of waves scattered from individual dislocations and no interference of their images unless the image is obtained far out of focus. Simple Fresnel diffraction theory gives this out-of-focus distance as approximately $A^2/2\lambda$ or, in this case, about 10 μm.

Calculations of this sort, involving the imposition of an artificial superlattice periodicity, have been carried out for dislocations by Fejes [1973] and for a completely non-periodic object chosen to simulate a rod-like protein molecule, negatively stained to increase the electron microscope contrast (Grinton and Cowley [1971]) with results which will be discussed in a later chapter. Calculations involving 12 000 beams have been used by O'Keefe and Iijima [1978] to simulate images of defects in oxide crystals, using a superlattice unit cell approximately 30 Å × 30 Å in size.

Problem

Derive the differential equation formulation of eqs. (10.35) and (10.36) from slice formulation, e.g. equation (11.49).

Section IV

APPLICATIONS TO SELECTED TOPICS

CHAPTER 12

Diffuse scattering and absorption effects

12.1. Thermal diffuse scattering

12.1.1. Phonons and vibrational waves

Apart from being scattered by the average periodic structure of a crystal, the incident radiation may interact with the crystal in a number of ways. The interaction may be described within the framework of the quantum mechanical theory. A quantum of radiation interacts to create or annihilate an excitation of the crystal such as a phonon, plasmon or exciton with consequent loss or gain of energy. The result of the interaction is, in general, the production of diffuse inelastically scattered radiation in the background of the diffraction pattern with an associated loss of intensity from the sharp Bragg reflections from the average periodic structure.

The main object of studying these effects is to investigate the nature and properties of the crystal excitations. A secondary, but still important aim, is to find the diffuse scattering and absorption effects due to these processes with sufficient accuracy so that they may be subtracted out from measurements of diffraction intensities required for crystallographic purposes.

So long as we stay within the kinematical or first Born approximation we may describe the scattering due to these excitations by use of the generalized Patterson function which we introduced in Chapter 7 and which we prefer to use because it forms a natural extension of our other considerations of kinematical scattering.

In Chapter 7 we used the simple Einstein model of thermal motion of atoms as an example for the use of the generalized Patterson function following Van Hove [1954]. Here we apply the same method to introduce a rather more realistic discussion of the atomic vibrations in crystals which constitute the most universal and important of crystal excitations.

The assumption of the Einstein model is that the motions of all atoms are independent. At the other extreme it is customary in rather more sophisticated treatments to consider the motion to be strongly correlated. The displacements of the atoms are considered to be given by the sums of displacements corresponding to the individual lattice waves which can occur in the crystal with a relationship between wavelength (or wave vector $\boldsymbol{k}$, or momentum $h\boldsymbol{k}$) and frequency ω (or energy $h\omega$) given by the dispersion relationship for the material. The simple model is that each individual lattice wave may be represented by a plane wave with correlation of atom displacements extending throughout the crystal and so, effectively, to infinity in space and time. This is an over-simplified picture since the range of correlations of atom displacements is limited in both space and time. One may say that a phonon has a finite extent in space and a finite lifetime, being not much more than one wavelength or one period, respectively, for high frequency phonons in many cases. However we stick to the simpler picture of infinite waves since this gives results which are usually not too far wrong.

12.1.2. Scattering for a longitudinal wave

For a longitudinal wave travelling in the x-direction, the displacements of the atoms from their mean positions may be written

$$\Delta = A \cos 2\pi(lx - \nu_l t) \tag{1}$$

where $l = 1/\Lambda$ and Λ is the wavelength and ν_l the frequency of the lattice wave. The electron density function for an atom displaced from the average lattice position at $x = X$ is

$$\varphi_0(x, t) = \rho_0(X + \Delta, t)$$

$$= \rho_0(X) + \Delta(X, t) \left.\frac{\mathrm{d}\rho_0}{\mathrm{d}x}\right|_{x=X} + \frac{\Delta^2(X, t)}{2!} \left.\frac{\partial^2 \rho_0}{\partial x^2}\right|_{x=X} + \dots . \tag{2}$$

The corresponding generalized Patterson function is then

$$P(r, t) = \left[\rho_0 * \sum_n \delta(x - x_n)\right] * \left[\rho_0(-r) * \sum_m \delta(x + x_m)\right]$$

or we can write

$$P(r, t) = \bar{\rho} * \bar{\rho} + (\bar{\rho} - \rho) * (\bar{\rho} - \rho), \tag{3}$$

where the convolution is with respect to x and t.

Here we have separated the Patterson into the Patterson of the spatially periodic, time invariant, averaged electron density $\bar{\rho}$ and the Patterson function for the deviations from this average, in analogy with equation (7.9).

From (2) it is readily seen that in averaging over space or time, $\langle\Delta(t)\rangle = 0$ so that

$$\bar{\rho}_0 = \rho_0(X) + \tfrac{1}{2}\langle\Delta^2\rangle \frac{\partial^2 \rho_0}{\partial x^2} + \dots .$$

Fourier transforming gives

$$\bar{F}(u) = F_0(u)\left[1 - \frac{4\pi^2}{2} u^2\langle\Delta^2\rangle + \dots\right]$$

$$\approx F_0(u) \exp\{-2\pi^2\langle\Delta^2\rangle u^2\} ,$$

and the average intensity function in reciprocal space is

$$\bar{F}\cdot\bar{F}^* = (F_0(u)|^2 \exp\{-4\pi^2\langle\Delta^2\rangle u^2\} , \tag{4}$$

where the exponential is the contribution to the Debye–Waller factor for this wave.

12.1.3. Diffuse scattering component

The second part of the Patterson, equation (3), giving the diffuse scattering is, to a first approximation for a small Δ,

$$\rho_0'(x) * \sum_n \delta(x - na) A \cos 2\pi(lna - \nu_l t) * \rho_0'(-x)$$

$$* \sum_m \delta(x + ma) A \cos 2\pi(lma - \nu_l t) ,$$

where the prime on $\rho_0'(x)$ denotes differentiation. Carrying out the convolution with respect to both x and t, we get

$$\begin{aligned}\Delta\rho * \Delta\rho &= \sum_n \sum_m A^2 \cos 2\pi \{l(n - m)a - \nu_l t\} \\ &\quad \times [\rho_0'(x) * \rho_0'(x) * \delta[x - (n - m)a\}] \\ &= NA^2 \sum_n [\rho_0'(x) * \rho_0'(-x) * \delta(x - na)] \cos 2\pi(lna - \nu_l t) .\end{aligned} \tag{5}$$

Hence the intensity distribution function in reciprocal space is

$$I_{th}(u) = A^2 \{2\pi u F(u)\}^2 \sum_h \delta\left(u - \frac{h}{a}, v\right) * \delta(u \pm l, v \pm \nu_l). \tag{6}$$

The two functions (5) and (6) are sketched in Fig. 12.1(a), (b) and (c). In (a) and (b) the functions $\rho_0'(x) * \rho_0'(-x)$ and $4\pi^2u^2F^2(u)$ have not been in-

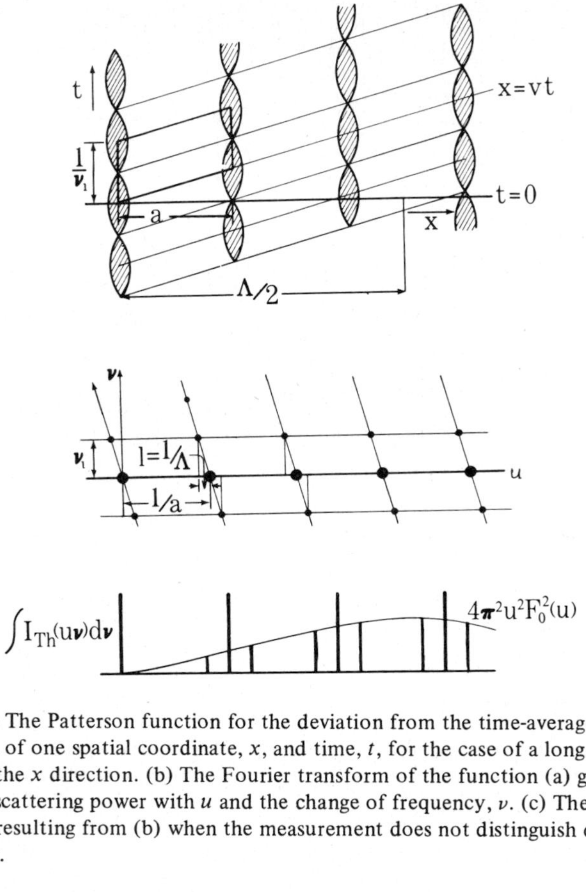

Fig. 12.1. (a) The Patterson function for the deviation from the time-averaged structure as a function of one spatial coordinate, x, and time, t, for the case of a longitudinal wave travelling in the x direction. (b) The Fourier transform of the function (a) giving the variation of scattering power with u and the change of frequency, ν. (c) The intensity distribution resulting from (b) when the measurement does not distinguish differences of frequency.

cluded in the diagram. The width of the lines parallel to the t axis represent the magnitudes (positive and negative) modulating the function $\rho_0'(x) * \rho_0'(-x)$ sketched on the left-hand side. The periodicity is obviously $1/\nu_l$ in the t direction. The other axis of the "unit cell" of the distribution is the constant-phase line $x = vt$ where v is the velocity of the lattice wave.

Correspondingly in reciprocal space, Fig. 12.1(b), one axis of the reciprocal lattice unit cell is the u direction, with a periodicity $1/a$, and the other axis, the ω axis, is perpendicular to the $x = vt$ line, with a periodicity corresponding to $\nu = \nu_l$. The average Patterson gives the strong scattering points at $u = h/a$, $\nu = 0$. The Patterson for the deviation from the average lattice adds to these the satellite spots for $u = \pm l$, $\nu = \pm \nu_l$.

From Fig. 12.1(b) we deduce that the purely elastic scattering $I(u, 0)$, the $\nu = 0$ section of $I(u, \nu)$, contains only the peaks corresponding to the average periodic lattice. The lattice wave introduces reflections with gain or loss of energy $h\nu_l$.

If, as in the case of X-ray or electron diffraction, no energy resolution is possible, what is measured is the total scattering, or the projection $\int I(u, \nu)\, d\nu$. From Fig. 12.1(c), this includes the strong diffraction spots due to the average lattice plus the weak satellite peaks due to the lattice wave. The former have intensity given by (4). The latter have intensity proportional to the square of the wave amplitude A, and also to $4\pi^2 u^2 F_0^2(u)$. This function is zero at the origin of reciprocal space, increases to a maximum usually for u of the order of 0.5 to 1 $Å^{-1}$ and then decreases slowly.

Each thermal wave occuring in the lattice will give a pair of satellite peaks around each reciprocal lattice point, with a displacement inversely proportional to the wavelength and the relative intensity depending on its amplitude squared, and so on the energy of the wave ($= h\nu_l$).

12.1.4. Dispersion curves

The relationship of the wave energy or frequency to the inverse wavelength, which is described by the dispersion curves, is linear for long wavelengths, the slope of the lines being given in the long wavelength limit by the elastic constants of the material. Hence the intensity of the satellite peaks will be inversely proportional to the square of the distance from the reciprocal lattice points. Then for a large number of possible wave with values of inverse wavelengths equally distributed, the satellites will merge into a diffuse intensity peaking sharply at the reciprocal lattice positions.

If the intensity distribution can be determined as a function of both the reciprocal distance, u, or momentum change and the change in frequency ν, or energy of the incident radiation, then the form of the dispersion curve may be derived. This may be done with neutron diffraction since the energy of the incident thermal neutron is of the order of 0.02 eV while the thermal waves have energies of the same order. The changes of energy of the incident neutrons are relatively large and can be deduced by using a diffracting crystal to analyse the energy (or wavelength) distribution of the scattered neutrons. For X-rays or electrons incident with energies in the range of tens of kilovolts or higher, the changes of energy due to the thermal waves are in general not detectable. On the assumption that the dispersion curves are linear in the long-wavelength range, measurements of the diffuse intensity allow the derivation of the elastic constants (Warren [1969]).

12.1.5. Three-dimensional generalizations

We have treated only the simplest case of longitudinal waves in one dimension. The three-dimensional case is further complicated by the existence of both longitudinal and transverse waves, each with different dispersion curves, and by the presence of both "optical" and "acoustic" modes of vibration. Thus the vector displacement $\boldsymbol{\Delta}$ may be at any angle with the diffraction vector $\boldsymbol{u}$ and we must consider the scalar product $\boldsymbol{u} \cdot \boldsymbol{\Delta}$. For a treatment including these complications we refer the reader to the various detailed discussions in the literature, including those by Hoppe [1956], Laval [1958], Warren [1969] for the X-ray diffraction case and Bacon [1962] for the description of neutron diffraction methods and measurements. See also Willis and Pryor [1975].

We have limited our discussion also to the first-order approximation valid for small displacements, Δ, only. Taking into account the higher order terms of the series (2) includes the effects of multiple scattering by single lattice waves. When more than one lattice wave must be present at a time it is necessary for completeness to consider also the successive scattering by different waves.

These effects are of importance in practice for accurate diffraction studies and particularly for studies of thermal diffuse scattering by crystals at high temperatures. It is easy to see how our treatment may be extended to include them. They have been discussed using other approaches by Paskin [1958, 1959], Borie [1970] and others.

For the general three-dimension case the expression (4) for the Debye–Waller factor remains valid if $\langle \Delta^2 \rangle$ is interpreted as the mean-square displacement of the atom in the direction of the diffraction vector. In general this value will not be isotropic and will vary with the type of atom and its environment. In accurate crystal structure analysis, three parameters defining a vibration ellipsoid are normally refined for each non-equivalent atom.

12.2. Static atom displacements

12.2.1. Relaxation around point defects

Diffuse scattering in diffraction patterns may arise from an irregular arrangement of atoms on the lattice sites so that there is only short range ordering of the site occupancies (Chapter 17) or from disordered arrangement of vacancies, interstitials or impurity atoms. The most important diffraction effect produced by small defects in crystals often comes from the strain field produced in the surrounding crystal. For an atom vacancy, an interstitial atom or pair of atoms, or an impurity atom, the neighboring atoms may be displaced from their average-lattice sites by appreciable fractions of the interatomic distance. The atom displacement usually decreases fairly slowly with distance from the point defect so that a very large number of atoms may be affected. Then the diffraction effect due to the atom displacements may well be much greater that the scattering by the impurity, interstitial or missing atom itself.

Simple models for the diffraction by vacancies and by the relaxation of atoms around a vacancy were discussed in Chapter 7 as examples for the use of the generalized Patterson function. We now treat this type of problem in a different context, providing a basis for the understanding of experimentally important situations.

For the static displacement of atoms there will be no essential difference from the dynamical case of thermal vibrations except that, since there is no time dependence there will be elastic scattering only and the practical distinction between the capabilities of X-rays or electrons and neutrons will removed. The generalized Patterson function is considered as a function of the three spatial coordinates only. As we shall see, there will be diffuse scattering rather similar to the thermal diffuse and a factor applied to the sharp Bragg reflections which is similar to the Debye–Waller factor.

It would be possible to resolve the static displacements into plane wave Fourier components and proceed as in the last section. Instead we illustrate another method of approach to diffuse scattering problems and, at the same time, provide an introduction to the treatment to be used later (Chapter 17) in the discussion of short-range order diffuse scattering.

12.2.2. Diffraction intensities for displaced atoms

For simplicity we assume the crystal lattice to be primitive with one atom site per unit cell. Then the structure amplitude is

$$F(\boldsymbol{u}) = \sum_i f_i \exp\{2\pi i \boldsymbol{u}\cdot(\boldsymbol{R}_i + \boldsymbol{\Delta}_i)\}$$

and

$$F(\boldsymbol{u})\,F^*(\boldsymbol{u}) = \sum_i \sum_j f_i f_j \exp\{2\pi i \boldsymbol{u}\cdot(\boldsymbol{R}_i - \boldsymbol{R}_j)\}\,\exp\{2\pi i \boldsymbol{u}\cdot(\boldsymbol{\Delta}_i - \boldsymbol{\Delta}_j)\}\,, \tag{7}$$

where the $\boldsymbol{R}_i$ vectors are lattice vectors for the average periodic lattice and $\boldsymbol{\Delta}_i$ is the displacement of the atom from the lattice site defined by $\boldsymbol{R}_i$.

For simplicity also we consider that the lattice sites are occupied by two types of atom, A and B, present in the crystal in fractional proportions m_{A} and m_{B} defined so that $m_{\mathrm{A}} + m_{\mathrm{B}} = 1$. The atom B will be considered as impurities (or vacancies) so that m_{B} is small and the B atoms may be assumed as a first approximation, to occur at random.

We now specify the occupancy of the sites by means of the occupational parameters (Flinn [1956])

$$\sigma_i^{\mathrm{A}} = \begin{cases} 1 & \text{for an A on site } i\,, \\ 0 & \text{for a B on site } i\,, \end{cases}$$

$$\sigma_i^{\mathrm{B}} = \begin{cases} 1 & \text{for a B on site } i\,, \\ 0 & \text{for an A on site } i\,. \end{cases} \tag{8}$$

If an isolated atom B is larger than the average atom size we would expect its neighboring atoms to be displaced outwards, away from it. We assume that we can describe these displacements in terms of a "displacement field" which is a function of the distances between lattice sites (which approximate to the distance between atoms). Assuming that the displacement caused by a B atom does not depend on the nature of the atom being displaced we can write the displacement of an atom on the j site due to the B atom on the i site as Δ_{ij}^{B}. If a A atom is considered to also give rise to a displacement field, the reasonable requirement that an average atom should provide no displacement gives the relationship

$$m_{\mathrm{A}} \boldsymbol{\Delta}_{ij}^{\mathrm{A}} + m_{\mathrm{B}} \boldsymbol{\Delta}_{ij}^{\mathrm{B}} = 0\,, \tag{9}$$

which implies that displacements due to A and B atoms are collinear and oppositely directed.

The displacement of the atom at i from its average lattice site defined by $\boldsymbol{R}_i$ is assumed to be given by the sum of the displacements due to all other atoms, i.e.

$$\boldsymbol{\Delta}_i = \sum_k \{\sigma_k^{\mathrm{A}} \boldsymbol{\Delta}_{ki}^{\mathrm{A}} + \sigma_k^{\mathrm{B}} \boldsymbol{\Delta}_{ki}^{\mathrm{B}}\}\,, \tag{10}$$

and, by definition

$$\langle \boldsymbol{\Delta}_i \rangle = 0\,, \tag{11}$$

where the angular bracket represents an averaging over all lattice sites of the average, periodic lattice.

12.2.3. The Bragg peaks

The sharp Bragg peaks correspond to diffraction from the average structure for which, from (7)

$$\bar{F}(\boldsymbol{u}) = \left\langle \sum_i f_i \exp\{2\pi i \boldsymbol{u}\cdot(\boldsymbol{R}_i + \boldsymbol{\Delta}_i)\}\right\rangle$$

$$= \sum_i \exp\{2\pi i \boldsymbol{u}\cdot\boldsymbol{R}_i\}\langle f_i \exp\{2\pi i \boldsymbol{u}\cdot\boldsymbol{\Delta}_i\}\rangle\,. \tag{12}$$

The average effective atomic scattering factor included as the last term of (12) is

$$\bar{f} = \langle(\sigma_i^A f_A + \sigma_i^B f_B)\exp\{2\pi i \boldsymbol{u}\cdot\boldsymbol{\Delta}_i\}\rangle\,. \tag{13}$$

If it can be assumed that the atomic displacements are small so that $\boldsymbol{u}\cdot\boldsymbol{\Delta}_i \ll 1$, the exponential can be expanded in a power series to give, using (10),

$$\bar{f} = f_A\Big[\langle\sigma_i^A\rangle + 2\pi i \sum_k \{\langle\sigma_i^A\sigma_k^A\rangle(\boldsymbol{u}\cdot\boldsymbol{\Delta}_{ki}^A) + \langle\sigma_i^A\sigma_k^B\rangle(\boldsymbol{u}\cdot\boldsymbol{\Delta}_{ki}^B)\}$$

$$-2\pi^2\sum_k\sum_l\{\langle\sigma_i^A\sigma_k^A\sigma_l^A\rangle(\boldsymbol{u}\cdot\boldsymbol{\Delta}_{ki}^A)(\boldsymbol{u}\cdot\boldsymbol{\Delta}_{li}^A)$$

$$+\langle\sigma_i^A\sigma_k^A\sigma_l^B\rangle(\boldsymbol{u}\cdot\boldsymbol{\Delta}_{ki}^A)(\boldsymbol{u}\cdot\boldsymbol{\Delta}_{li}^B) + \ldots\} + \ldots\Big] + f_B\Big[\langle\sigma_i^B\rangle + \ldots\Big]\,. \tag{14}$$

From the definition of σ_i^A we have $\langle\sigma_i^A\rangle = m_A$. If the B atoms are distributed at random in the lattice we have, for example,

$$\langle\sigma_i^A\sigma_k^B\rangle = \langle\sigma_i^A\rangle\langle\sigma_k^B\rangle = m_A m_B\,,$$

$$\langle\sigma_i^A\sigma_k^B\sigma_l^B\rangle = m_A m_B^2\,. \tag{15}$$

Then in (14) the first order terms in $(\boldsymbol{u}\cdot\boldsymbol{\Delta})$ are zero because of the relationship (9) and the second order term is similarly zero unless $k = l$ so that we obtain

$$\bar{f} = f_A\Big[m_A - 2\pi^2 m_A \sum_k \{m_A(\boldsymbol{u}\cdot\boldsymbol{\Delta}_{ki}^A)^2 + m_B(\boldsymbol{u}\cdot\boldsymbol{\Delta}_{ki}^B)^2\} + \ldots\Big]$$

$$+ f_B\Big[m_B - 2\pi^2 m_B \sum_k \{m_A(\boldsymbol{u}\cdot\boldsymbol{\Delta}_{ki}^A)^2 + m_B(\boldsymbol{u}\cdot\boldsymbol{\Delta}_{ki}^B)^2\} + \ldots\Big]$$

or

$$\bar{f} \approx (m_A f_A + m_B f_B) \exp\left\{-2\pi^2 \frac{m_B}{m_A} \sum_k (\boldsymbol{u}\cdot\boldsymbol{\Delta}_{ki}^{B})^2\right\}$$

$$\approx (m_A f_A + m_B f_B) \exp\{-2\pi^2 \langle(\boldsymbol{u}\cdot\boldsymbol{\Delta}_i)^2\rangle\}\,. \tag{16}$$

Thus, at least to the second order approximation, the effect of the atom displacements on the Bragg peaks is to multiply the structure amplitudes by an exponential factor which is of exactly the same form as the Debye–Waller factor due to thermal motion. The fact that this pseudo Debye–Waller factor is the same for both types of atom is a result of our assumption that the displacement fields act on all atoms equally.

Detection of this factor depends on separation from the thermal Debye–Waller factor by making use of either its lack of dependence on temperature or its linear dependence on impurity concentration as evidenced by the first part of (16).

12.2.4. The diffuse scattering

To calculate the diffuse scattering, we may evaluate the total scattering intensity by use of equation (7) and then subtract out the Bragg scattering terms. The double summation in (7) can be interpreted as indicating that we may take each site j in turn as origin and consider the scattering amplitude and relative phases for atoms separated from it by vectors $\boldsymbol{r}_i - \boldsymbol{r}_j$. Thus (7) may be written, for real f values,

$$|F(\boldsymbol{u})|^2 = N \sum_i \langle f_0 f_i \exp\{2\pi\mathrm{i}\boldsymbol{u}\cdot\boldsymbol{R}_i\} \cdot \exp\{2\pi\mathrm{i}\boldsymbol{u}\cdot(\boldsymbol{\Delta}_i - \boldsymbol{\Delta}_0)\}\rangle\,. \tag{17}$$

Then making the same sort of substitutions and expanding the exponential as in (13) and (14)

$$|F(u)|^2 = N \sum_i \exp\{2\pi\mathrm{i}\boldsymbol{u}\cdot\boldsymbol{R}_i\}\Big\langle(\sigma_0^A f_A + \sigma_0^B f_B)(\sigma_i^A f_A + \sigma_i^B f_B)$$

$$\times\left[1 - 2\pi\mathrm{i}\sum_k \{\sigma_k^A(\boldsymbol{u}\cdot\boldsymbol{\Delta}_{k0}^{A}) + \sigma_k^B(\boldsymbol{u}\cdot\boldsymbol{\Delta}_{k0}^{B})\} + \ldots\right]$$

$$\times\left[1 - 2\pi\mathrm{i}\sum_k \{\sigma_k^A(\boldsymbol{u}\cdot\boldsymbol{\Delta}_{ki}^{A}) + \sigma_k^B(\boldsymbol{u}\cdot\boldsymbol{\Delta}_{ki}^{B})\} + \ldots\right]\Big\rangle. \tag{18}$$

We isolate first the case $i = 0$ and use the relations $\sigma_0^A\sigma_0^A = \sigma_0^A$ and $\sigma_0^A\sigma_0^B = 0$ to get the contribution to the diffuse part as

$$N(m_A f_A^2 + m_B f_B^2) - N(m_A f_A + m_B f_B)^2 = N m_A m_B (f_A - f_B)^2 \, . \quad (19)$$

Continuing with the assumption $i \neq 0$, we see that the zero order term in $(\boldsymbol{u} \cdot \boldsymbol{\Delta})$ in (18) is part of the Bragg scattering. In the first order term the contributions for $k \neq 0, i$ are zero from (10) and (11). For $k = 0$ and i this is not so since $\Delta_{ii}^B = \Delta_{00}^B = 0$. Then using (9) to express the result in terms of Δ_{ij}^B only, we obtain

$$2N(m_A f_A + m_B f_B)(f_B - f_A) m_B \, 2\pi \mathrm{i} \sum_i (\boldsymbol{u} \cdot \boldsymbol{\Delta}_{0i}^B) \exp\{2\pi \mathrm{i} \boldsymbol{u} \cdot \boldsymbol{R}_i\}$$

$$= 2\pi N m_B (m_A f_A + m_B f_B)(f_B - f_A) \sum_i (\boldsymbol{u} \cdot \boldsymbol{\Delta}_{0i}^B) \sin 2\pi(\boldsymbol{u} \cdot \boldsymbol{R}_i) \, . \quad (20)$$

The second order terms of the series in the square brackets of (18) and the second order term obtained by multiplying these series give summations of the two sets of indices indicated by k and l. Again using (9) or (11) we can set all terms equal to zero unless $k = l$, when the expressions simplify again for $k = 0$, or else for the special cases $k = 0, l = 1$ or $k = 1, l = 0$. Then we derive the second order contribution in the form

$$4\pi^2 N \frac{m_B}{m_A} (m_A f_A + m_B f_B)^2 \sum_i \left\{ \sum_{k \neq 0, i} (\boldsymbol{u} \cdot \boldsymbol{\Delta}_{k0}^B)(\boldsymbol{u} \cdot \boldsymbol{\Delta}_{ki}^B) \right\} \cos 2\pi \boldsymbol{u} \cdot \boldsymbol{R}_i$$

$$- 4\pi^2 N \left[\frac{m_B}{m_A} (m_A f_A + m_B f_B)(m_B f_A + m_A f_B) + m_B^2 (f_A - f_B)^2 \right]$$

$$\times \sum_i (\boldsymbol{u} \cdot \boldsymbol{\Delta}_{i0}^B)^2 \cos 2\pi \boldsymbol{u} \cdot \boldsymbol{R}_i \, . \quad (21)$$

The forms of these contributions to the diffuse scattering are suggested in Fig. 12.2. The term (19) sketched in 12.2(a) is symmetric about the reciprocal space origin and falls off monotonically. The first order term in $(\boldsymbol{u} \cdot \boldsymbol{\Delta})$, given by (20) is symmetric about the origin but anti-symmetric about each reciprocal lattice point. Then the second order term (21) gives a contribution symmetric about each reciprocal lattice point. When these are added together we see that there will be no peak around the origin, but around each reciprocal lattice point there will be a diffuse peak with its maximum somewhat displaced from the reciprocal lattice point because of the influence of the asymmetrical contribution from (20). The direction of this displacement will depend on the relative signs of $(f_B - f_A)$ and $(\boldsymbol{u} \cdot \boldsymbol{\Delta}_{0i}^B)$. If the "impurity" atom B is larger in size than the host atom A and also has a larger scattering factor, these two quanti-

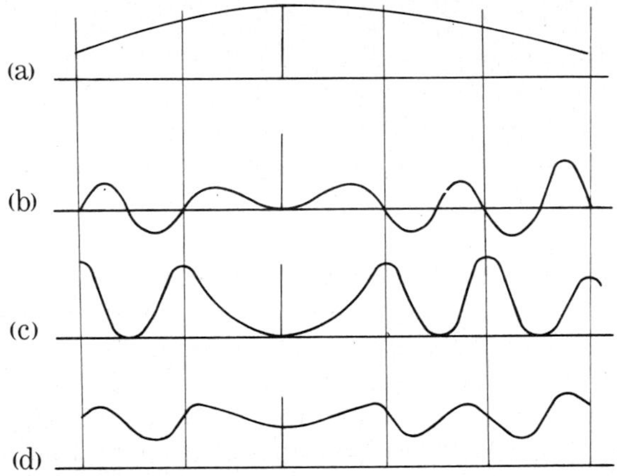

Fig. 12.2. The contributions to the intensity distribution of the diffuse scattering due to size-effect diaplacements of atoms in the presence of impurities or defects: (a) is the "Laue monatomic" scattering, (b) is the term asymmetric about the reciprocal lattice points and (c) is the part symmetric about the reciprocal lattice points: (d) is the summation of these components.

ties will be of the same sign and the displacement of the diffuse peaks will be towards the origin, but for a small, heavy impurity the displacement will tend to be outwards.

In order to calculate the distributions of diffuse scattering in detail, the exact form of the displacement vectors $\boldsymbol{\Delta}_{0i}^{\mathrm{B}}$ must be known, but there are few materials for which these vectors are known with any confidence. A reasonable first approximation, used by Huang [1947] and Borie [1957, 1959, 1961] is to make use of the formula derived for the macroscopic case of a center of dilation in a uniform isotropic solid, giving

$$\boldsymbol{\Delta}_{0i}^{\mathrm{B}} = \frac{c_{\mathrm{B}}\boldsymbol{r}_{0i}}{|\boldsymbol{r}_{0i}|^3} \,. \tag{22}$$

Then some of the summations in the above expressions may be performed analytically. For example

$$\sum_k (\boldsymbol{u}\cdot\boldsymbol{\Delta}_{ki}^{\mathrm{B}}) = c_{\mathrm{B}}^2 \sum_k \frac{(\boldsymbol{u}\cdot\boldsymbol{r}_{ki})^2}{|\boldsymbol{r}_{ki}|^6} \,, \tag{23}$$

and for a simple face-centered cubic lattice (Born and Misra [1940]) this is equal to $c_B^2 \cdot 33.7 \cdot |\boldsymbol{u}|^2/a_0^4$.

These solutions assuming the strain field to have the asymptotic form given by continuum elastic theory give reasonable results for the scattering very close to the Bragg peaks, but are less appropriate for the description of atom displacements near the defect and for the diffuse scattering. Calculations of the displacements of the near neighbors of a point defect in solid argon by Kanzaki [1957] and computer simulations of the surroundings of point defects in copper by Tewordt [1958] have shown behaviour very different from this. Along some directions, such as cube axes for argon, the displacements may actually change sign with increasing distance from the defect. Flocken and Hardy [1970] have deduced that the asymptotic solution is valid only for distances from the defect greater than several times the unit cell dimensions.

Calculations of the diffuse scattering distributions far from Bragg reflections given by various point defects have been made recently (e.g. Benedek and Ho [1973]) and observations of this diffuse scattering for X-ray diffraction from large concentrations of point defects has been reported by Haubold [1974]. Results of investigations of the structure and agglomeration of interstitials in metals irradiated at low temperatures have been reported by Ehrhart [1978].

Meanwhile a general approach to the scattering from defects which allows large displacements of atoms to be treated has been formulated by Krivoglaz [1969] and this has been further developed by many others.

12.3. Electron excitations

12.3.1. *Inelastic X-ray scattering*

For both X-rays and electrons an important source of diffuse background scattering and of absorption of energy from the sharp Bragg reflections is the inelastic scattering of the incident radiation by electrons in the crystal. The addition of real and imaginary parts to the X-ray atomic scattering factor as a result of excitation of electrons of the inner electron shells has been discussed in Chapter 4. The imaginary part of the scattering factors provides an absorption coefficient which may be very large for incident wavelengths just smaller than the absorption edges, i.e., when the incident quanta have just enough energy to eject an electron from one of the inner shells. Diffuse background is then produced in the diffraction pattern by the generation of characteristic radiation from the atoms of the specimen.

The collision of X-ray quanta with nearly-free electrons in the specimen gives the well-known Compton scattering. In Chapter 5 we considered the one electron case as an example of the use of the generalized Patterson function $P(\boldsymbol{r}, t)$, giving correlations in both space and time. The total scattering from one electron is 1 electron unit. The elastic scattering is

$$\mathcal{F}\int P(r, t)\, \mathrm{d}t = |F(u, 0)|^2 = |f_e(u)|^2 ,$$

so that the inelastic scattering, given by the difference, is

$$I_{\text{inel}} = 1 - |f_e(u)|^2 . \tag{24}$$

For an atom containing many electrons, the total time-average electron density-distribution $\rho(\boldsymbol{r})$ is the sum of the electron density functions for all electrons so that, Fourier transforming,

$$\begin{aligned} I_{\text{el}} &= \left| \sum_k f_k(u) \right|^2 \\ &= \sum_k |f_k|^2 + \sum_{j \neq k}\sum f_j f_k^* . \end{aligned} \tag{25}$$

A first approximation to the inelastic scattering is given by assuming all electrons to scatter independently and write, from (24),

$$I_{\text{inel}} = \sum_k \{1 - |f_k(u)|^2\} .$$

However at this stage we must take into account the quantum-mechanical properties of electrons in atoms including the effects of the Pauli Exclusion Principle (James [1948]). Then the simple classical generalized Patterson is no longer sufficient. The result obtained is

$$I_{\text{inel}} = \sum_k (1 - |f_k|^2) - \sum_{j \neq k}\sum |f_{jk}|^2 \tag{26}$$

where f_{jk} is the transition matrix element involving exchange.

The Compton scattering intensity is seen to be practically zero at small scattering angles for which the f_k for each electron is near unity. Then it increases with increasing angle until, particularly for light elements, it may be comparable with the elastic scattering given by near-amorphous materials. (James [1948]).

12.3.2. Electron excitation by electrons – plasmons

Incident electron beams may excite electrons from the inner shells of atoms.

This, of course, provides the means for X-ray generation in X-ray tubes. Apart from the characteristic X-rays and white radiation, or "brehmstrahlung", this may give rise to the emission of "Auger" electrons having energies characteristic of the atoms. While both the characteristic X-rays and Auger electrons provide important means for analysis of the chemical constitution of materials, neither contributes appreciably to background in electron diffraction patterns or to the absorption coefficients of electrons in solids. In these respects the excitations of the outer-shell or valence electrons is much more important.

For some metals, such as Al and Mg, the predominant electron excitation is the excitation of collective oscillations, or plasmons. The sea of nearly free electrons may be set vibrating with a characteristic plasmon frequency by the impact of a fast incident electron. The energy of the plasmon and the energy loss of the incident electron are typically of the order of 10 to 20 eV. The wavelength of the plasmon oscillation is normally many unit cells and the angle of scattering of the incident electron is very small compared with Bragg angles for crystalline reflections.

The theory of the excitation of plasmons has been given by Pines and Bohm (see Pines [1955, 1956, 1964]; Ferrell, R.A. [1956]). Ritchie [1957] introduced the concept of the surface plasmon, generated at the boundaries of solids, lower in energy and important for diffraction from very small crystals. A survey of the theory and experimental results has been given by Raether [1980].

The characteristic frequency for bulk plasmons is given by elementary considerations as $\omega_p = (ne^2/m\epsilon_0)^{1/2}$. The slight dependence of the frequency on k is given by the dispersion relationship,

$$\omega(k) = \omega_p \left(1 + \tfrac{3}{5}\left(\frac{E_F}{\hbar\omega_p}\right)\frac{\hbar}{m\omega_p} k^2 + \ldots\right) \tag{27}$$

where E_F is the Fermi energy.

The intensity distribution of the electrons which have suffered one plasmon loss is given by the differential cross-section relationship of Ferrell [1956] as

$$\sigma_B(\theta) = \frac{\theta_E}{2\pi a_0 n} \frac{1}{\theta^2 + \theta_E^2} G^{-1}(\theta) \tag{28}$$

where $\theta_E = \Delta E/2E_0$, a_0 is the Bohr radius, n is the electron density and $G^{-1}(\theta)$ is a function which is unity for small angles but falls to zero at $\theta = \theta_c$, the cut-off angle given by

$$\theta_c = 0.74(E_F/E_0)^{1/2} \ .$$

The scattering intensity for surface plasmons depends on the thickness of the specimen foil or size of the specimen particles (Fujimoto and Komaki [1968]) but for the limiting case of a thick foil is

$$\sigma_s(\theta) = \frac{e^2}{\pi\hbar v}\,\frac{2}{1+\epsilon}\,\frac{\theta\cdot\theta_E}{(\theta^2+\theta_E^2)^2}\,f \tag{29}$$

where v is the incident electron velocity and f is a geometric factor. For a foil of thickness D there are two values for the surface plasmon frequency,

$$\omega_{\pm}(k) = (1+\epsilon)^{-1/2}\,\omega_p(1 \pm \exp\{-kD\})^{1/2}\,, \tag{30}$$

where ϵ is the dielectric constant. Experiments have verified the existence of these ω_+ and ω_- frequencies for very thin films.

From (28) it is seen that for bulk plasmons the half width of the angular distribution is given by $\theta = \theta_E$. For 50–100 kV electrons and characteristic energy losses ΔE of 10 to 20 eV this half width is of the order of 10^{-4} radians. This may be compared with the scattering angle of about 10^{-2} radians for the first Bragg reflection from simple metals. For surface plasmons the distribution is even narrower. Thus plasmon scattering may be considered as essentially small angle scattering, tending to broaden diffraction spots slightly. With increasing specimen thickness multiple pasmon losses occur with increasing frequency and the angular distribution is progressively broadened.

The path length for plasmon scattering is typically of the order of 1000 Å. For Al, for example, Spence and Spargo [1970] found a value of 740 ± 40 Å for 58.5 kV. For very thin films surface plasmons predominate but usually these losses become less important for thicknesses of the order of 100 Å.

It is only for relatively few metals, including Al and Mg that well defined plasmon losses dominate the energy loss spectrum. For most metals, semiconductors and insulators the energy-loss spectrum is complicated, with occasional peaks which can be assigned to plasmon excitation but with other strong, sharp or diffuse bands usually described as being due to single electron excitations although the associations with particular excitation processes of the crystal electrons is at best rather vague.

12.3.3. Single-electron excitations

For nearly-free, or valence, electrons in the crystal the excitations may be considered in terms of the energy-band picture as corresponding to transitions from one energy level to another within the same band (an intra-band transitions, possible when the band is not filled, as in metals) or to a level within a higher unfilled band (and inter-band transition).

In general it may be said that the wave function for an inelastically scattered electron derives from the wave equation

$$\left(-\frac{\hbar^2}{2m}\nabla^2 + H_{nn} - E_n\right)\psi_n = -\sum_{m\neq n} H_{mn}\psi_n\,, \tag{31}$$

where

$$H_{nm}(r) = \int a_n^*(r_1 \ldots r_N)\, H(r, r_1 \ldots r_N)\, a_m(r_1 \ldots r_N)\, \mathrm{d}r_1 \ldots \mathrm{d}r_N\,. \tag{32}$$

Here $a_n(r_1 \ldots r_N)$ is the wave function for the electrons in the crystal having positions $\boldsymbol{r}_1 \ldots \boldsymbol{r}_N$ and $H(\boldsymbol{r}, \boldsymbol{r}_1 \ldots \boldsymbol{r}_N)$ is the interaction energy of the incident electron (position $\boldsymbol{r}$) with the crystal electrons.

For electrons in a crystal the interaction energy must show the periodicity of the crystal lattice and it was shown by Howie [1963] that

$$H_{nm}(\boldsymbol{r}) = \exp\{-2\pi\mathrm{i}\boldsymbol{q}_{nm}\cdot\boldsymbol{r}\} \sum_g H_g^{nm} \exp\{2\pi\mathrm{i}\boldsymbol{g}\cdot\boldsymbol{r}\} \tag{33}$$

where $\boldsymbol{q}_{nm}$ is the wave vector for the excitation in the crystal and $\boldsymbol{g}$ is the reciprocal lattice vector.

Assuming a simple Coulomb interaction of the electrons the interaction energy is of the form $H(\boldsymbol{r}) = \Sigma_n\, e^2/(\boldsymbol{r} - \boldsymbol{r}_n)$, which allows the intergration over $\boldsymbol{r}$ in (32) to be carried out directly. Then, following Cundy et al. [1966] the kinematical amplitude of the scattering involving a single electron excitation from a state m to a state n is proportional to

$$H_g^{nm} = \frac{e^2}{\pi V\epsilon(\boldsymbol{q}-\boldsymbol{g}, \omega)\cdot|\boldsymbol{q}-\boldsymbol{g}|^2}\int a_n^*(\boldsymbol{r})\exp\{2\pi\mathrm{i}(\boldsymbol{q}-\boldsymbol{g})\cdot\boldsymbol{r}\}\, a_m(\boldsymbol{r})\,\mathrm{d}\tau_{\boldsymbol{r}} \tag{34}$$

where $\epsilon(\boldsymbol{q}-\boldsymbol{g}, \omega)$ is the value of the dielectric constant for a change of momentum $h(\boldsymbol{q}-\boldsymbol{g})$ and a change of energy $h\omega$; V is the volume of normalization. As a special case of this we note that H_g^{nn} terms relate to elastic scattering. In particular H_g^{00} is the scattering for all electrons remaining in the ground state, and this depends on the Fourier transform of $a_0^*(\boldsymbol{r})\,a_0(\boldsymbol{r})$ which is just the electron density distribution $\rho(\boldsymbol{r})$ for the ground state of the crystal. Taking the g^{-2} term into account relates this to the Fourier transform of the unperturbed potential, $\varphi_0(\boldsymbol{r})$.

For the inelastic processes the amplitude of scattering is given by the Fourier transform of the product of wave functions $a_n^* a_0$. For nearly free electrons both of these functions may be Bloch waves of crystal electrons and therefore periodic. The diffracted amplitudes will be sharply peaked at the

Bragg reflection positions, just as for elastic scattering. However, in contrast to the elastic scattering case, only the outer electron shells will be involved. Hence the effective scattering distribution will have a broader spread around the lattice points: the effective atomic scattering factor will fall off more rapidly with angle than for elastic scattering. This is the effect observed by Kuwabara in measurements of the relative intensities of diffraction rings obtained from thin metal and other films when an energy analyser was used to select electrons having only particular values of energy loss (Kuwabara and Cowley [1973]).

The greatest contribution of diffuse scattering intensity due to electron excitations will come from the $g = 0$ terms of equations such as (34). For these the intensity will be proportional to $|q|^{-2}$ and so will fall in much the same way as for plasmon scattering, with a comparable half-width. (Humphreys and Whelan [1969].) The width of the scattering distribution can be taken as inversely dependent on the localization of the inelastic scattering process. The inelastic scattering is localized to a region of width approximately $\lambda E/\Delta E$ for an energy loss of ΔE. The effect of the localization or delocalization of the process on the contrast of electron microscope images of defects in crystals has been discussed, for example, by Craven et al. [1978].

12.4. Dynamical effects in diffuse scattering

12.4.1. Scattering and re-scattering

Under the conditions of diffraction for which dynamical diffraction of the sharp Bragg reflections is important, the dynamical diffraction will also affect the diffuse scattering intensity. Firstly it must be taken into account that the incident beam is not the only strong beam in the crystal. Each of the diffracted beams will also act as a source of diffuse scattering. Secondly the diffusely scattered radiation will undergo diffraction as it passes through the crystal. Diffuse scattering in two directions separated by twice the Bragg angle for a Bragg reflection may be coupled dynamically, giving rise, among other things, to the appearance of Kossel or Kikuchi lines (see Chapter 14). Finally, diffusely scattered radiation may be diffusely scattered a second time or more times so that for a thick crystal the observed diffuse intensity may be the sum of many multiply scattered components, all modified by dynamical interactions through Bragg reflections.

For X-ray diffraction in a perfect crystal a two beam dynamical theory is normally sufficient. In the case of thermal diffuse scattering for example (see

O'Connor [1967]), both the incident and diffracted beams can be considered to generate diffuse scattering in proportion to their intensities. In general the diffusely scattered radiation will be transmitted through the crystal with the average absorption coefficient. But where it meets a diffracting plane at the Bragg angle it will be diffracted giving rise to sharp Kossel or Kikuchi lines.

For electron diffraction the situation is normally complicated by n-beam diffraction effects, although some useful results were obtained for thermal diffuse scattering by Takagi [1958] for a two beam case and by Fujimoto and Kainuma [1963], Fujimoto and Howie [1966] and Ishida [1970] who extended this approach and included other types of diffuse scattering. A useful approach to the general n-beam dynamical treatment was made by Gjønnes [1965, 1966] and applied by Gjønnes and Watanabe [1966] for cases involving relatively few beams and by Fisher [1965] to the case of short-range order diffuse scattering. In extension of this approach, Cowley and Pogany [1968] gave a general theory and outlined computational methods which were subsequently used by Doyle [1969, 1971] for detailed calculations on thermal diffuse and plasmon scattering and by Cowley and Murray [1968] for consideration of short-range order scattering.

If a crystal is considered to be divided into many thin slices almost perpendicular to the incident beam, the total single-diffuse scattering can be consid-

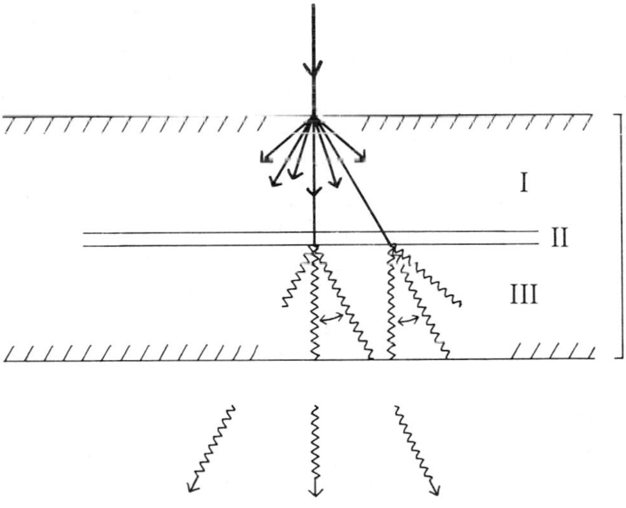

Fig. 12.3. Diagram representing the scheme for dynamical electron diffraction calculation of diffuse scattering.

ered as the sum of contributions from each slice taken separately. For diffuse scattering from a particular slice the crystal can be considered as divided into three parts as suggested in Fig. 12.3. The incident beam enters region I and undergoes n-beam dynamical diffraction as for a perfect crystal. Region II is the slice in which each Bragg beam gives rise to diffuse scattering. Then in region III the diffusely scattered radiation undergoes n-beam dynamical interactions through the Bragg reflections.

12.4.2. Coherent and incoherent scattering

Even though the electron wave may have lost energy or otherwise have been rendered incoherent with the elastically scattered beam it will maintain coherence, or the ability to interfere, with itself. As suggested in Fig. 12.4, if the diffuse scattering process corresponds to a change $\boldsymbol{q}$ in the scattering vector, n-beam dynamical diffraction will take place between the points $\boldsymbol{h} + \boldsymbol{q}$, where $\boldsymbol{h}$ is a reciprocal lattice vector, with interactions depending on the structure amplitudes $\Phi(\boldsymbol{h}_1 - \boldsymbol{h}_2)$ and the appropriate excitation errors. The calculation for region III must be made for each vector within the Brillouin zone (or basic reciprocal lattice unit cell).

If the amplitude of diffuse scattering from the m slice of the crystal is $\Psi_m(u, v)$ the total single-scattering diffuse intensity will be given by adding either the amplitudes or intensities of diffuse scattering from each slice, depending on whether there is correlation between the diffuse scattering processes of separate slices or not. Thus in the case of plasmon diffuse scattering it may be assumed that the plasmon wave extends through the whole thickness of a crystal. This provides a definite relationship between the amplitudes of scattering for all slices and the diffuse scattering intensity will be given by

$$I_{\text{coh}}(u, v) = \left| \sum_m \Psi_m(u, v) \right|^2 . \tag{35}$$

If the diffuse scattering comes from a random distribution of point defects

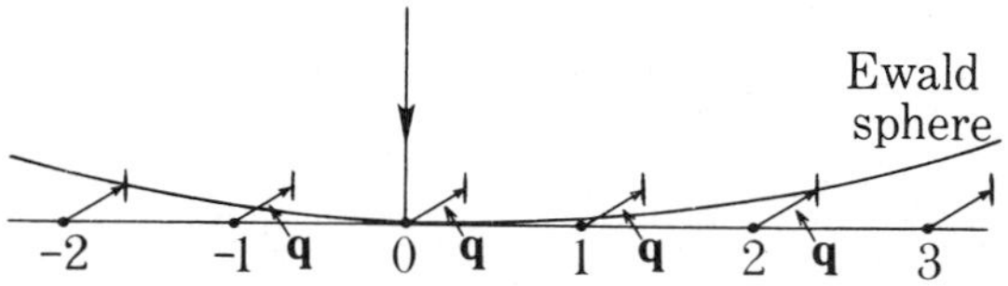

Fig. 12.4. The intersection of the Ewald sphere with the regions of scattering power corresponding to a phonon having momentum q.

there will be no definite phase relationship between the scattering from successive slices and when the average is taken over a large number of defects the diffuse intensities will be added incoherently so that

$$I_{\mathrm{incoh}}(u, v) = \sum_m |\Psi_m(u, v)|^2 . \tag{36}$$

There are intermediate cases for which there is a phase relationship in the diffuse scattering over a limited distance. This is the case for some phonon scattering for which it is suggested, for example by neutron diffraction observations, that the correlation in the movements of the atoms extends over perhaps 10 to 100 Å. Then the amplitudes of diffuse scattering must be added over a limited range. The total intensity may then be approximated by an expression such as

$$I(u, v) = \sum_m \left| \frac{1}{n} \sum_{p=m-n/2}^{m+n/2} \Psi_p(u, v) \right|^2 . \tag{37}$$

The addition of amplitudes as in (35) was the basis for the detailed n-beam calculations of Doyle [1971] of the intensity of plasmon scattering given by thin crystals of Al oriented so that only the hhh systematic set of reflections are excited. In this way he showed agreement with the experimental observa tion that when the strong 111 reflection is excited the diffuse scattering tends to the excluded from the region between the strong 000 and 111 beams (see also Howie [1963]). Also he derived the form of the thickness fringes obtained with the 000 and 111 beams for electrons which had suffered a single plasmon energy loss (15 eV), in agreement with experiment.

The summation of intensities as in (36) may be expected to be appropriate for the diffuse scattering well away from the strong Bragg reflections since this will be produced mostly by scattering involving short wavelength phonons, small defects, or the excitation of electrons tightly bound to atoms. For each of these cases the excitation is limited to a small region of the crystal.

The assumption of the addition of intensities was used by Gjønnes [1966] and has since been extended (Gjønnes and Høier [1971]; Høier [1973]) to provide a general description of dynamical diffuse scattering in terms of a six-dimensional real-space distribution.

The case of an intermediate range of correlation of the excitation giving diffuse scattering, treated by use of (33), was the basis for the study of thermal diffuse scattering in thin crystal of gold, carried out by Doyle [1969]. One result was that the need to average at each point over the diffuse scattering due to a variety of phonons having the same components of q perpendicular to the beam has the effect of greatly reducing the dependence of the diffuse intensity on phonon correlation length, so that all assumptions as to correlation length tended to give much the same result as the use of the incoherent addition of intensity given by (36). For this case the oscillations of intensity with thickness, characteristic of the Bragg beams, are almost eliminated. This is in agreement with the experimental observation that dark-field images of wedge-shaped crystals obtained from the pseudo-elastic diffuse scattering (mostly thermal) show thickness fringes of very low contrast (Watanabe [1966]; Castaing [1966]; Cundy, Metherall and Whelan [1966]).

The calculations by Doyle also showed the generation of Kikuchi lines and bands in the thermal diffuse scattering, including the "zero-order" Kikuchi line, often seen in the center of a Kikuchi band, which can only arise from n-beam dynamical scattering (Kainuma and Kogiso [1968]).

12.4.3. Analysis of diffuse scattering

The possibility of obtaining useful information on crystal defects, disorders or perturbations from diffuse scattering in electron diffraction patterns has been considered by several authors. There are obvious limitations because of the formation of Kikuchi line patterns in any diffuse-scattering distribution, but in practice these effects can be largely eliminated by averaging over a small range of angles of incidence (or crystal orientation) since the Kikuchi lines are strongly orientation-dependent. The initial calculations by Fisher [1965] suggested that the intensity of short-range order diffuse scattering in electron diffraction patterns from Cu-Au alloys could be related to the kinematical scattering intensities by a smoothly-varying "dynamical factor". However, it was found that the "size-effect" modification of the diffuse scattering from such alloys could be eliminated by the strong two-dimensional dynamical interactions occuring near principle orientations (see Chapter 16). This, and related observations on thermal diffuse scattering, plus arguments based on phase-grating approximations led Cowley [1965] to propose that the dynamical effects may prove useful in their ability to distinguish between different sources of diffuse scattering. Strong dynamical interactions tend to suppress diffuse scattering due to displacements of atoms but not the diffuse scattering due to atom replacements (short-range ordering, vacancies, substitutional impurities, etc.).

The subsequent analysis of dynamical diffuse scattering by Gjønnes and Høier [1971] has provided a basis for a more complete interpretation of experimental observations, particularly for relatively thick crystals for which the parts of intensity expressions which oscillate with thickness may be ignored.

The application to the study of the defect distribution in non-stoichiometric vanadium oxide is given by Anderson, Gjønnes and Tafto [1974].

A fundamental assumption made in most of the treatments of dynamical diffuse scattering is difficult to justify. This is the assumption that each layer of the structure gives diffuse scattering which is the same as that given by averaging over the whole structure. Since, in the scheme of fig. 12.3, the contribution to the exist wave at any point (and hence the image intensity or the contribution to the diffraction pattern) will depend only on the atom positions in a rather narrow column through the crystal, it is assumed that the displacements or replacements of the few atoms of each layer of a column will give scattering equal to that from a statistical averaging of the whole crystal. Thus, to calculate the diffraction pattern the averaging is applied to the structure before dynamical scattering is considered, rather than being applied to the exit wave function.

A method for avoiding this assumption has been developed recently by Cowley and Fields [1979] (also Fields and Cowley [1978]) and appropriate computing methods have been outlines. The concept of a dynamical factor or more approprately, a "multiplicative dynamical scattering function", is still valid within the approximations of single and weak scattering for the deviations from the average structure. An evaluation of the errors inherent in the assumption made in earlier treatments must rely on detailed calculations for particular cases but these have yet to be made.

12.5. Absorption effects

12.5.1. The nature of absorption parameters

We will emphasize repeatedly that the effects to be considered as constituting "absorption" are strongly dependent on the nature of the experimental observation being made. A number of elastic and inelastic scattering processes take place when a beam of radiation interacts with matter. The degree to which the scattered radiation is included in the experimental measurement determines whether a particular scattering process contributes to the measured intensity directly or through the application of an absorption function. For example in neutron diffraction with energy analysis, measurement of the sharp Bragg reflections from a crystal will exclude thermal diffuse scattering. The loss of energy from the incident beam and Bragg reflections to give the thermal diffuse intensity will provide a small absorption effect.

In the case of X-ray and electron diffraction, much of the thermal diffuse scattering may be included in some measurements of the Bragg reflection intensity. Then this does not provide any absorption effect. However if the experiment is made in such a way as to separate the sharp Bragg peaks from the attendant thermal diffuse, then an absorption function must be applied in the calculation of the Bragg reflection intensity.

For X-ray or neutron diffraction the absorption function due to thermal diffuse scattering is very small since it enters first in second-order scattering terms and so, unlike the Debye–Waller factor, is negligible under kinematical scattering conditions. Under dynamical scattering conditions for X-rays the probability of double diffuse scattering with appreciable amplitude is also negligible. But, as we shall see, under the conditions for dynamical diffraction of electrons the absorption coefficients due to thermal diffuse scattering may be important.

The influence of the experimental conditions of the measurement of intensities on the nature of the absorption function is probably most severe in electron microscopy of crystals, where the objective aperture transmits a combination of the direct beam with inelastically and elastically scattered beams both in the Bragg spots and the diffuse background. Then these various components are influenced differently by the lens aberrations. This case will be discussed in the next chapter. Here we confine our attention to the case of most universal significance; the absorption functions relevant to the sharp elastic Bragg reflections given by a crystal.

12.5.2. Absorption of X-rays and neutrons

For X-rays and neutrons the major absorption effect is one which usually provides no contribution to the diffraction pattern. Incident X-rays may excite inner-shell electrons from the specimen atoms, losing most of their energy in the process. The characteristic radiation emitted by the excited atoms is normally filtered out. As has been discussed in Chapter 4, the atomic scattering factors for the specimen atoms are thereby made complex by the addition of a real and an imaginary part: $f = f_0 + f' + if''$. The imaginary part is associated with absorption. For the incident beam direction for example, the scattered radiation is added $\pi/2$ out of the phase to give an amplitude, in electron units, $\psi_0 + if(0)$. Hence $f''(0)$ is subtracted from ψ_0 and so reduces the incident intensity.

For neutron scattering the absorption component of the scattering factor comes from the imaginary part of equation (4.27). This arises from inelastic interactions of the neutron with the nucleus and results in capture of the neutron by the nucleus with excitation of the nucleus and often emission of secondary radiations which are not normally detected in the diffraction experiment.

Of the other scattering processes which subtract energy from the incident and Bragg beams and so contribute absorption effects, the most important for X-rays are the Compton and thermal diffuse scatterings. The relative contributions from these sources to the average attenuation coefficients for X-rays in crystals have been calculated and compared with experiment by DeMarco and Suortti [1971]. They find that for a variety of elements and X-ray wavelengths, these effects contribute 1 to 3 per cent of the absorption coefficient due to inner electron excitation.

12.5.3. "Absorption" for electrons

For high energy incident electron beams the situation is different in that the processes contributing to the diffuse elastic and inelastic scattering observed in the diffraction patterns are the major contributors to the absorption effects. These are the excitation of plasmons, single electron excitations and phonons, with appreciable contributions from short-range order or defect scattering in particular cases.

By an extension of the Schrödinger equation formulation to include excited states of the scattering atoms as in (31), Yoshioka [1957] showed that the effect of inelastic scattering on the elastic scattering amplitudes could be represented by the addition of an imaginary component to the scattering potential and so to the structure factors for centrosymmetric crystals. A number of authors have subsequently estimated or derived from experiment the contributions to this imaginary, absorption component due to the various scattering processes.

The added imaginary parts of the structure factor can be regarded as the Fourier coefficients of a three-dimensional "absorption function", $\mu(\boldsymbol{r})$. In general, $\mathcal{F}\{\Phi(\boldsymbol{u}) + \mathrm{i}\Phi'(\boldsymbol{u})\} = \varphi(\boldsymbol{r}) + \mathrm{i}\varphi'(\boldsymbol{r})$.

For a thin object we may then write in the phase object approximation for the transmitted wave

$$\psi(x,y) = \exp\{-\mathrm{i}\sigma[\varphi(x,y) - \mathrm{i}\varphi'(x,y)]\}$$

$$= \exp\{-\mathrm{i}\sigma\varphi(x,y) - \mu(x,y)\}\,, \qquad (38)$$

where the absorption function $\mu(x,y)$ is the projection in the beam direction of the absorption function $\mu(\boldsymbol{r})$.

We may use this phase object formulation as a basis for an approximate treatment of the general consideration of absorption in dynamical diffraction because, in the slice methods used for developing dynamical scattering, the basic form of the interaction is introduced in this way.

We deduce, for example, the form of the absorption coefficients to be applied to sharp Bragg reflections when deviations from the perfectly periodic structure give rise to diffuse scattering. As in Chapter 7 we write the projection of the potential distribution as

$$\varphi(x,y) = \bar{\varphi}(x,y) + \Delta\varphi(x,y)\,,$$

where $\bar{\varphi}(x,y)$ is the average periodic structure. For a thin phase object the transmission function is

$$q(x,y) = \exp\{-\mathrm{i}\sigma[\bar{\varphi}(x,y) - \Delta\varphi(x,y)]\}\,.$$

The sharp Bragg reflections are then given by the Fourier transform of

$$\langle\exp\{-\mathrm{i}\sigma[\bar{\varphi}(x,y) + \Delta\varphi(x,y)]\}\rangle$$

where the brackets $\langle\ \rangle$ indicate a periodic ensemble average over all unit cells. Since $\bar{\varphi}$ is periodic, this may be written

$$\exp\{-\mathrm{i}\sigma\bar{\varphi}(x,y)\}\ \{1 - \mathrm{i}\sigma\langle\Delta\varphi\rangle - \tfrac{1}{2}\sigma^2\langle\Delta\varphi^2\rangle + \ldots\}\,. \tag{39}$$

By definition, $\langle\Delta\varphi\rangle$ is zero. Hence to a first approximation this is

$$\exp\{-\mathrm{i}\sigma\bar{\varphi}(x,y)\}\ \exp\{-\tfrac{1}{2}\sigma^2\langle\Delta\varphi^2\rangle\} = \exp\{-\mathrm{i}\sigma\bar{\varphi}(x,y) - \mu(x,y)\}\,, \tag{40}$$

where $\mu(x,y) = \tfrac{1}{2}\sigma^2\langle\Delta\varphi^2\rangle$.

Thus the effective absorption coefficient is proportional to the mean square deviation from the average potential. The effective value of the structure factor to be included in n-beam dynamical calculations is $\bar{\Phi}(u,v) - \mathrm{i}M(u,v)$ where $\bar{\Phi}(u,v)$ is the Fourier transform of $\bar{\varphi}(x,y)$ and $M(u,v)$ is the Fourier transform of $\sigma^{-1}\mu(x,y)$.

12.5.4. Absorption due to thermal vibrations

For the case of thermal vibrations of the lattice, $\bar{\varphi}(x,y)$ includes the averaging over all displaced atoms so that $\bar{\Phi}(u,v)$ includes the Debye–Waller factor. For each atom the contribution to $\Delta\varphi$ is written using a Taylor series as

$$\varphi_0(x+\epsilon) = \varphi_0(x) + \epsilon\frac{\partial\varphi}{\partial x} + \ldots$$

where ϵ is the displacement from the mean position. Then

$$\langle\Delta\varphi^2\rangle = \langle\epsilon^2 4\pi^2\,\{u f(u) * u f(u)\}\rangle\,, \tag{41}$$

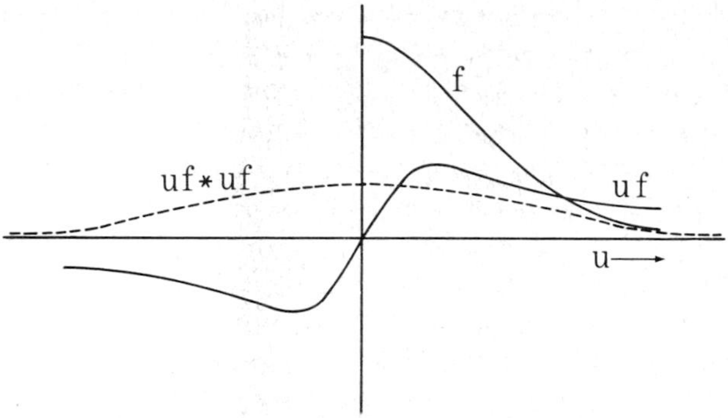

Fig. 12.5. Sketch suggesting the relation between the functions $f(u)$, $u\,f(u)$ and the self convolution of $u\,f(u)$, where $f(u)$ is an atomic scattering factor.

In Fig. 12.5, $u\,f(u)$ and $u\,f(u) * u\,f(u)$ are compared with $f(u)$. It is seen that the imaginary component of the structure factor, $M(u, v)$, will in general fall off much more slowly than the elastic part $\overline{\Phi}(u, v)$ with $|\boldsymbol{u}|$, at least for the angles of scattering usually considered.

The values for the absorption coefficients calculated from (41) are in good agreement with those derived by Hall and Hirsch [1965] using a rather different set of approximations and the Einstein model for thermal vibrations and by Hall [1965] using a many-phonon Debye model. Doyle [1970] derived the

Table 12.1

Calculated absorption coefficients due to plasmon scattering (P), thermal diffuse scattering (TDS) and single electron excitations (SE) for 111 systematic reflextions of Al at 40 keV (from Doyle [1970]).

Units: $\times 10^{-3}$ Å^{-1}				
g	μ^{P}	μ^{TDS}	μ^{SE}	μ^{total}
000	1.448	0.667	0.235	2.350
111		0.591	0.053	0.644
222		0.440	0.047	0.487
333		0.288	0.038	0.326
444		0.165	0.029	0.194
555		0.081	0.020	0.101
666		0.032	0.013	0.045
777		0.007	0.008	0.015
888		−0.003	0.004	0.001

values for Al (111) reflections for 40 kV electrons at 300 K as given in Table 12.1.

For static displacements of atoms due to defects or impurities, the same derivation as for thermal vibrations will be appropriate. It can usually be assumed that the density of defects is sufficiently small to make this contribution negligible (see Hall et al. [1966]) but the possibility should be born in mind that situations may arise when this is not so, as in cases where the incident electron beam produces intense irradiation damage.

12.5.5. Absorption from electron excitations

Although the argument of equations (38) to (41) may not be strictly appropriate for quantum processes, it does suggest the form of the absorption coefficients to be expected for the important cases. For plasmon excitations the deviations from the average potential have a wavelength of hundreds of Å with little or no modulation with the periodicity of the lattice. Correspondingly the contribution to $M(\boldsymbol{u})$ is confined to the $|\boldsymbol{u}| = 0$ Bragg peak. This, like more rigorous theories gives a uniform absorption, $\mu_0^{\mathrm{P}} (\equiv M^{\mathrm{P}}(0))$ inversely proportional to the mean free path for plasmon excitation and given by Ferrell [1957] as

$$\mu_0^{\mathrm{P}} = \frac{1}{\Lambda} = \frac{\theta_E}{a_0} \int_0^1 \frac{x\, G^{-1}(x)}{x^2 + (\theta_E/\theta_{\mathrm{c}})^2}\, \mathrm{d}x\,, \tag{42}$$

where the notation is as for (28) except that the cut-off angle θ_{c} is equal to $0.74\,(E_{\mathrm{F}}/E)^{1/2}$. A similar term due to surface plasmons is relatively small for the thick specimens for which absorption terms become important.

Single electron excitations involving inner shells are highly localized, while those involving outer electron shells will be influenced by the periodicity of the lattice. Hence we might expect the associated part of $\mu(x, y)$ to show the lattice periodicity and the Fourier transform to give $M(\boldsymbol{u})$ values for all Bragg reflections. The values given in Table 12.1 for μ_h^{SE} (values of $M(h)$) by Doyle were obtained by use of a simple approximation due to Heidenreich [1962]. More sophisticated treatments by Whelan [1965], Ohtsuki [1967], Pogany [1968], Cundy et al. [1969] and Humphreys and Whelan [1969] give much the same sort of result.

12.5.6. Values of absorption coefficients

The figures of Table 12.1 and other calculations show that for the zero beam the ratio of the imaginary part of the structure factor due to absorption to the real part is usually of the order of 0.05 for light elements. The ratio is about 0.03 for inner reflections, increases with increasing scattering angle and then falls away rapidly for larger angles. However the very small or negative values for μ_h due to thermal diffuse at large angles are probably not reliable and presumably are subject to large corrections due to neglect of higher order terms in the expansion of (39) and other effects.

For heavier atoms the ratios of the imaginary to the real parts of the scattering factor are larger, being about 0.10 for the inner reflections of gold crystals, for example. Data on a number of elements is given by Humphreys and Hirsch [1968].

A number of further complications may become important when greater accuracy of calculation of absorption coefficients is required. For example, the assumption that any absorption coefficient derived as in (40) for a thin slice can be used in thick-crystal calculations (or equivalent assumptions in other treatments) implies that the deviation from the average periodic lattice is uncorrelated for successive slices. This is not usually the case. Arguments may be made which suggest that the error may not be too great but an adequate treatment taking into account the considerable extent of phonon or lattice defects would be a very complicated business and the effect may not be describable in terms of a simple absorption function picture.

Other sources of absorption may be important for particular samples. It was shown by Cowley and Murray [1968] that short-range ordering of atoms in binary solid solutions, giving rise to short-range-order diffuse scattering, may produce appreciable absorption coefficients in some cases (see Chapter 17).

We have been concerned so far with the one idealized case of the effect of diffuse scattering on sharp Bragg reflections. For comparison with experiment it must be taken into account that a measurement of a Bragg reflection maximum usually includes some thermal diffuse scattering and, unless energy filtering is used, most of the plasmon scattering. For different experimental conditions, the appropriate absorption coefficients will be different. This is important for the understanding of observations in electron microscopes when the aperture sizes and aberrations of the lenses have a strong influence on the apparent absorption effects.

CHAPTER 13

Electron microscope imaging

13.1. Optics of the electron microscope

13.1.1. Conventional transmission electron microscopy

In the conventional transmission electron microscope using electromagnetic lenses, a beam of electrons of energy between about 20 keV and several MeV is generated by an electron gun. The most commonly used voltage is 100 keV. The illumination of the specimen is controlled by, usually, two condenser lenses. The effective source size is normally of the order of a few microns. The beam divergence at the specimen may be made as small as 10^{-6} radians but for the high intensities of illumination required for high magnification work it may approach 10^{-2} radians, especially if the specimen is immersed in the field of the objective lens to the extent that the fore-field of the objective acts as a short-focal-length condenser.

The initial stage of magnification by the objective lens is the critical stage for the determination of resolution and contrast of the image and it is the aberrations and aperture of this lens which are considered in detail. The two or more following lenses, the intermediate and projector lenses, serve to provide the desirable range of magnification from zero up to perhaps 1,000,000 times.

While an electromagnetic lens differs in a fundamental way from a light-optical glass lens because of the vector nature of the field, the differences which arise in imaging properties, such as the rotation and rotational aberration of the image, are usually minimized by the electron optical design. For the paraxial imaging properties with which we are mostly concerned, the considerations are much the same as in light optics.

The most important aberration is the third-order spherical aberration. Lenses of focal length 2 or 3 mm usually have a spherical aberration constant of a few mm, limiting the resolution in principle to a few Å although, until recently, the resolution limitation has most commonly come from the instabilities of high voltage supplies for the electron gun and current supplies for the electron lenses, or mechanical instabilities. Astigmatism of the objective lens is normally corrected in operation by the adjustment of the magnetic field of the lens by the use of electromagnetic stigmators.

One very useful feature of the optics of the electron microscope is the combination of diffraction with imaging by use of the selected-area diffraction technique, possible because the focal lengths of electromagnetic lenses may be varied simply and rapidly. Fig. 13.1(a) is a geometric optics representation of the action of a typical 3 lens magnifying system with objective, intermediate and projector lenses, each in turn providing magnification of the image. The Fraunhofer diffraction pattern of the object is formed in the back focal plane

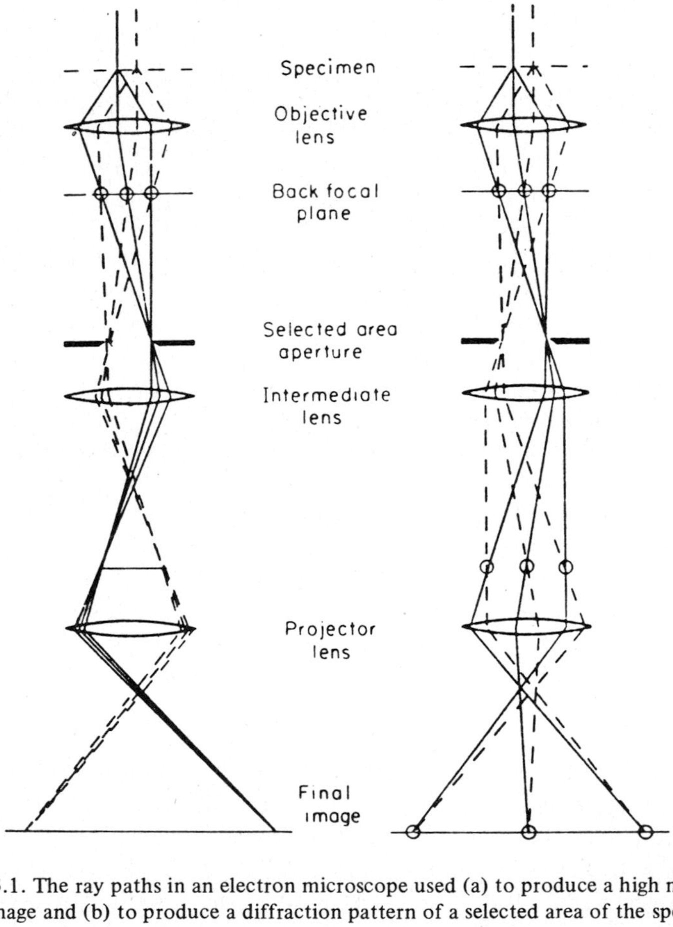

Fig. 13.1. The ray paths in an electron microscope used (a) to produce a high magnification image and (b) to produce a diffraction pattern of a selected area of the specimen.

of the objective as in Fig. 3.2. In Fig. 13.1(b) the focal length of the intermediate lens has been increased so that, instead of the image plane, the back focal plane of the objective is imaged at the object plane of the projector lens and so is projected, magnified, on the fluorescent viewing screen or recording photographic plate.

If a small aperture is placed in the image plane of the objective lens it will have the effect of blocking out all but a selected part of the image in the magnifying mode of Fig. 13.1(a). Since switching of the intermediate lens will not change the selection of electrons to pass through the aperture, the diffraction pattern produced in the diffraction mode of Fig. 13.1(b) will be that of the selected area of the specimen. This "selected-area electron diffraction" possibility is of great value in enabling image contrast to be correlated with diffraction conditions, especially for crystalline specimens. In practice the minimum area which can be usefully selected in this way is limited to about 0.5 μm at 100 keV and perhaps 500 Å at 1 MeV by the spherical aberration of the lens.

13.1.2. Scanning transmission electron microscopy

An alternative to the conventional "fixed-beam" transmission electron microscope, providing much the same information, is the transmission scanning electron microscope (STEM) illustrated in principle in Fig. 13.2. Here the short-focus objective type lens is used to form an electron probe of small diameter at the specimen level by demagnification of a small bright electron source, preferably a cold field-emission tip. This fine electron probe is scanned across the specimen in a television-type raster and a selected part of the transmitted beam is detected to provide a signal which is then displayed on a cathode ray tube to form a magnified image of the object.

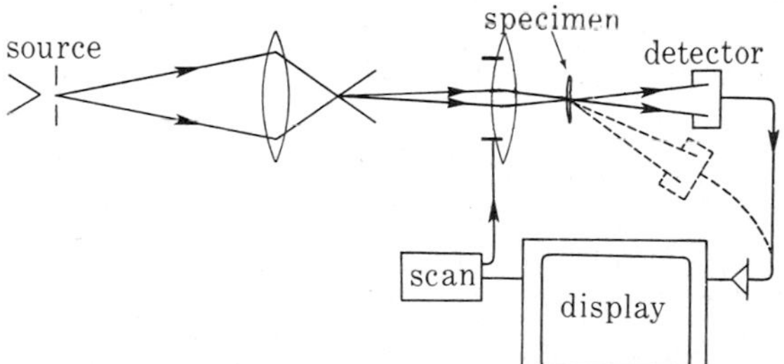

Fig. 13.2. Diagram suggesting the mode of operation of a scanning transmission electron microscope.

A straight-forward application of the reciprocity theorem (Cowley [1969]) shows that, for equivalent geometry, the image contrast and resolution for the STEM should, in principle, be the same as in the conventional microscope. This has been shown to be the case in practise by Crewe and Wall [1970] who obtained a resolution of about 5 Å and observed the familiar types of contrast features due to Fresnel and Fraunhofer diffraction in a scanning microscope operating at about 20 keV. The possibilities for the STEM at voltages up to 500 keV or more have been explored by Cowley and Strojnik [1968] and Cowley [1970].

The STEM has a number of potential advantages, notably in the ease with which the transmitted and diffracted beams can be detected, analysed and manipulated and in the flexibility possible in control of the image by manipulation of the electrical signal which constitutes the immediate output of the microscope. Furthermore, if the scan of the incident electron probe is stopped, a convergent beam diffraction pattern is obtained from a region of diameter much the same as the resolution limit of the microscope. A detailed comparison of the STEM and conventional transmission microscopes in terms of the efficiency in the collection of information from the electron scattering has been made by Langmore, Wall and Isaacson [1973].

An important advantage for thick specimens is that the resolution of the STEM is not limited in the same way as that of fixed beam instruments by the effects of chromatic aberration of the imaging lenses on the electrons which have suffered multiple inelastic scattering and hence have a large energy spread. (Sellar and Cowley [1973].) Also it is possible to take advantage of the flexibility of the detection system, possible with STEM, to obtain the greatest possible contrast and signal strength for each particular value of the specimen thickness. There is an optimum detector aperture angle, up to 10^{-1} radians for very thick specimens, which gives the best image. Also if an energy filter is used to enhance the contrast by cutting out electrons which have lost energy through inelastic collisions, there is an optimum cut-off value for the energy loss which gives the greatest contrast enhancement. (See Cowley [1978b].)

Since, in the STEM technique, a very small, intense electron probe is focussed on the specimen, it has become feasible to combine STEM imaging with other techniques involving the use of finely focussed electron beams. If the probe is held stationary on the specimen a microdiffraction pattern can be observed on the detector plane, given by the illuminated region of the specimen which may be as small as 5 Å in diameter (Cowley [1980]). The transmitted electrons may be passed through an energy analyzer to receal the inelastic scattering due to the plasmons or single-electron excitations discussed in Section 12.3. The excitation of inner-shell electrons with characteristic energy losses is the basis for identification of chemical species by energy loss spectroscopy (ELS). An alternative method for microanalysis is to detect the characteristic X-ray lines emitted from the irradiated portion of the specimen using energy dispersive spectroscopy (EDS). Either of these techniques for microanalysis may be applied to regions of the specimen 100 Å or less in diameter.

The combination of these techniques is possible by the use of either dedicated STEM instruments or by the addition of attachments to conventional transmission electron microscopes. For convenience, and to recognize the important increase in the power of electron optical methods represented by this combination, the term Analytical Electron Microscopy (AEM) has become current (Hren et al. [1979]).

13.2. Image formation

The fact that both image and diffraction pattern may be observed and recorded and both are of importance for the study of many materials suggests that the most useful approach to imaging is through the Abbe theory as outlined in Chapter 3. For the range of accelerating voltages used, the small angle approximation is adequate for almost all applications and so will be used here because of its simplicity and clarity.

If an object has a transmission function $q(x, y)$ we may assume that, since the incident beam can be approximated as a plane wave of amplitude unity, the wave leaving the object has amplitude $q(x, y)$. Then in the back-focal plane of the objective the amplitude distribution is $Q(u, v)$ where $u = x/f\lambda$ and $v = y/f\lambda$. The amplitude in the image plane is given as $q(-x/M, -y/M)$ where M is the magnification. Often, for convenience, we ignore the factor $-M$ and, referring the image back to the object plane, assume that the image amplitude given by an ideal lens would be $q(x, y)$, and the intensity $I(x, y) = |q(x, y)|^2$.

Then the effects of aperture limitations and aberrations of the lens are represented by changes in amplitude and phase of the wave in the back-focal plane: the changes may be described as the result of multiplication by an optical transfer function. The effect of the insertion of an objective lens aperture is represented by multiplying $Q(u, v)$ by an aperture function $A(u, v)$ which is unity for $(u^2 + v^2) < (u_0/2)^2$. The image is then affected by convoluting the amplitude with $J_1(\pi u_0 r)/\pi r$, as in equation (3.5).

A defocus of the lens by an amount Δ is represented as in equation (3.20) by multiplying $Q(u, v)$ by a phase factor $\exp\{\pi i \Delta\lambda(u^2 + v^2)\}$. The effect of spherical aberration of the lens may be simulated by introducing a phase change on the back focal plane proportional to $(u^2 + v^2)^2$ so that $Q(u, v)$ is multiplied by $\exp\{+\frac{1}{2}\pi C_s \lambda^3 (u^2 + v^2)^2\}$. Hence, introducing these, the most important limitations of the imaging process, the back focal plane distribution becomes

$$Q(u, v) A(u, v) \exp\{i\chi\} \tag{1}$$

where the phase factor is

$$\chi = \pi\Delta\lambda(u^2 + v^2) + \tfrac{1}{2}\pi C_s \lambda^3 (u^3 + v^2)^2 \, . \tag{2}$$

The amplitude distribution of the image is then

$$\psi(x, y) = q(x, y) * J_1(\pi u_0 r)/\pi r * \mathcal{F} \exp\{i\chi\} \tag{3}$$

and the intensity is then $\psi\psi^*$. This is the result for perfectly coherent radiation. It is difficult to appreciate the effects of the convolutions except for some specially simple cases which will be treated later.

For electron microscopy in the low- or medium-resolution range which prevailed until recently, or for most thick specimens, it is often sufficient to assume the classical case of incoherent imaging, for which the phase relationships of the various terms of (3) are cancelled out by summation of intensities for the various incident beam directions. Then we may replace (3) by the expression for the intensity

$$I(x, y) = |q(x, y)|^2 * (J_1(\pi u_0 r)/(\pi r))^2 * |\mathcal{F} \exp\{i\chi\}|^2 \, . \tag{4}$$

The limitation of resolution due to the aperture effect is then judged by the Rayleigh criterion, as in equation (3.8). The image intensity function is successively convoluted by the spread functions due to the aperture effect and the aberration effect as in (4), and then by a spread function due to a chromatic aberration effect (arising from fluctuations of the accelarating voltage or the objective lens current or from energy spread due to inelastic scattering). If these spread functions are assumed to be approximately Gaussian, with half-widths d_A, d_s and d_c, we see from the equation (2.62) that the total spread function will have half width d given by

$$d^2 = d_A^2 + d_s^2 + d_c^2 \, . \tag{5a}$$

Ignoring the chromatic aberration effect, the optimum resolution is obtained for $d_A = d_s$ and is given approximately by

$$d = \tfrac{1}{2}\lambda^{3/4} C_s^{1/4} \, . \tag{5b}$$

13.3. Contrast for thin specimens

13.3.1. Phase object approximation

High resolution electron microscopy normally implies the use of thin specimens since for thick specimens too much fine-scale detail is usually superimposed and also the resolution tends to be degraded by the effects of multiple elastic and inelastic scattering.

Thin specimens may be described as thin phase objects with a small amount of aborption. The transmission function may be written

$$q(x,y) = \exp\{-i\sigma\varphi(x,y) - \mu(x,y)\}\,. \tag{6}$$

The use of the projections $\varphi(x, y)$ and $\mu(x,y)$ depends on the ignoring of the spread of the electron wave by Fresnel diffraction. In order to resolve the distance, d, the maximum thickness must not then exceed, approximately

$$H_{max} = d^2/2\lambda\,. \tag{7}$$

For 100 keV electrons this gives H_{max} equal to 1200 Å for 10 Å resolution and 100 Å for 3 Å resolution.

The transmission function (6) is largely that of phase object and, as discussed in Chapter 3, contrast in the image may be given by defocus, by aperture limitation, by the lens aberrations or by application of special techniques such as Zernike phase contrast. Although the Zernike and related methods have been used in electron microscopy with apparent success (see Unwin [1971]) the most usual and general methods involve defocus.

Equation (3) can be used as a basis for calculations with the appropriate phase factor χ. However, there are several special cases which allow a much more immediate appreciation of the nature of the image contrast.

In Chapter 3 we saw that if we consider the effect of a small defocus distance and ignore aberrations we may write

$$I(x,y) = 1 - \frac{\Delta\lambda}{2\pi}\,\sigma\left(\frac{\partial^2\varphi}{\partial x^2} + \frac{\partial^2\varphi}{\partial y^2}\right) - \mu(x,y)\,,$$

where we have assumed $\mu(x,y)$ but not $\sigma\varphi(x,y)$ to be very small.

The second differential term in this expression is the two dimensional form of $\nabla^2\varphi$. Using Poisson's equation,

$$\nabla^2\varphi = -4\pi\rho(x,y)\,,$$

where $\rho(x,y)$ is the projection of the charge density (including both positive and negative charges), we obtain

$$I(x,y) = 1 + 2\Delta\lambda\sigma\,\rho(x,y) - \mu(x,y)\,. \tag{8}$$

This is the "projected charge density" approximation. Its application to the interpretation of the images of thin crystals has been investigated and discussed by Allpress et al. [1972], Anstis et al. [1973], and Moodie and Warble [1967]. It has an important range of validity but cannot deal with the higher ranges of u, v values where, from (2), the spherical aberration term dominates the phase factor.

For thin biological specimens, it has been common practise for many years to underfocus the objective lens to obtain maximum contrast and then to interpret the image as if it was due to a pure absorption process. A partial justification for this approach, and the basis for a technique for obtaining optimum resolution and contrast from weak phase objects, is provided by a body of theory introduced by Scherzer [1949] and developed by Eisenhandler and Siegel [1965], Heidenreich and Hamming [1965] and Erickson and Klug [1970] (see also Cowley [1974]). To rederive their results we start from the expression (1). The function $Q(u, v)$ for the object given by (6) is

$$[\delta(u,v) + C(u,v) - \mathrm{i}S(u,v)] * \mathcal{F}\,[\exp\{-\mu(x,y)\}\,, \tag{9}$$

where

$$C(u,v) = \mathcal{F}[\cos\sigma\varphi(x,y) - 1]$$

and

$$S(u,v) = \mathcal{F}[\sin\sigma\varphi(x,y)]\ .$$

If $\mu(x,y)$ is assumed to be small we may write $M(u,v) = \mathcal{F}\mu(x,y)$ and (9) becomes

$$Q(u,v) = \delta(u,v) + C(u,v) - M(u,v) - \mathrm{i}S(u,v)$$

$$- C(u,v) * M(u,v) + \mathrm{i}S(u,v) * M(u,v)\,. \tag{10}$$

13.3.2. The absorption function

The absorption function $\mu(x,y)$ in this case arises because some electrons are scattered in such a way that they cannot contribute to the image. Some lose so much energy in inelastic scattering processes that they cannot be properly focussed in the image plane. For electrons which have lost 10 to 20 eV as a results of plasmon or single electron excitation, the chromatic aberration of the objective lens will defocus the image so that for these electrons the best resolution attainable in a 100 keV electron microscope will be about 10 to 20 Å. This defocussed image will be added to the in-focus image due to elastic electrons. Thus for the imaging of detail on a scale much larger than 20 Å the inelastic scattering will have no effect. For imaging of detail on a scale less than 10 Å the inelastically scattered electrons will give a slowly varying background, decreasing the contrast, and an appropriate absorption function will be needed for the interpretation of the image due to the elastically scattered electrons.

A major contribution to the absorption comes from the electrons which are scattered through sufficiently large angles to be either stopped by the objective aperture or else defocussed and displaced by spherical aberration. These will be both elastically scattered electrons and electrons inelastically scattered by thermal vibrations of the atoms or by excitation of atomic electrons. For thin specimens and apertures which are not too large the major contribution to absorption will be the exclusion of elastically scattered electrons. On the basis of the derivation of absorption functions for disordered materials by Cowley and Pogany [1968] it was shown by Grinton and Cowley [1971] that if the defocus is not too great the absorption function could be written

$$\exp\{-\tfrac{1}{2}\sigma^2\langle(\Delta\varphi)^2\rangle\} \tag{11}$$

where $\langle(\Delta\varphi)^2\rangle$ is the mean square deviation from the value of $\varphi(x,y)$ averaged over the resolution distance. This can be shown to be equivalent to the simple concept that the energy lost by application of the absorption function should be equal to the energy cut off by the aperture. The average value of $\mu(x,y)$ so obtained is then of the order of 10^{-3} Å^{-1}, considerably less than the average value of $\sigma\varphi(x,y)$.

By introducing the absorption function into $q(x,y)$ we have thus eliminated the need to consider the aperture function in (1) explicitly and have also redefined $\varphi(x,y)$ to be an average over the resolvable distances.

13.3.3. *Weak-phase object approximation*

For sufficiently thin specimens involving sufficiently light atoms we may now make a further simplifying assumption that $\sigma\varphi(x,y) \ll 1$. If as usual atoms are not resolved, $\varphi(x,y)$ is the projection of the mean inner potential of the object and so is equal to $\Phi_0 H$, where H is the specimen thickness. The condition then becomes

$$\frac{\pi}{\lambda E}\,\Phi_0 H \ll 1\,. \tag{12}$$

This implies that for 100 keV electrons and a mean potential of 10 volts the specimen thickness must be much less than about 150 Å. But this would be the requirement if specimens of this thickness were compared with specimens of zero thickness. Since an average phase shift can always be ignored, we must amend the condition to read that, in (12), H should represent the derivation from the average thickness (or $\Phi_0 H$ should represent the deviation in the product).

If this condition is met, we may simplify (10) by omitting all powers of $\sigma\varphi$ beyond the first and (1) becomes

$$\begin{aligned} Q'(u, v) &= [\delta(u, v) - M(u, v) - \mathrm{i}\sigma\Phi(u, v)] \; \exp\{\mathrm{i}\chi\} \\ &= \delta(u, v) - M(u, v) \cos\chi + \sigma\Phi(u, v) \sin\chi - \mathrm{i}\{\sigma\Phi(u, v) \cos\chi\} \,. \end{aligned} \tag{13}$$

The intensity in the image plane will be, for a centro-symmetric structure,

$$\begin{aligned} &[1 - \{\mu(x, y) * \mathcal{F} \cos\chi\} + \{\sigma\varphi(x, y) * \mathcal{F} \sin\chi\}]^2 \\ &\quad + [\sigma\varphi(x, y) * \mathcal{F} \cos\chi]^2 \,. \end{aligned} \tag{14}$$

In this, the terms of second power in $\sigma\varphi$ are considered negligible so that the main contribution to the image intensity will come from

$$Q'(u, v) = \delta(u, v) - M(u, v) \cos\chi + \sigma\Phi(u, v) \sin\chi \,. \tag{15}$$

The phase factor χ is given by (2). Since the system has cylindrical symmetry we may use u as a single radial coordinate and put

$$\chi(u) = \pi\lambda\Delta u^2 + \tfrac{1}{2}\pi C_s \lambda^3 u^4 \,. \tag{16}$$

The second-power defocus term is either added to or subtracted from the constant aberration term depending on the direction of defocus. Forms of the function $\sin\chi$ for various Δ values are illustrated in Fig. 13.3. From (15) it is seen that the contribution to the image contrast of diffracted beam amplitudes $\Phi(u)$ for a particular range of u, depends on the value of $\sin\chi$. For Δ values near to that for Fig. 13.3(c) the χ value is near $\pi/2$ and the $\sin\chi$ value is close to unity over a considerable range of u. Over this range the contributions from the potential distribution to the image intensity will be the same as if an ideal in-focus lens were used to image an object of transmission function.

$$q(x, y) = 1 - \sigma\varphi(x, y) \,.$$

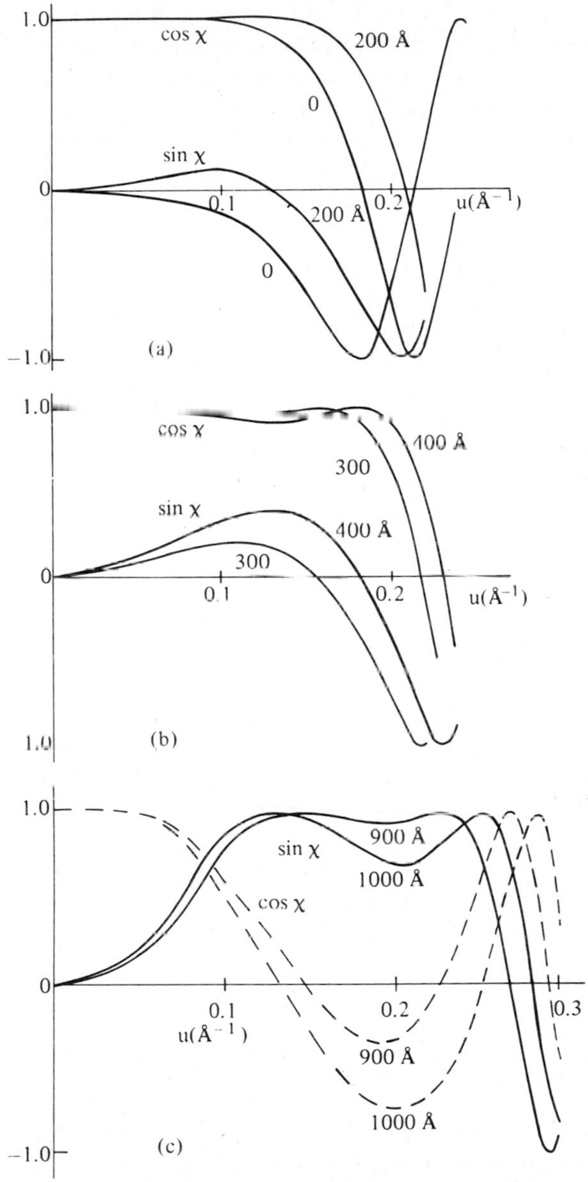

Fig. 13.3. Values of the functions $\sin \chi(u)$ and $\cos \chi(u)$ for $\chi(u)$ defined by equation (13.16) for defocus values R of (a) 0 and −100 Å, (b) −300 and −400 Å, and (c) −900 and −1000 Å. The curves are drawn for 100 keV electrons and C_s = 1.8 mm.

To these contributions must be added those of the absorption effects in (15). For small values of u where contributions of the $\sin \chi$ term are lacking, $\cos \chi$ will be approximately unity and there will be added contributions from the absorption term in equation (15).

Hence for a small range of under-focus Δ values there will be an optimum contribution from the whole of the diffraction beam amplitudes of interest and for this optimum defocus the image intensity will approximate to $1 - 2\sigma\varphi(x, y)$. From (3.28) this is seen to be similar to that for ideal Zernike phase contrast. The optimum defocus distance, given by putting $d\chi/du = 0$ for χ approximately equal to $-2\pi/3$, is

$$\Delta = -(\tfrac{4}{3}C_s\lambda)^{1/2} . \tag{17}$$

The optimum aperture size is then

$$u_{max} = 1.51\, C_s^{-1/4} \lambda^{-3/4} , \tag{18}$$

corresponding to a least resolvable distance in the image of

$$\Delta x = 0.66\, C_s^{1/4} \lambda^{3/4} \tag{19}$$

or $\Delta x = 3.5$ Å for 100 keV electrons and $C_s = 1.8$ mm.

The form of the intensity function for this optimum defocus provides some justification for the interpretation of high-resolution images of large molecules of biological interest (proteins, viruses) in terms of a simple "absorption" function. For most observations of biological samples the resolution is much poorer than that given by (19), being limited by the severe radiation damage of the specimen by the incident beam. The contrast arises mostly from the use of a relatively small objective. It appears that there is some limited area of validity for the image interpretation which forms the basis for the three dimensional reconstruction of the configuration of small objects using computerized Fourier transforms of series of micrographs obtained with differing angles of incidence of the electron beam (see deRosier and Klug [1968]).

The rough estimates for the range of validity of the weak-phase-object approximation, equations (6) and (12) have been confirmed by Grinton and Cowley [1971] who made detailed n-beam calculations of image intensities for models of stained biological objects using the computer techniques outlined in the final section of Chapter 11.

When the weak-phase-object approximation no longer applies, there is little option to the calculation of the image contrast from the full expression (10) or from a full n-beam dynamical treatment. Some further approximate expressions which may be valid for somewhat greater phase changes than allowed by (12) have been derived by Cowley [1974] and these may serve to indicate the nature of the failure of the simpler treatment.

The use of a small objective aperture, with little or no defocus, to achieve the best images of biological samples has been a standard practice of biologists for many years. From the above treatment we see that, since for small values of u $\sin \chi(u)$ is very small, the main contribution to the contrast must come from the term in (15) involving $\cos \chi$ which is very nearly equal to unity for small u. From Subsection 13.3.2, this contribution involves the effective absorption function which arises because the outer part of the diffraction pattern is cut off by the objective aperture.

If the objective aperture used is larger than that required to give the useful resolution (usually 10–20 Å for biological specimens) the image will be confused by the addition of a great amount of unwanted detail arising from phase-contrast contributions from the outer parts of the diffraction pattern. Detailed computer calculations by Cowley and Bridges [1979] for a model system representing a negatively stained protein molecule, using the periodic continuation method of calculation of Section 11.5 in two dimensions, are in agreement with the experimental observations of Massover [1978]. With the optimum defocus and aperture size for high resolution imaging, the image of the protein molecule is submerged in granular phase-contrast noise and is not recognizable. If the objective aperture is made smaller this noise disappears and the protein molecule appears much more clearly.

13.3.4. Dark field images

For dark-field images, when the contribution of the directly transmitted incident beam to the image is removed, it has usually been considered that the image intensity is given simply by the amount of scattering matter present at each point of the object. This is a reasonable approximation for low-resolution images but cannot be valid for high resolution pictures of thin specimens. We demonstrate this for the case of a weak phase object, following Cowley [1973].

In the ideal case that only the directly transmitted beam is removed, the effect is to subtract the delta-function, $\delta(u, v)$, from the expression (13). In order to obtain the best directly-interpretable image we need to make both $\cos \chi$ and $\sin \chi$ constant over a large range of u, v. From Fig. 13.3, this can best be done by making $\cos \chi \approx 1$ and $\sin \chi \approx 0$. If we allow a maximum value of $|\sin \chi| = 0.3$ we obtain an optimum dark field imaging for a defocus given by $\Delta_{\mathrm{opt}}^2 = (0.61/\pi) C_s \lambda$ or $\Delta \approx -300$ Å for 100 keV electrons and $C_s = 1.8$ mm.

Equation (13) then becomes

$$Q'(u, v) = -\mathrm{i}\{\sigma(\Phi(u, v) - \Phi(0, 0)) - \mathrm{i}M(u, v)\} ,$$

where the $\Phi(0, 0)$ is subtracted since this part of the scattered amplitude must be removed when we block out the transmitted beam. The zero order Fourier coefficient $\Phi(0, 0)$ must be equal to $\bar{\varphi}$, the average potential, taken over the region of coherent illumination. The image intensity is therefore given by

$$I_{\mathrm{DF}}(x, y) = \sigma^2 [\varphi(x, y) - \bar{\varphi}]^2 \tag{20}$$

and depends on the square of the deviation from the average projected potential. Thus a small bright spot in the image may represent either a concentration of atoms or a hole in the specimen: a sinusoidal variation of projected potential will produce an image having half periodicity.

As a matter of experimental convenience, the idealized case, in which only the transmitted beam is stopped, is not normally used. More commonly a considerable portion of the scattered radiation is also blocked off. Then the dark-field image intensity cannot be expressed by a simple formula such as (20) and detailed calculations are required for each case (e.g. Hashimoto et al. [1973]). When the weak-phase-object approximation fails, the calculation becomes correspondingly more complicated.

13.4. The imaging of crystal structures

13.4.1. Imaging of thin crystals; structure images

For crystals, as for amorphous material, the phase-object approximation may be used for sufficiently small thicknesses. The most interesting cases will be those for which the incident beam is parallel to a principal axis of the crystal because then the projected potential will be a multiple of the projected potential for one unit cell and, with sufficient resolution, the crystal structure may be directly visualized.

In this case, of course, the variation in the projected potential function, and in the phase change of the electron wave, will be much greater than for amorphous materials since the beam will pass either through a whole row of atoms or else will not traverse any atoms. Hence we may expect that the weak-phase-object approximation will fail for very small thicknesses.

Nevertheless, images obtained under the optimum defocus imaging conditions derived for weak phase objects have shown very good representations of the atom arrangements in crystals. Uyeda et al. [1970, 1972] obtained images of thin crystals of Cu-hexadecachlorophthalocyanine showing the shapes of the individual molecules in projection. Even more striking is the image, Fig. 13.4(a) of a thin crystal of $Ti_2Nb_{10}O_{29}$ obtained with an underfocus of about 900 Å and with the incident beam in the direction of the short *b*-axis of the orthorhombic unit cell (Iijima [1971]). Comparison with the diagram, Fig. 13.4(b) shows an excellent correlation between dark spots in the image and the metal atom positions. At the position where there is one metal atom per unit cell there is a gray spot. Where two metals in each unit cell are close together in projection there is a much darker spot. The clearly defined

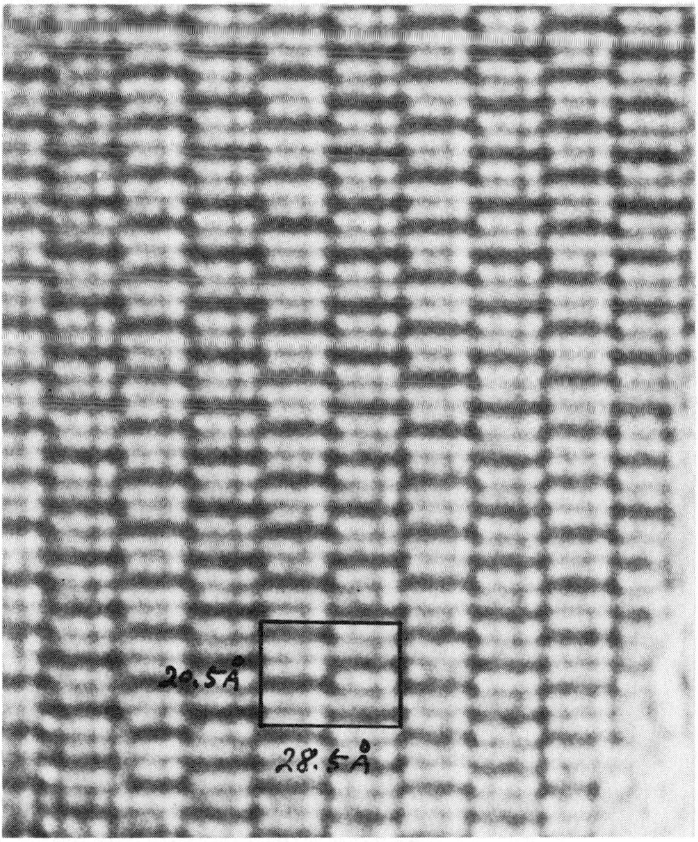

Fig. 13.4(a)

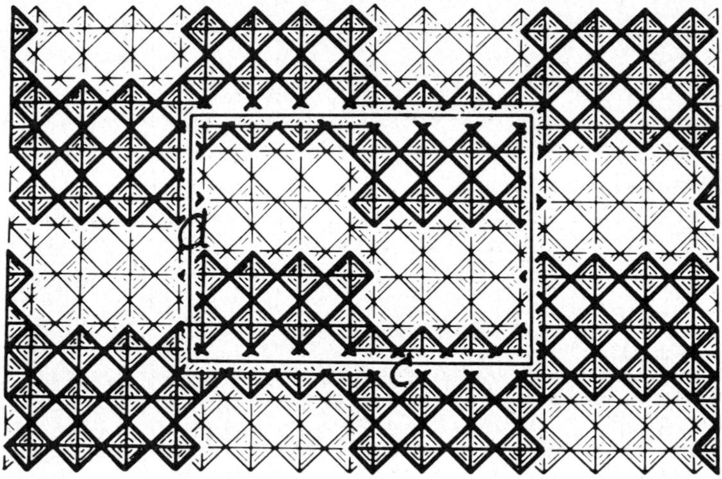

Fig. 13.4(b)

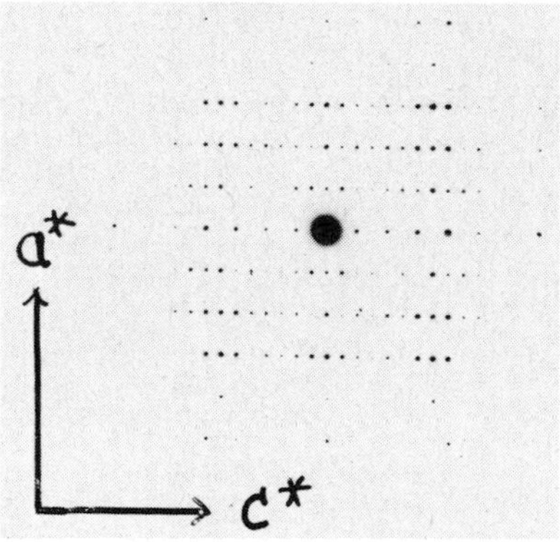

Fig. 13.4(c)

Fig. 13.4. (a) High resolution electron micrograph of a thin portion of a crystal of $Ti_2Nb_{10}O_{29}$ with the incident parallel the *b*-axis. In this projection the axial lengths are $a = 20.5$, $c = 28.5$ Å (Iijima [1971]). (b) Diagram of the idealized structure of $Ti_2Nb_{10}O_{29}$ (Wadsley [1961]). The shaded squares represent octahedra of oxygen atoms containing metal atoms, stacked on two levels indicated by dark and light shading. (c) Electron diffraction pattern from a crystal of $Ti_2Nb_{10}O_{29}$ in the same orientation as for (a).

configuration of two rows of three white spots corresponds to the characteristic arrays of six sites in the lattice where no metal atoms exist.

In order to obtain such pictures which are readily interpreted in terms of the crystal structure several conditions must be met (Cowley and Iijima [1972]). The incident beam must be aligned parallel to a principal axis of the crystal with an accuracy of a fraction of a degree by use of a high-quality goniometer specimen stage. The crystal thickness must be no more than 50 to 100 Å (with some exceptions – see below). The microscope must be underfocussed by an amount approximately equal to the optimum defocus for a weak phase object (about 900 Å in this case) for very thin crystals but decreasing as the crystal thickness increases. Deviation by more than about 200 Å in focal length from this optimum value gives images having little or no apparent relationship with the structure.

Similar results have been obtained for a number of related oxide systems based on the juxtaposition of rectangular blocks of metal-oxygen octahedra in the configuration of the ReO_3 structure, including various titanium-niobium oxides and Nb_2O_5 itself (Iijima [1973]) (see Fig. 13.5), and also some niobium

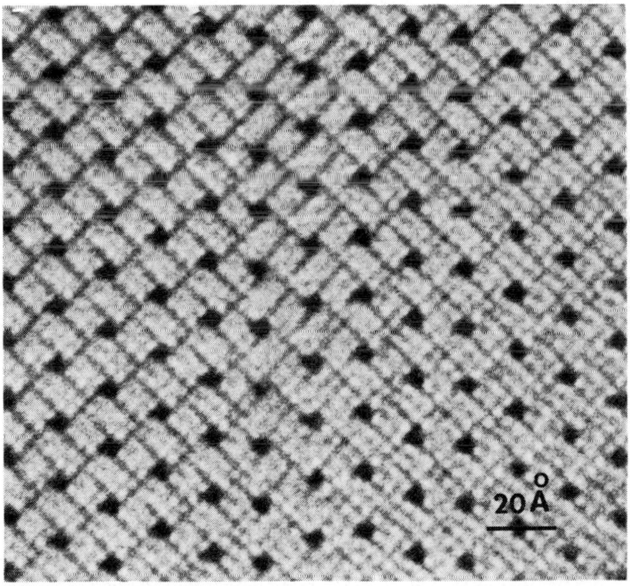

Fig. 13.5. High resolution image of a portion of a crystal of $H-Nb_2O_5$, of thickness about 800 Å, showing a clear representation of the structure in both perfect-crystal regions and in a defective region.

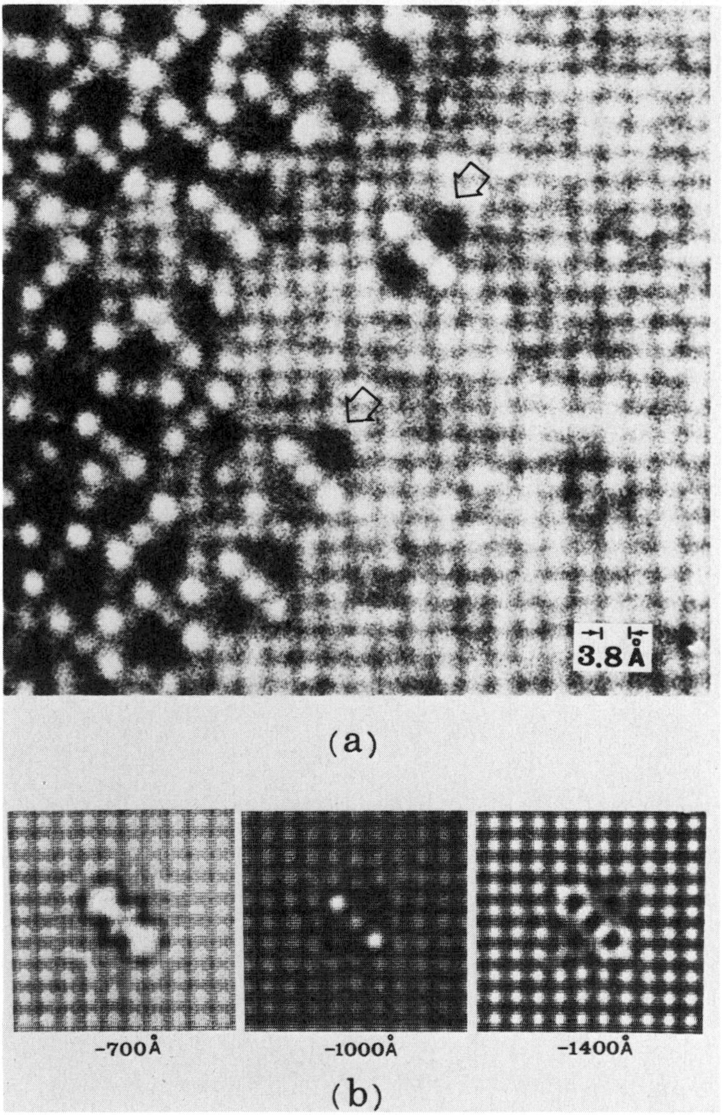

Fig. 13.6. (a) A high-resolution electron micrograph showing a disordered arrangement of the atoms in a crystal having a nominal composition 19 $Nb_2O_5 \cdot 63 WO_3$. The region on the left of the images has a tetragonal tungsten bronze-type structure. The square patterns shown by arrows are described by coherent intergrowth of a single unit cell of the TTB-type structure with the ReO_3 type. The dark contrast appearing inside the region of the TTB element, corresponding to the pentagonal tunnels, indicates that they are occupied by metal atoms. (b) A through-focus series of images calculated for the model structure of the TTB element with two pentagonal tunnels occupied. The "optimum focus" image (−1000 Å) resembles the experimental images and thus the assumed structure for the TTB element is shown to be reasonable (O'Keefe and Iijima [1976]).

tungsten oxides and others having superlattices based on the tungsten bronze type of structure (Iijima and Allpress [1974 a, b]) (see Fig. 13.6).

The technique has also been successfully applied to the study of a variety of other oxides and various minerals (Buseck and Iijima [1974]; Buseck [1980] and to an increasingly wide range of other inorganic materials. More recent reviews of the accomplishments and possibilities of the method are given by Anderson [1979], Cowley [1979] and other contributors to the proceedings of the 1979 Nobel symposium.

In the Fig. 13.5 some defects in the crystal structure are seen to be clearly resolved. This observation tends to emphasize the fact that the imaging of individual features within the structure of a crystal having a large unit cell is almost exactly the same as the imaging of any isolated group of atoms not included in a crystal.

For these specimens, electrons emerging from a region of the specimen of width equal to the resolution limit will have been influenced only by the column of atoms immediately above it. Also the imaging of these electrons on the photographic plate does not involve interference with electrons from any other part of the image. Hence the atoms in the regular crystal lattice and in imperfections of the structure will be imaged with equal clarity provided that the separation of the atoms concerned exceeds the resolution limit of the microscope. For detail on the same scale as the resolution limit or less, the representation of atomic arrangements in the images becomes much more complicated (see, for example, Skarnulis, Iijima and Cowley [1976]) and a direct intuitive interpretation of the images is not possible. The only reliable method of image interpretation is through comparison with detailed calculations.

The high resolution imaging of crystals thus offers important possibilities for the study of the structures and defects of the more complicated stable crystals. Extensive observations with more limited resolution, particularly by Allpress and his collaborators (see Allpress and Sanders [1973]), have provided important information on the ordering principles and processes in non-stoichiometric oxide systems. The improved resolution of about 4 Å illustrated in Figs. 13.4 to 13.6 allowed a considerable extension of the possibilities. More recently, the improvement of the resolution to 3 Å or better with 100 keV and 200 keV instruments, and to something approaching 2 Å with the high resolution 1 MeV instruments has extended further the range of materials which may be usefully imaged and the accuracy of the structural information obtainable.

The problem remains of providing a ready means of accounting for the observed contrast of these images. The requirement that the Fresnel diffraction spread of the wave should be less than the resolvable distance is satisfied in that the crystal thicknesses conform with the requirements (2). However, the condition that $\sigma\varphi(x, y) \ll 1$ is clearly not satisfied since a calculation shows that for the crystals giving Figs. 13.4 to 13.6 the differences in $\sigma\varphi$ for points a few Å apart may be 5π or more.

13.4.2. Calculation of images of crystals

To examine the imaging process in detail, the intensity distribution in the image of a crystal lattice may be calculated with any desired degree of precision by making n-beam dynamical calculations as outlined in Chapters 10 and 11 to obtain the amplitudes of the diffracted beams and then using these as coefficients in the Fourier series, modified by the phase and amplitude factors giving the effects of defocus, spherical aberration and aperture limitation. For the large unit cells involved in the studies mentioned above, the number of simultaneous diffracted beams to be considered may be very large, amounting to several hundreds, as suggested by Fig. 13.4(c).

Allpress et al. [1972] and Anstis et al. [1973] found detailed agreement between calculations and measurements of the intensity distributions of one-dimensional fringes obtained from crystals of $W_4Nb_{26}O_{77}$ tilted so that the main contributions to the image came from a single systematic row of reflections, although in this study up to 435 beams were included in the calculations.

The extension of the calculations to two-dimensional images was made by Lynch and O'Keefe [1973], O'Keefe [1973] and Fejes [1977] and further developments of the computing methods have been provided, for example, by Van Dyck [1978] and Ishizuka [1980]. Excellent agreement has been obtained between computed and observed images, including through-focus series, for a number of materials. The need to include into the computations the effects of the finite convergence angle of the incident electron beam and the effects of chromatic aberration on the finite spread of electron energies has been emphasized by Anstis and O'Keefe [1976] and Ishizuka [1980].

Successful calculations have also been made of the intensities of images of defects in crystals, making use of the periodic continuation method of Section 11.2 with a large artificial superlattice unit cell. The calculated images of defects shown in fig. 13.6, for example, were made by O'Keefe and Iijima [1978] with the superlattice unit cell sizes shown, using 12 000 beams in the calculation.

Such calculations are necessarily large and time consuming. The prospect of using such calculations on a trial-and-error basis for interpreting the contrast given by crystal structures or defects, for which the atom positions are imperfectly known, is rather formidable. There is thus considerable point in the development of relatively simple approximations which may serve as a rough guide to the nature of the contrast to be expected from a particular structure.

One such approach was suggested by Cowley and Iijima [1972] and the approximate expressions of Cowley [1973, 1974] may also be applied. These suggest that for a high peak of projected potential the image intensity will no longer be linearly proportional to $\sigma\varphi(x, y)$ but will tend, through possible weak fluctuations to a constant limiting value as the maximum value of $\sigma\varphi(x, y)$ increases.

Two conclusions follow which are contrary to the predictions of the weak phase-object approximation. The contrast of the dark spots due to rows of atoms parallel to the beam will first increase and then become constant (apart from an overall absorption effect) as the crystal thickness increases. Also an unresolved pair of such rows will give approximately twice the contrast of one. Both of these predictions are in agreement with observations.

There are some observations on thicker crystals of oxides, however, for which the phase-grating approximation is clearly irrelevant. In some cases where a good "amplitude object" type of image is produced with optimum defocus for the thin regions (up to 150 Å) of a crystal oriented so that the beam is very closely parallel to a crystal axis, a good amplitude image is also seen for thickness in the range of about 700 to 1000 Å, although not for thickness between 150 and 700 Å (Fejes, Iijima and Cowley [1973]). The Fig. 13.5 was, in fact, obtained in this thick-crystal range.

The explanation for this thick-crystal contrast seems to lie in the special nature of the dynamical effects under diffraction conditions of high symmetry and in the existence of special relationships between the structure amplitudes and excitation errors for particular crystals.

13.4.3. Lattice fringe imaging

The earliest (Menter [1956]) and for many years the predominating, observation on crystal lattice periodicities was that of parallel sets of near-sinusoidal fringes corresponding to the interference of the transmitted beam with one diffracted beam. These may be obtained under diffraction conditions approximating the classical two-beam case, although similar fringes are obtained in n-beam situations when the contributions of other beams to the image are excluded by the objective aperture or by lack of resolution.

If three or more, non-collinear strong reflections contribute to the image, two or more intersecting sets of parallel fringes will be observed, giving a two-dimensional, periodic image which may in some cases resemble the images of very thin crystals discussed in the previous paragraphs. Unless the special imaging conditions for these "structure images" apply, there is, in general, no direct connection between the intensities given by these intersecting fringe patterns and the crystal structure except that some of the lattice periodicities will be represented. We therefore refer to these more general images, periodic in one or two dimensions, as "lattice fringe images". Their nature and lack of relationship to the structural images is best discussed initially in terms of the simple two-beam approximation.

For the ideal two-beam situation, the amplitudes of the transmitted and diffracted beams at the exit face of a plane-parallel crystal slab are given from equations (8.27), (8.28), (8.29) or as by Cowley [1959] for the symmetrical case, $\theta_0 = \theta_h$, as

$$\Psi_0 = \exp\{-\mathrm{i}\sigma H\Phi_0 - H\mu_0 + \tfrac{1}{2}\mathrm{i}\zeta_h H\}\left[\cos WH - \frac{\mathrm{i}\pi\zeta_h}{W}\sin WH\right] \tag{21}$$

$$\Psi_h = \exp\{-\mathrm{i}\sigma H\Phi_0 - H\mu_0 - \tfrac{1}{2}\mathrm{i}\zeta_h H\}\left[2\pi\mathrm{i}\sigma\Phi_h \frac{\sin WH}{W}\right] \tag{22}$$

where we have put

$$\Phi_0 = \Phi(0,0,0)\,, \qquad \mu_0 = \mu(0,0,0)$$

and

$$W = 2\pi\sigma\Phi_h(1+w^2)^{1/2} \tag{23}$$

The intensity distribution for the in-focus fringes is given by

$$I = \psi\psi^* \quad \text{where} \quad \psi(x) = \Psi_0 + \Psi_h \exp\left\{\frac{2\pi\mathrm{i}hx}{a}\right\}. \tag{24}$$

Then

$$I(x) = \exp\{-2\mu_0 H\}\left[1 + \left(\frac{2w}{1+w^2}\right)\sin^2(2\pi\Phi_h H\sigma\sqrt{1+w^2})\cos\frac{2\pi hx}{a}\right.$$

$$\left. + \frac{1}{(1+w^2)^{1/2}}\sin(4\pi\sigma\Phi_h H\sqrt{1+w^2})\sin\frac{2\pi hx}{a}\right]. \tag{25}$$

At the Bragg angle, this reduces to

$$I(x) = \exp\{-2\mu_0 H\}\left[1 + \sin(4\pi\sigma\Phi_h H)\sin\frac{2\pi hx}{a}\right]$$

or, taking the absorption into account explicitly,

$$I(x) = \exp\{-2\mu_0 H\}\left[1 - \sin(4\pi\sigma\,\Phi_h H)\cos 2\pi\left(\frac{hx}{a} - \alpha\right)\right], \tag{26}$$

where $2\pi\alpha = \tan^{-1}(\sigma\Phi_h/\mu_h)$.

If we assume that $\sigma\Phi_h \gg \mu_h$, the phase angle $2\pi\alpha$ is approximately $\pi/2$. Thus the image shows sinusoidal fringes having their maximum intensity at one quarter of the way between atomic planes. The contrast of the fringes varies sinusoidally with the thickness H. As the value of $4\pi\sigma\Phi_h H$ passes through π the contrast goes to zero and then reverses.

The more general situations involving deviations from the exact Bragg condition have been studied in detail by Hashimoto et al. [1961]. They showed that for varying ζ_h and H the fringes are no longer parallel to the lattice planes giving, on occasion, the false impression that there may be defects in the crystal structure. From (25) it is seen for example that for w large the intensity distribution is

$$1 + \frac{2}{w}\sin\zeta_h H \sin 2\pi\left(\frac{hx}{a} + \frac{\zeta_h H}{2\pi}\right),$$

so that a phase factor is introduced which shifts the fringes by an amount depending on both ζ_h and H.

For the two-beam symmetrical case, defect of focus or spherical aberration can make no difference to the image since the phase factor χ will be the same for both beams. This makes the detection of the fringes easier but also provides the basis for the belief, of limited validity, that imaging of crystal lattices does not impose the same requirements upon the quality of the electron microscope since a "fringe resolution" is always better than "point-to-point" resolution. In fact electron microscopes having a point-to-point resolution of more than 3 Å have given lattice fringe spacings of less than 1 Å.

Inclusion of more than one beam in a systematic row of reflections introduces more complicated intensity profiles of the fringes and a more complicated variation of the profiles with crystal thickness and orientation and with defocus. Early calculations related to such fringe profiles were made by Niehrs [1962] and by Miyake et al. [1964]. More sophisticated n-beam calculations related to high-resolution imaging of crystals are reported by Allpress et al. [1973], Anstis et al. [1973].

The perturbations of a crystal structure associated with dislocations, stacking faults and other defects will produce variations of the excitation errors and effective thickness and so will produce variations of relative fringe positions and spacings. These effects have been observed, for example, in high resolution imaging of fringes corresponding to (111) planes of silicon and germanium (Phillips and Hugo [1968, 1970]; Parsons and Hoelke [1969, 1970]). The interpretation of the images in terms of two-beam diffraction theory by Cockayne et al. [1971] confirmed the implications of the earlier treatments (Hashimoto et al. [1961], Cowley [1959]) that, particularly under the symmetrical two-beam conditions, the relationship between the positions of fringes and of lattice planes is not straightforward as can be seen from equations (25) and (26) and the direct interpretation of fringe configurations in terms of defect structure is rarely possible.

Usually it is considered that in a symmetrical situation with the incident beam and diffracted beam making equal angles with the objective lens axis, the transfer function of the lens has no effect on the two beam lattice fringes. This is not true in the cases of defects involving variations of lattice plane spacings. With the incident beam inclined to the axis at the Bragg angle for the host lattice planes, a local variation of lattice plane spacing will result in a diffracted beam inclined at some greater or lesser angle and therefore subject to a phase change relative to the incident beam corresponding to the variation of the function $\chi(u)$ [equation (16)]. For the case of alloys having a periodic variation of lattice constant, resulting from a spinodal decomposition mechanism, it has been shown by Spence et al. [1979a] at the apparent variation of the lattice constant can be greatly exaggerated by this effect.

For two-dimensional fringe patterns given by crystals of simple structure in principal orientations some special imaging conditions can be achieved. The relatively small number of sharp Bragg reflections sample the transfer function at sharply defined points and the values of the transfer function in between those points are irrelevant. For example for a crystal with a unit cell which has a square projection in the beam direction the Bragg reflections will occur for $u, v = h/a, k/a$ and the phase factor $\chi(u, v)$ given by equation (2) can be made equal to a multiple of 2π for all h, k by choosing particular values for the defocus, Δ, and the spherical aberration constant C_s. For these values the image will then be the same as for an ideal lens with Δ, $C_s = 0$. The fineness of the image detail will be limited only by the chromatic and lateral coherence effects and by high order aberrations. The image intensities will not necessarily show any relationship to a projection of the structure, since the relative phases of the diffracted beams will be strongly influenced by dynamical scattering effects, but for particular crystal thicknesses, some resemblance may be fortuitously possible.

In this way it was possible for Hashimoto et al. [1977] to obtain images of thin gold crystals, viewed along the [100] axis, showing clearly resolved intensity maxima, 2 Å apart, at the gold atom row positions, each with a central dark spot, suggesting a fine structure with an image resolution of 1 Å or better even though the microscope they used had a resolution of 3 Å or more, as defined by the Scherzer limit of equation (19). Also Izui et al. [1977] produced pictures of silicon crystals, viewed in a [110] direction, in which the rows of silicon atoms 1.3 Å apart appeared to be resolved (as white spots about 1.5 Å apart) with a similar instrument.

Such special cases can occur only for simple structures in principle orientations which give favorable numerical relationship of the u, v parameters. They can occur only for near-perfect crystals since when imperfections are present the images of the defects are produced from diffuse scattering between the Bragg reflections so that the selection of special values of the transfer function does not apply. A discussion of the degree of crystal perfection or the limits of crystal size which allow these special imaging conditions to apply has been given by Cowley [1978a]. It has been shown by Spence et al. [1978] that it is possible to use the concepts involved to obtain images of defects in crystals, to a limited but important extent, with resolution well beyond the Scherzer limit.

A further range of phenomena is introduced when crystals are superimposed. Moire fringes are produced, as when larger periodic objects are superimposed and viewed with visible light, when the two crystals have the same periodicity but differ slightly in orientation, or when the two crystals have slightly different periodicities. Treatment of the moire pattern intensities has been given for the two-beam case by Hashimoto et al. [1961] and by Cowley [1959]. An n-beam treatment for the limiting case that both crystals are very thin was provided by Cowley and Moodie [1959].

13.4.4. *Crystal imaging without lattice resolution*

If the interference of the diffracted beams with the transmitted beam, or among themselves, is prevented by limitation of the objective aperture or because the microscope does not have sufficient resolution under the conditions of operation, the bright-field image will show the variation of the intensity of the transmitted beam as a function of position. Dark-field images obtained from individual diffracted beams will show the variation of intensity of the diffracted beams with position. Perfect, unbent crystals of uniform thickness will show no contrast, but contrast may be produced by any crystal defect, or by bending or thickness variation.

By use of the n-beam diffraction theory described in previous chapters it is possible to calculate in detail the intensity variation in the image corresponding to any one or any combination of these perturbations. The column approximation may be used where that seems valid or else the more general methods proposed above (Section 11.5) for non-periodic objects. However for most purposes a very rough indication of the type and scale of intensity variations is sufficient to characterize the feature of interest, and for this purpose it is sufficient to use the nearest simple approximation. This is usually a two-beam approximation for cases where it is not obviously inapplicable.

An account of the dependence of the transmitted and diffracted beams on crystal thickness and orientation in the two beam approximation has been given in Chapter 9. This offers a first-approximation explanation for many of the observations on thickness fringes and extinction contours in bright- and dark field electron microscopy.

The contrast produced by crystal defects is found by application of the column approximation as described in Chapter 10. The progressive changes of lattice displacement in each column are fed into the equations (10.32), (10.33) or for n-beam situations (10.35), in order to trace the effect of the strain fields or local discontinuities on the intensities.

There is a great deal of experience and empirical knowledge in the subject of electron microscopy of extended crystal defects and a theoretical basis derived from two-beam and occasionally n-beam dynamical theory. Because of the very wide range of application of this type of observation and the consequent interest in the topic, it has been the subject of numerous extended treatments in books (Hirsch et al. [1965]; Heidenreich [1966]; Amelinckx [1964]) and review articles. We do not attempt to reproduce any of this material here but instead confine ourselves to a few topics of more general and theoretical interest. Some results for extended defects will be mentioned in Chapter 18.

13.5. High voltage microscopy

As the accelerating voltage, or the size of the unit cell of the crystal increases, the two-beam approximation has decreasing relevance. In either case the number of diffracted beam with comparable intensity becomes larger. Even in the absence of non-systematic interactions, the systematic interactions become increasingly important.

An analysis of this situation for simple metals with increasing voltages has been made by Humphreys et al. [1971] in terms of the transfer of energy from one Bloch wave to another. Their calculations showed that for lower voltages the strongest diffraction effects are observed when the Bragg condition for a strong reflection is satisfied giving a good approximation to pure two beam effects. With increasing thickness the anomalous transmission effect described in Chapter 9 becomes apparent. The extinction contour patterns show maximum transmission at the Bragg reflection conditions with energy transmitted preferentially in Bloch wave 2. With increasing voltage, about 250 keV for Au(111) reflection or 600 keV for Cu, the increasing strength of other reflections increases the importance of other Bloch waves. The maximum transmitted intensity may then be found for the symmetrical orientation, with the incident beam parallel to the lattice planes. For voltages in the megavolt range the orientation for maximum interaction and maximum transmission may switch to any one of a number of possibilities as the relative structure factors and excitation errors vary.

In general for very high voltages the number of simultaneous reflections is so large that the discussion in terms of Bloch waves is not profitable and the calculation of intensities is very laborious. The phase-object approximation is valid for the limiting case of infinite voltage but is still limited to very thin crystals in the megavolt range. A development from the phase object approximation due to Berry [1971] and Berry and Mount [1972] takes this idea further and allows a connection with the semi-classical channelling theory (Chapter 14). It is possible to formulate the scattering theory on the basis that the incident electrons travel almost parallel to planes or rows of atoms and, considering motions in directions perpendicular to the beam only, one can envisage oscillations or spiral paths of the electrons in the potential fields of these planes or row of atoms. The various quantized, tightly or loosely bound orbitals of the electrons can then be considered and the effects of absorption on each can be envisaged (Kambe et al. [1974]).

Increasingly high-voltage microscopes, specially designed to have the best possible resolution, are being used to produce structure images of thin crystals along the lines of Subsections 13.4.1 and 13.4.2. To a reasonable approximation, the ability of the microscope to produce images readily interpretable in terms of a projection of the structure is indicated by the Scherzer optimum resolution, given by equation (19), provided that the effects of beam divergence and chromatic aberration are not too severe. The dependence of Δx in (19) on λ suggests that an increase in electron energy from the usual 100 keV to 1 MeV will give a considerable improvement on the resolution. In practice the spherical aberration constant C_s will increase with electron energy, but even if we make the somewhat pessimistic assumption that $C_s\lambda$ = constant we find a dependence on $\lambda^{1/2}$, or an improvement by a factor of 2.1, suggesting a resolution limit of less 2 Å at 1 MeV.

To date the performance of the high-voltage, high-resolution microscopes has been limited mostly by chromatic aberration effects, resulting from fluctuations of the high voltage or of the lens current supplies, but these limitations are gradually being overcome and resolution limits approaching 2 Å have been attained and used to great effect in studies of inorganic oxide phases by Horiuchi and colleagues (Horiuchi [1979]), of phthalocyanines and some inorganic compounds by Uyeda et al. [1979] and of various alloys by Hiraga et al. [1977].

In addition to the possibility of better resolution, the use of higher voltages offers other advantages. Because the wavelength is shorter the spread of the electron waves by Fresnel diffraction is less and the phase object approximation may be used for thicker crystals. It has been verified by detailed calculations that the thickness of specimen which can be used to obtain images directly interpretable in terms of projection of the structure (qualitatively but non-linearly) may be greater by a factor of 2 or 3 at 1 MeV as compared with 100 keV. Also the contrast obtainable from individual atoms or small groups of atoms should improve with increasing voltage. The interaction constant σ decreases very little above about 200 keV (see Section 4.16) because of relativistic effect but the relativistic effects enhance the decrease of the wavelength, which determines the resolution (Cowley [1975]).

Problems

1. From the expressions (13.14) and (13.15), derive the variation of the contrast and the position of the two-beam lattice fringes as a function of crystal thickness and excitation error for the case of no absorption. (c.f. Hashimoto et al. [1961]).

2. Inelastic scattering effects will tend to decrease the contrast of thickness fringes. Phonon scattering gives little contrast and the plasmon contribution, although showing strong dynamical effects, will be defocussed by the chromatic aberration of the lens. Assuming an average energy loss of 20 eV, a chromatic aberration constant C_c of 2 mm and a loss of resolution given by $\Delta r = C_c \alpha \Delta E/E$, ($E$ = 100 kV) find the loss of contrast of thickness fringes of 200 Å periodicity given by the 200 reflection from MgO. How will this affect the derivation of absorption coefficients from thickness fringe intensity variations? What other factors will cause errors in absorption coefficients derived in this way?

3. A perfect crystal has flat faces and uniform thickness except that there is a small rectangular protuberance 20 Å high on one face. Assuming 100 keV electrons, a mean inner potential Φ_0 of 20 volts and that the amplitude for the h reflection is 10 volts, find the contrast in the image due to the protuberance,

when

(a) no strong diffracted beams occur and the microscope is suitably defocussed;

(b) the Bragg angle for the reflection h is exactly satisfied? (Use a column approximation and find the bright- and dark-field images assuming that the lattice spacing is not resolved.)

CHAPTER 14

K-line patterns and channelling

14.1. Kossel lines

14.1.1. Geometry of Kossel lines

Diffraction patterns from single crystals produced by divergent radiation and consisting of continuous sets of lines have been observed with X-rays, electrons and other radiations under a great variety of experimental conditions. Kossel lines are formed in diffraction patterns produced by X-rays generated within a single crystal. The somewhat similar "Kikuchi lines" are produced when electrons diffusely and inelastically scattered within a crystal are diffracted by the crystal. These and all the other similar types of patterns to be discussed in this chapter we refer to collectively as "K-line patterns".

In the initial observations by Kossel et al. [1934] and the earlier rather questionable observations of Clarke and Duane [1922], a single crystal was made the anticathode in an X ray tube. The characteristic X-rays excited by the incident electron beam were then diffracted in the crystal to produce the Kossel line pattern on a photographic plate. A pattern recorded by Voges [1936] from a copper single crystal is reproduced in James [1950]. Similar patterns were obtained by Borrmann [1936] using X-rays rather than electrons to excite the fluorescent X-rays in the crystal. The patterns in these cases were observed on the same side of the crystal as the incident electron or X-ray beam but a transmission geometry is also possible using a thin crystal.

The geometry of the line patterns can be understood by considering the diffraction of radiation emitted by a point source in a crystal as in Fig. 14.1. For a particular set of planes, radiation from the point P will make the Bragg angle θ_h for a particular direction of emission giving the transmitted beam T_h and diffracted beam D_h. The sum of all directions of T_h, where intensity is removed by the diffraction process will be a cone of half angle $\pi/2 - \theta_h$ and axis perpendicular to the diffracting planes. The sum of all diffracted beams D_h will form a similar cone of directions having increased intensity with oppositely directed axis. The intersections of these cones with a flat photographic

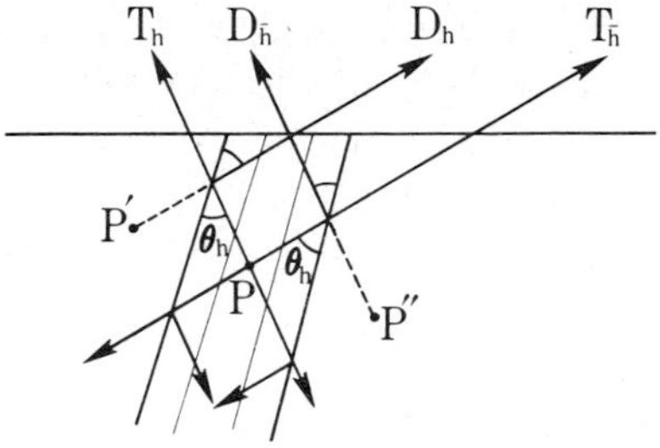

Fig. 14.1. Diagram illustrating the formation of Kossel line patterns with radiation produced at a point P in a crystal.

plate or film will be conic sections: ellipses, parabolas or hyperbolas. A recent account of the geometry of the patterns has been given by Tixier and Wache [1970].

However from Fig. 14.1 we see that there will be a reflection $\bar{h}$, from the opposite side of the atomic planes also, giving a transmitted beam $T_{\bar{h}}$ parallel to D_h and a diffracted beam $D_{\bar{h}}$ parallel to T_h, and the excess or deficit of intensity observed due to diffraction from the h planes will tend to be cancelled out. In fact, if X-rays are generated at all points in the crystal with equal intensity, we see that the intensity excess due to D_h will be cancelled directly by transmitted radiation from a point such as P′. According to simple kinematical diffraction ideas or a two beam dynamical treatment without absorption, the sum of transmitted and diffracted beams in any direction will be always the same and the Kossel lines will show no contrast. To predict anything like the observed black-white contrast of the lines we must use a dynamical theory with absorption.

14.1.2. Dynamical theory of Kossel intensities

At first sight the dynamical theory for a point source in a crystal emitting a spherical wave seems formidable. However von Laue [1935] showed how to simplify the problem by use of the reciprocity theorem (Chapter 1). From this theorem, the amplitude at a point A far distant from a crystal due to a point source B within the crystal will be equal to the amplitude at the same position B within the crystal due to a point source at A.

Thus the problem of calculating the Kossel pattern is just that of calculating the wave field in the crystal due to a plane incident wave. Since the Kossel pattern is given by the independent emission of characteristic radiation from atoms distributed throughout a volume of crystal, the intensity of the wave-field in the crystal due to a plane incident wave is summed for all positions of emitting atoms. This can be done making use of the theory of diffraction as developed by von Laue [1935], and later by Zachariasen [1945], Hirsch [1952], and Kato [1955], among others.

The polarizability of the crystal may be written

$$\psi(\boldsymbol{r}) = \psi'(\boldsymbol{r}) + \mathrm{i}\psi''(\boldsymbol{r}) \,, \tag{1}$$

where the imaginary part is added to take account of absorption. Both real and imaginary parts are periodic and have Fourier coefficients denoted by Ψ'_h and Ψ''_h respectively. For a crystal having more than one type of atom these are not necessarily similar because the ratio of absorption factors for the various atoms may be very different from the ratio of their elastic scattering factors. The distribution of atoms emitting the characteristic radiation will be different again and may be denoted by

$$\xi(\boldsymbol{r}) = \sum_h \xi_h \exp\{-2\pi\mathrm{i}\boldsymbol{h}\cdot\boldsymbol{r}\} \,. \tag{2}$$

The expression for the Kossel line intensity K_α as a function of the deviation from the Bragg angle, $\alpha = 2(\theta_B - \theta) \sin 2\theta_B$, is complicated. It is simplified if we consider that the Kossel lines are generated in a crystal which is thin compared to X-ray extinction distances and also if we consider the symmetric case that h and $\bar{h}$ reflections make equal angles with the exit surface. Then, following Cowley [1964] we obtain

$$K_\alpha \propto \left[\xi_0 + \frac{K^2 a_0 \Psi'_h \Psi''_h}{2(K^2 \Psi'^2_h + \alpha^2/4)} \{\xi_0 \alpha/2 + K\Psi'_h \xi_h\} \right] . \tag{3}$$

Here K is the polarization factor equal to 1 for the normal component and $|\cos 2\theta|$ for the parallel component and a_0 is the maximum value of a thickness parameter $a = \pi t_0/\lambda\gamma_0$, where t_0 is the actual thickness of crystal and γ_0 is the direction cosine of the incident beam relative to the surface normal.

Thus the intensity is made up of a term symmetric in α, depending on the product $\Psi'_h \xi_h$ and an antisymmetric term with α in the numerator, proportional to the product $\Psi'_h \Psi''_h$. The sum of these terms as suggested in Fig. 14.2 will show a black-white contrast. The antisymmetric part of this contrast will be reversed if $\Psi'_h \Psi''_h$ changes sign and so can be used as an indicator of the relative positions within the unit cell of the atoms which are emitting the characteristic radiation giving the Kossel lines. For crystals containing only

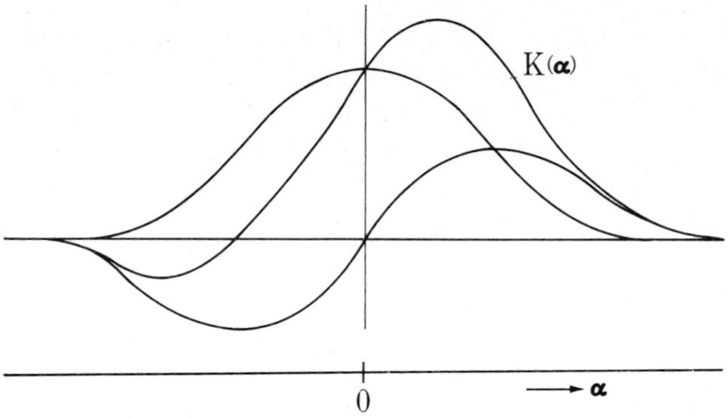

Fig. 14.2. The intensity distribution K_α across a Kossel line shown as the sum of symmetric and antisymmetric parts.

one type of atom Ψ'_h, Ψ''_h and ξ_h will be of the same sign and the asymmetry will be such that the white line on a photographic print (high intensity) will always be on the same side of the line, namely on the outside of the cones of diffracted beams.

From equation (3) it is seen that the half width of the symmetric part of the intensity peak is given by $\alpha = 2K\Psi'_h$, which gives values of the order of 10^{-5} radians. The separation of black and white (defect and excess intensity) lines will be much the same i.e. of the order of 10^{-3} mm on a film placed 10 cm from the crystal.

14.1.3. Kossel lines with limited resolution

For many observations such as those of Borrmann [1936] the lack of resolution results in an integration over the intensities of the black-white pairs to give a single line which may be either defect or excess, depending on geometric factors. But for many other observations the lines show black-white contrast with an angular separation of the maximum and minimum of 10^{-3} to 10^{-4} radians. The contrast then is not explainable purely in terms of this idealized two-beam dynamical theory. We must add a more realistic consideration of the geometry of the experiments including the finite width of the incident electron beam.

Referring to Fig. 14.1, suppose that the electron beam which excites the characteristic X-rays is centered on P but does not extend to P′ or P″. Then using an argument based on simple geometric optics ray diagrams (justifiable because the crystal dimensions involved are sufficient to define directions of propagation and beam locations with the necessary accuracy) we see that the diffracted beam D_h due to radiation from P is not compensated by a reduced transmission of radiation from P′. The reduced transmission of the beam $T_{\bar{h}}$ from P is not compensated by a diffracted beam of radiation from P″. Hence there will be a dark line where $T_{\bar{h}}$ intercepts a photographic plate and a white line due to D_h separated by a distance which depends on the separation of P′ and P″ from P. As a refinement of this kinematical treatment we may take into account the attenuation of both transmitted and diffracted beams by absorption and by diffraction processes (the "extinction" effect). Then the relative intensities of these beams will depend on the distances they travel through the crystal which are not, in general, equal.

If we then take into account the finite width of the incident electron beam and the generation of X-rays from the volume of crystal defined by the penetration and spread of the electron beam, we see that we will get a separation and contrast of the black-white pairs of lines which depends in a rather complicated way on the dimensions of this region of X-ray generation, the dimensions of the crystal, the Bragg angle, the absorption of the X-rays in the crystal and the angle between the diffracting planes and the crystal surface.

We can see why some of the clearest Kossel line patterns have been obtained by use of the configuration suggested in Fig. 14.3(a), where an electron beam is focussed to a small spot on one side of a thin crystal and the Kossel pattern is observed in the transmitted X-rays. From the geometry of this case it is seen that the white (excess intensity) line tends to be on the outside of the cones of radiation and so on the side of the curve on the photographic plate away from the center of curvature. The contrast then depends very little on the positions of the emitting atoms in the unit cell.

The diagram 14.3(a) is appropriate for a near-kinematic case of diffraction from an imperfect crystal. For a thick perfect crystal, Fig. 14.3(b) is more ap-

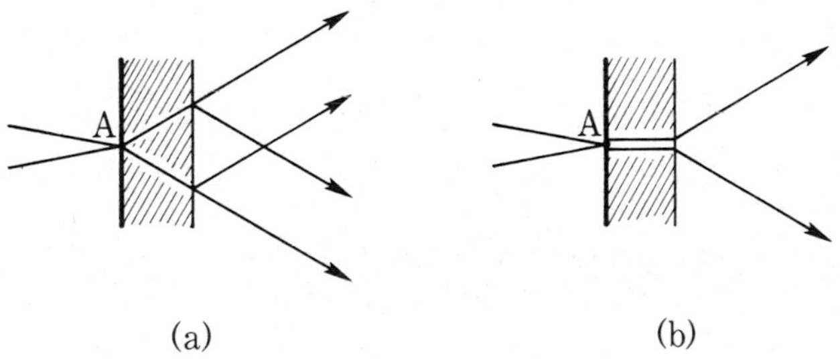

Fig. 14.3. Production of a Kossel-line pattern from a source of radiation, A, outside the crystal for the cases of (a) a thin crystal and (b) a thick crystal.

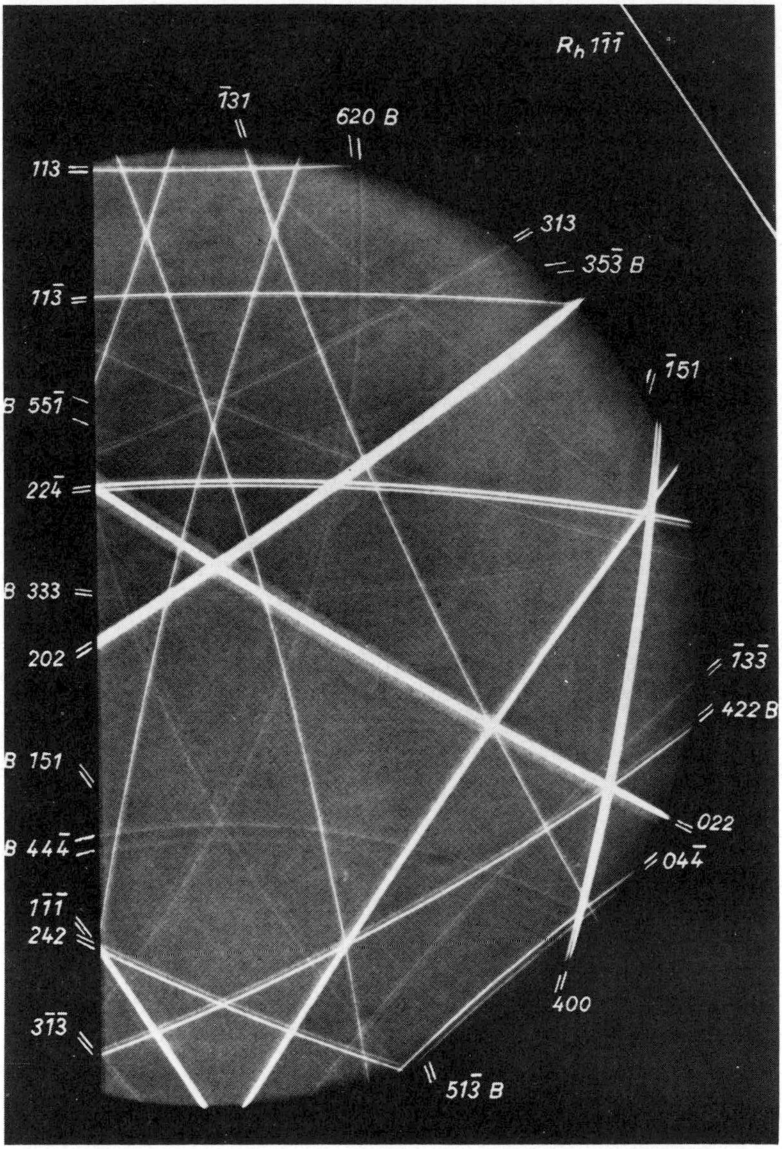

Fig. 14.4. Kossel line pattern from a crystal of germanium, thickness 0.23 mm, produced by placing a source of CuK_α radiation 3 cm from the crystal. A number of transmission lines and one reflection line, $1\bar{1}\bar{1}$, are visible. (From Borrmann [1964].)

propriate. As we saw in discussion of the Borrmann effect (Chapter 9) the energy will be conveyed through the crystal preferentially along strongly diffracting planes. The transmitted and diffracted beams are equally strong, narrow intensity peaks. Thus in both directions making angles $\pm\theta_h$ with the diffracting planes sharply defined bright lines will be produced. The background to the pattern is very low because of the much higher absorption coefficient in non-diffracting directions. A Kossel line pattern produced by a source of radiation considerably removed from a near-perfect crystal of germanium is shown in Fig. 14.4.

Where Kossel lines intersect, two diffracted beams occur simultaneously. For the perfect crystal cases the intensities of the lines are not additive but are given by the complicated 3-beam X-ray dynamical theory. (Ewald and Heno [1968].) In particular for the thick-crystal case of Borrmann effect transmission the 3-beam situation can give an even smaller absorption coefficient than the 2-beam case so that bright spots are seen at the intersection points (Borrmann and Hartwig [1965]; Balter et al. [1971]).

14.2. Kikuchi lines

The line patterns first observed by Kikuchi [1928] in the diffuse background of electron diffraction patterns from single crystals are of much the same form as Kossel lines except that, because of the much smaller wavelengths and diffraction angles in the case of electrons, the cones of diffracted rays are very shallow and intersect a photographic plate in what appear to be straight lines (see Fig. 14.5). The other obvious difference is that, while in Kossel lines the $+\theta_h$ and $-\theta_h$ lines are equivalent, each with a close black-white contrast, in the Kikuchi pattern one of the $+\theta_h$ and $-\theta_h$ lines will be white and the other black, with the black line always nearest to the central beam of the diffraction pattern. For those symmetrical cases where the $+\theta_h$ and $-\theta_h$ lines are almost the same distance from the central beam, there is often a more complicated "Kikuchi band" structure with an excess or defect of intensity over the whole area between the line pairs and an asymmetry of intensity about the actual line positions.

The diffuse scattering in single-crystal electron diffraction patterns is made up of pseudo-elastic scattering due to thermal vibrations of the atoms, atomic disorder or crystal defects plus inelastic scattering due to electron excitations. For thick crystals it becomes largely multiple diffuse scattering with a broader distribution in angle and in energy. Since the scattering processes render the electron beams incoherent with the incident beam and with each other the

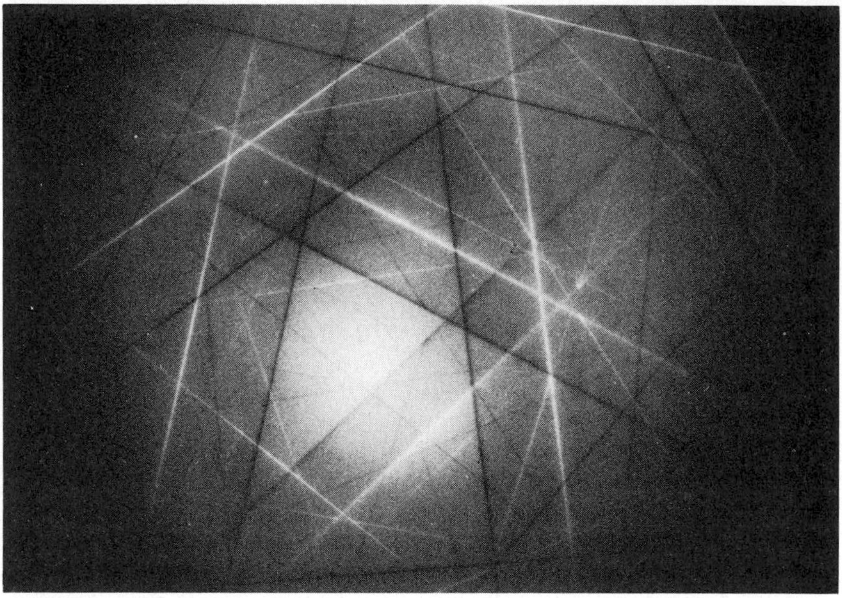

Fig. 14.5. Kikuchi line pattern obtained by transmission of 483 keV electrons through a crystal of aluminum. (From Beauvillain [1970].)

diffuse scattering can be considered to be generated within the crystal. However in contrast to the case of Kossel lines where the emission of X-rays is isotropic, the scattering is strongly peaked in the forward direction of the incident beam. Then a simple kinematical treatment gives a reasonable account of the line-pair intensities. For the case sketched in Fig. 14.6 for example, there is more scattered intensity in the $-\theta_h$ than in the $+\theta_h$ direction. Hence in the $-\theta_h$ direction there will be more energy lost by the Bragg reflection than will be gained from the beam diffracted from the $+\theta_h$ direction. There will be a defect of intensity (black line) in the $-\theta_h$ direction and an excess in the $+\theta_h$ direction (white line).

This simple treatment fails to predict any contrast for the symmetrical case of the Kikuchi bands. For these cases a two-beam dynamical theory with absorption gives an asymmetry of the lines as in the X-ray case. For electrons the magnitude of the structure factors is much greater because of the much stronger interaction of the radiation with atoms. For strong reflections the width of the reflections or the distance between maxima and minima of the asymmetric part of the intensity profile is an appreciable fraction of the distance between $+\theta_h$ and $-\theta_h$ lines and the long tails on the intensity profiles

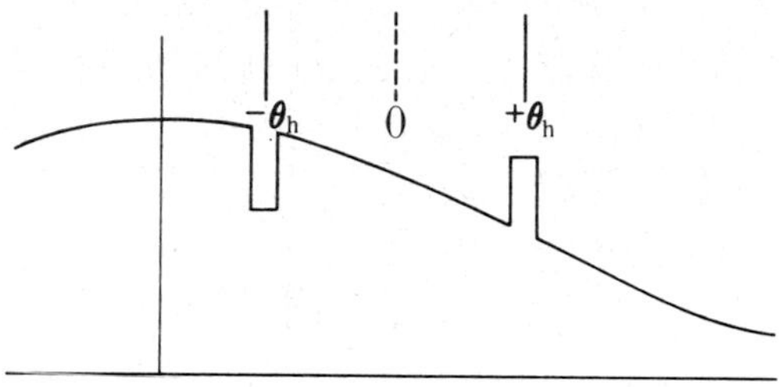

Fig. 14.6. Diagram illustrating the formation of a black-white pair of Kikuchi lines in an electron diffraction pattern.

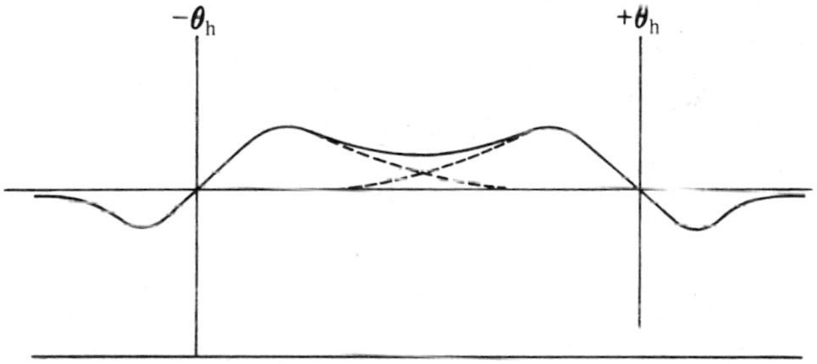

Fig. 14.7. Diagram suggesting the origin of a Kikuchi band with excess intensity between the lines.

can add up to give a band of excess or defect intensity as suggested in Fig. 14.7. However for a complete account of the intensity profiles an n-beam treatment is necessary, particularly for higher voltage electrons (> 100 keV) for which the intensity maximum may actually occur in the middle of the band rather than near a Bragg angle, and the whole intensity distribution due to a systematic row of reflections may be quite complicated. Then when bands or lines intersect, very complicated configurations may result from strong non-systematic dynamical interactions, as seen in the extinction lines of Fig. 9.3.

The simplest case of dynamical interaction, and the first n-beam dynamical diffraction effect to be investigated (Shinohara [1932]) is given by the crossing of two well defined isolated Kikuchi lines. The 3-beam interactions in this case may cause the lines to avoid the point of intersection, turning the intersection of two straight lines into the two branches of a hyperbola as sketched in Fig. 14.8.

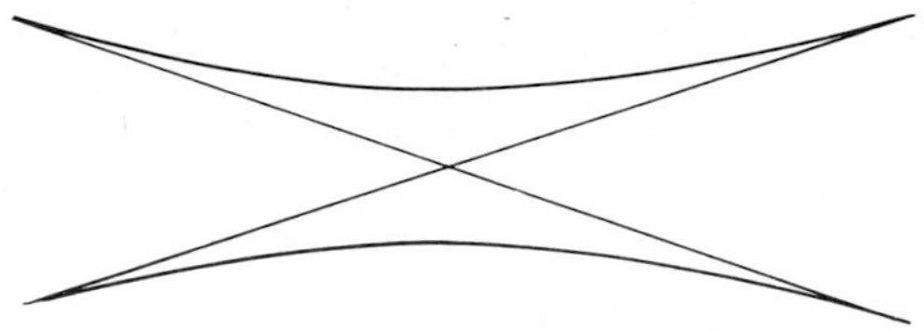

Fig. 14.8. The interaction of intersecting Kikuchi lines due to 3-beam dynamical interactions.

The complete dynamical calculation of the intensities of Kikuchi patterns is complicated both for formulation and for computation and has been attempted only with simplifying assumptions of very restricted validity such as the assumption of single inelastic scattering only. The problem has been formulated using an extension of the Schrödinger equation method by Pfister [1953] and Kainuma [1953] and extended by Fujimoto and Kainuma [1963] and Okamoto et al. [1971]. A somewhat different, and conceptually simpler approach was initiated by Gjϕnnes [1966]. On the basis of the ideas of Gjϕnnes, Doyle has made detailed calculations of both thermal diffuse scattering (Doyle [1969]) and plasmon scattering (Doyle [1971]) using the three-stage calculations outlined in Chapter 12 and showing the generation of Kikuchi lines at least in the thermal diffuse scattering for thin crystals of gold. His results showed clearly the zero-order line, the line appearing in the center of a set of parallel lines due to a set of systematic reflections (Kainuma and Kogiso [1968]). This is a specifically n-beam phenomenon.

From the early days of electron diffraction until the present day the Kikuchi line patterns have been used as a testing ground for electron diffraction theory, as in the studies of n-beam diffraction effects (Shinohara [1932]; Kambe [1957]), the observation and theoretical treatment of the failure of Friedel's Law under conditions of dynamical scattering (Miyake and Uyeda [1955]; Kohra [1954]) and the relativistic variations of n-beam dynamical effects (see Chapter 15). These studies have been made without reference to the fact that the Kikuchi pattern arises from multiple inelastic and incoherent scattering processes. However, so long as quantitative intensity values are not being considered it seems to be sufficient to take into account that all electrons in the crystal, scattered by any number of processes, have almost the same wavelengths and undergo much the same n-beam dynamical scattering process, so that consideration of elastic scattering of electrons from a single source is sufficient.

14.3. External sources of divergent radiation

For both X-rays and electrons the same sorts of K-line patterns can be generated using sources of divergent radiation outside of the crystal. A small source of X-rays may be formed in a thin metal foil by focussing a beam of electrons on it. Then if the foil forms the wall of a vacuum system a thin crystal of any material may be placed against the foil and a K-line pattern is generated by transmission with the geometry of Fig. 14.3. Or if a crystal, necessarily of larger size, is held a few cm away from the X-ray source a K-line pattern is obtained by back reflection. Patterns of this sort, generated by the use of a converging cone of X-rays with angle restricted to about 60° were obtained first by Seemann [1919, 1926].

In general X-ray line patterns given by an external source, as in Fig. 14.4, are usually known as "pseudo-Kossel patterns". They may be obtained from crystals in which no useful characteristic radiation can be generated or crystals which may deteriorate in vacuum or under electron beam irradiation. For example, Lonsdale [1947] made extensive studies of organic crystals in this way.

For a source of radiation which is not integral with the crystal lattice giving the diffraction effects there will be no dependence of the diffracted intensities on the source atom position. The intensity is no longer calculated by use of Laue's method employing reciprocity. The X-ray sources are usually sufficiently far from the diffracting crystal to allow the incident radiation from each point source to be considered as the sum of plane waves for which the intensities are calculated separately and added, using the geometric considerations for either kinematic scattering (Fig. 14.3(a)) or dynamical perfect crystal scattering (Fig. 14.3(b)) or something between these limiting cases.

An interesting observation by Lonsdale [1947] and others is that the contrast of the pseudo-Kossel line patterns is much better for imperfect crystals than for near-perfect ones. This can be understood intuitively on the basis of several combinations of poorly-defined ideas. It may be said, for example, that for perfect crystals the width of the black-white lines will be as determined from the two-beam dynamical theory and so will be so small that the contrast is lost when the line is broadened by finite source size effects. Alternatively one could say that for near-perfect crystals there is very strong absorption of the transmitted and diffracted beams by Bragg reflection (extinction) and this reduces the contrast of the diffraction effects.

On the other hand, if a crystal is too imperfect, the range of angular orientations of the individual small regions of the crystal will cause a spreading of the Kossel line intensity profile which comes from the geometric effects previously described and a consequent loss of contrast. Hence for best pseudo-Kossel patterns it is necessary to use a crystal having just the right amount of imperfection.

In recent years the development of the electron probe micro-analyser has provided an almost ideal means for the observation of Kossel and pseudo-Kossel lines. In this instrument an intense beam of electrons is focussed to form a spot a few μm in diameter on a specimen to produce fluorescent X-rays which may be analysed to show the composition of the small irradiated area. But by using this small intense X-ray source for the production of K-line diffraction patterns some excellent results have been obtained (Castaing and Guinier [1951]).

For electrons, small sources of radiation outside the crystal can be obtained readily by using a lens to focus an electron beam to a small cross-over. In this way Kossel and Mollenstedt [1939] produced convergent beam patterns (see Chapter 9 and Figs. 9.4 and 9.5) in which the round discs of intensity produced for each diffracted spot are crossed by dark or white lines. These differ from Kikuchi lines (which, incidentally, are produced in the background inelastic scattering of convergent beam patterns) in that there is not the same overlapping of transmitted and diffracted beams.

Such overlapping occurs if the angle of divergence of the incident radiation is increased until the incident beam cone is great enough to include the diffracted beam directions for many reflections. Then the line patterns are somewhat similar to pseudo-Kossel patterns in that within limits the intensity distribution of an incident radiation is uniform. The simple kinematical explanation of contrast for Kikuchi lines based on the rapid fall-off of intensity with angle does not apply.

The kinematical type of explanation of Kossel line contrast as in Fig. 14.3(a) applies to some extent but for electrons the directions of the diffracted beams are not defined with such precision and the dimensions of the crystals are very much smaller. Thus the black-white contrast of the K-lines is seen if the pattern very close to the crystal is greatly magnified by use of an electron microscope, as in the observations of Dupuoy and Beauvillian [1970]. But if the pattern is observed at a large distance (20 to 50 cm) the spread of the diffracted beams due to their natural half-width is sufficient to smear out this black-white structure. The contrast is then due purely to dynamical interference effects which give a fine structure which has a greater angular width.

A further complication arises if the incident electron beam is not focussed exactly on the crystal. Then any bending, faulting or defects of the crystal will modify the K-line patterns. It was shown by Smith and Cowley [1971] that bending of the crystal causes the separation of a K-line into one strong black line and one white line with a separation proportional to the amount of defocus and the curvature.

14.4. Information from K-line patterns

The useful information concerning crystalline specimens which has been derived from K-line patterns has come almost entirely from considerations of their geometry. While there is obviously some relationship between the contrast of K-lines and the structure amplitudes for reflections from the planes involved, very little has been done toward quantitative interpretation of observed contrast, partly because there are complications due to dynamical diffraction effects and dependence on the experimental conditions, and partly because K line patterns are most often obtained from simple substances of well-known structure.

The form of the intensity distribution around the intersection of two K-lines depends on the relative phases of the reflections from the two sets of planes involved. Hence, as pointed out by Fues [1949] and others, observations of all available K-line intersections should in principle provide an unambiguous assignment of phases of structure amplitudes and hence a solution to the phase problem of crystal structure analysis. However for a number of reasons, mostly relating to experimental difficulties, no practical use has been made of this idea.

The geometry of K-lines obviously reflects the orientation of the crystal and shows the symmetry of the equivalent sets of atomic planes. When the pattern is produced by a fine electron beam giving X-rays or scattered electrons from a small region of a crystalline sample, it offers a sensitive indication of changes of orientation or of the degree of perfection of an imperfect crystal.

A further application of importance is the derivation of accurate lattice parameter values from observations of the details of the K-line geometry, preferably in regions where n-beam dynamical diffraction effects are small. The initial idea of Lonsdale [1947] for the use of pseudo-Kossel lines was elegant. Occasionally in K-line patterns three lines meet at a point. This is sometimes the inevitable result of crystal symmetry, as when the three planes concerned belong to one cone and the coincidence is independent of wavelength. For other cases it results because the line of intersection of two K-line cones lies on the reflection cone for a third set of planes for one particular wavelength.

For example in the sketch of Fig. 14.9 the lines marked 1, 2 and 3 are the intercepts of three sets of planes on a film. Diffraction with a particular wave-

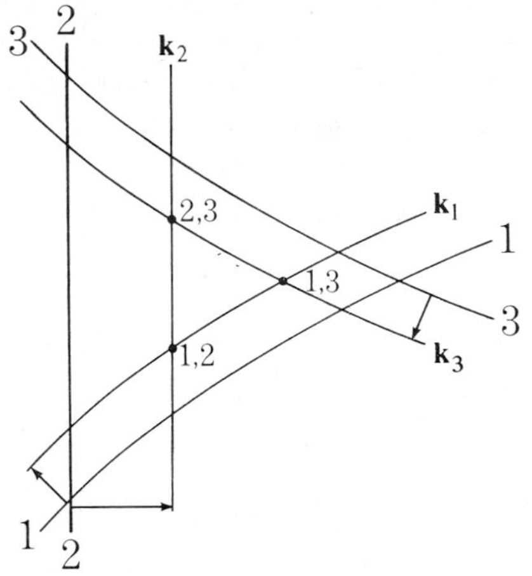

Fig. 14.9. The intersection of three Kossel lines to form a triangle. Lines 1, 2 and 3 are intersections of three crystal planes with the film. Lines K_1, K_2 and K_3 are the corresponding Kossel lines for one particular wavelength.

length λ gives the K-lines K_1, K_2 and K_3. As the wavelengths and hence the diffraction angles increase the points of intersection of K_1, K_2 and K_3 will move closer together until for a particular wavelength, λ_c, they coincide. Then the observation of a coincidence provides a unique relationship between the angles between the three planes, their interplanar spacings and the wavelength λ_c. If the crystal is cubic and the three sets of planes can be indexed, the one remaining unknown, the lattice constant, a_0, can be found in terms of the wavelength with an accuracy which Lonsdale estimated as about $\pm 5 \times 10^{-5}$ Å. For a unit cell with more than one parameter to be found, a greater number of coincidences must be observed.

The beauty of this method lies in its simplicity. The accuracy of the results does not depend at all on the geometry of the experimental arrangement. The shape, bending and distance from the crystal of the recording film is arbitrary. The main difficulty lies in the need for the accidental occurrence of at least one 3-fold coincidence for each crystal.

This difficulty is largely overcome by an extension of the method used by Schwarzenberger [1959] and Mackay [1966]. This takes advantage of the fact that pseudo-Kossel line patterns often include K-lines from the K_{α_1}, K_{α_2} and K_β characteristic X-ray emission lines. Hence instead of the set of three intersecting lines as in Fig. 14.9, there will be three parallel sets. If the critical wavelength λ_c for exact coincidence of the intersections comes close to these characteristic wavelengths it may be found with considerable accuracy by simple linear interpolation or extrapolation.

Other related methods for obtaining accurate lattice constants from Kossel or pseudo-Kossel patterns have been summarized by Tixier and Wache [1970]. In a similar way, the intersections of Kikuchi lines in electron diffraction patterns have been used to establish a relation between the lattice constant and the wavelength and so determine one in terms of the other. The ease with which the electron wavelength can be changed by changing the accelerating voltage makes the coincidence method of Lonsdale easy to apply (Høier [1969]).

14.5. Channelling

In recent years considerable interest has been aroused by the observation that fast charged particles show preferential penetration through crystals when the direction of incidence is almost parallel to prominent planes or axes of the lattice. The particles concerned range from protons to the ions of elements with quite high atomic number. Their energies may be as high as many MeV or as low as a few kilovolts. In the preferred directions of transmission through the crystal the number of particles transmitted is normally greater and the average energy loss is less.

Those concerned with the initial observations of this type were accustomed to thinking in terms of collisions of particles. The explanations evolved and developed (J. Lindhard [1965], Lervig et al. [1967]) to explain the observations invoked a picture of the incident particles travelling along channels in the crystals between planes or rows of atoms, oscillating from side to side of the channels as if reflected back and forth by the potential walls created by the atoms of the crystal. For reviews and discussion of channelling phenomena see Datz et al. [1967] and Chadderton [1970, 1973].

From the first, the similarity of channelling patterns recorded for these fast heavy particles with the Kikuchi line patterns given by electrons was apparent and, with time, the observations showed analogies with the whole variety of K-line effects which we have discussed in this chapter. For example, channelling patterns observed on a fluorescent screen when a near-parallel beam of protons is transmitted through a thin single crystal are very much reminiscent of transmission Kikuchi line and band patterns.

Other observations have been made by firing a beam of fast ions at a thin single crystal and measuring the total transmitted intensity as a function of incident angle. A simple application of the reciprocity theorem suggests that this is equivalent to a wide-angle convergent beam experiment in which a beam of high convergence angle strikes the specimen and the intensity transmitted is observed as a function of the angle of observation. There are similarities also in the production of secondary radiations as will be discussed in the next section.

Various attempts have been made to describe channelling in terms of diffraction phenomena. The observation of anomalous transmission in lattice plane directions is reminiscent of the Borrmann effect. But a moment's reflection will show that the two-beam dynamical theory usually used to discuss the Borrmann effect even for electrons, is quite out of place here. For protons the wavelength will be approximately 1/40 that for electrons of the same energy. At the same time the strength of the elastic interaction with matter, indicated by the value for $\sigma = \pi/\lambda E$, will be approximately 40 times as great and the strength of the inelastic scattering will be relatively much greater than this. Hence for proton diffraction the thickness of crystal over which coherent diffraction takes place will be tens of Å, the number of simultaneous reflections will be very great and the Ewald sphere will be very nearly planar. Under these circumstances the phase-grating approximation including absorption is very good indeed and of sufficient accuracy to deal with any conceivable observation of diffraction with protons or heavier ions.

The observations which have been reported are for crystal thicknesses very much greater than the range for elastic scattering. The channelling process essentially involves a large number of incoherent inelastic scattering processes. Hence a wave diffraction treatment of channelling would necessarily involve the consideration of elastic scattering processes, treated by use of a phase- and amplitude-grating approximation, combined with multiple inelastic processes.

Also there are important considerations of geometry to be considered. For protons having a wavelength of the order of 5×10^{-4} Å the diffraction angles for simple metals will be around 10^{-4} radians. The available sources of protons are not sufficiently bright to allow beams of divergence less than 10^{-4} radians to be obtained with appreciable intensity. Hence for protons or heavier charged particles diffraction effects are usually not observed.

This point may be restated in terms of the lateral coherence of the incident beam (Cowley [1968]). Diffraction from a crystal lattice will be observable only if the incident radiation is coherent in the direction perpendicular to the beam over a distance larger than the inter-planar spacing. The lateral coherence is approximately equal to λ/α where α is the beam divergence. For the proton beams usually employed this is a small fraction of 1 Å.

The lateral coherence of a particle beam may also be considered as a measure of the "size" of an individual particle. Since quantum mechanically it must be said that a quantum of radiation can only interfere with itself, radiation scattered by two atoms can interfere only if the wave packet which represents the particle overlaps the two atoms at the same time.

Hence for most channelling observations it may be assumed that the incident particles are small compared with the distance between lattice planes. Hence a description of channelling in terms of a series of collisions of the incident particles with individual atoms, resulting in preferred transmission by bouncing of the particles along channels in the crystal, is fully justified. The treatment of the phenomena in terms of waves is equally valid and possible in principle, but would be complicated without simplifying assumptions.

The case of very high voltage electrons is an intermediate one since the effects observed are essentially diffraction phenomena, describable in terms of n-beam dynamical diffraction but a semi-classical particle approach gives a reasonable approximation in some cases. A further special consideration is that for electrons the preferred minimum-potential paths through the crystal are close to rows or planes of atoms whereas for positive particles the preferred paths lie well away from the atoms. For electrons there is thus a minimum transmission near principal directions, although small maxima have been observed in the middle of these minima (Kumm et al. [1972]).

The interpretation of these small maxima in terms of the particle channelling model is that electrons travel along paths confined to the potential troughs around rows or planes of atoms, oscillating about the potential minima.

In the wave mechanical treatment, one can separate out the component of the momentum in the forward direction (almost identical for all waves) and then consider the components of the momentum at right angles interacting with, effectively, a two-dimensionally modulated potential (Howie [1966]). The preferential transmission of electrons near atom positions is correlated with the formation of Bloch waves having maxima there (Berry and Mount [1972]). These Bloch waves can be identified with bound states of the electrons in the potential minima, describable in terms of the tight-binding approximation to solution of the wave equation (Kambe et al. [1974]). This type of approach offers a potentially useful means for the interpretation of observations in the high voltage microscopy of crystals.

This type of approach has been further extended by Fujimoto and associates (Fujimoto [1977]; Sumida, et al. [1977]) who have used it to characterize a large range of high voltage observations of diffraction patterns and images. The Kikuchi and Kossel patterns given in the range of electron energies up to 2 MeV for monatomic crystals of simple structure show characteristic changes at voltages which vary systematically with the atomic number of the elements present and the radius of the channel through the crystal between the atoms.

14.6. Secondary radiations

In most cases the intensity variations in K-line patterns can be considered to result from variations of the effective absorption coefficient for a dynamical diffraction condition. The absorption process frequently involves the emission of secondary radiation and the intensity of the secondary radiation will then also fluctuate with the incident beam direction.

A striking example of this may be seen in scanning electron micrographs of single crystal faces as first seen by Coates [1967] and discussed by Booker et al. [1967]. (See also Booker [1970].) The scan of the beam is normally arranged so that the beam passes through the center point of the final lens. Then as the beam sweeps over the surface the angle of incidence on the crystal surface changes. When the beam is near the Bragg angle for a strong reflection the effective absorption coefficient is decreased and the primary beam penetrates further into the crystal. There is a reduction of the number of secondary electrons produced near enough to the surface to allow them to escape and be detected. Hence the image formed by displaying the signal obtained by collecting secondary electrons shows a pattern of lines which is very close to being the inverse of a wide angle convergent beam diffraction pattern obtained by transmission through a single crystal or, at one stage removed, a Kikuchi line pattern.

The intensity of secondary electrons emitted will depend on the way in which the primary electrons penetrate the crystal and the way in which they are scattered back towards the surface. A rough theory in terms of the behaviour of individual Bloch waves was proposed by Humphreys et al. [1972] and refined by Humphreys et al. [1973] to give reasonable agreement with experimental observations. A point of particular interest is the possibility that crystal defects at the surfaces of bulk crystals might be detectable through their perturbation of this scattering process. With other combinations of incident and secondary radiations a considerable variety of interesting results is possible. For example with incident electrons or X-rays the characteristic X-rays due to particular atoms could in principle be detected (see 9.3). The intensity distribution of the incident beam in the crystal is determined by the dynamical scattering from all atoms of the crystal. The secondary radiation is produced by one type of atom only and its intensity will depend on the position of this type of atom in the unit cell. Thus the observations could be used as an aid for structure analysis or to locate defects or impurity atoms in the unit cell (Cowley [1964]).

Analogous observations have been made in relation to the channelling of fast charged particles. The secondary radiations in this case may be X-rays or the γ-rays or particles produced in nuclear reactions. The reactions are then very much characteristic of particular types of nuclei and the intensity of emission depends on the sharply-defined positions of inner electron shells or nuclei.

If the incident particles are strongly channelled between planes of atoms, an impurity atom lying in the planes of atoms will be shielded from the incident radiation and so have a small probability of emitting secondary radiation. But if it lies between the planes, the probability of emission is increased. Hence the secondary radiation intensities may be used to detect and locate impurity atoms with considerable sensitivity.

For example, Cairns and Nelson [1968] detected the K_α radiation of Si and L radiation from Sb when a proton beam was incident on a silicon crystal doped with 1 per cent antimony. The fact that both radiations detected gave the same dip in intensity when the protons were channelled confirmed that the Sb atoms form a substitutional solid solution. For this case it is predicted that one impurity atom in 10^6 could be detected and located.

CHAPTER 15

Applications of dynamical effects in single crystals

15.1. Dependence of dynamical effects on crystal parameters

As can be seen from the discussions of dynamical scattering effects in Chapters 8 to 11, the intensity of a diffracted beam given by a near-perfect single crystal may be a strongly-varying function of the structure amplitude for the reflection, the crystal thickness in the beam direction, the orientation of the crystal relative to the incident beam and the form and magnitude or frequency of recurrence of deviations from the crystal periodicity. Also, less directly, the intensity will depend on the other conditions of the scattering process including the temperature and the presence of absorption or inelastic scattering processes. It follows that observations on dynamical diffraction intensities may be used to obtain measures of any of these quantities or effects with high precision provided that the others are sufficiently well controlled. Recently a number of techniques have been devised whereby dynamical effects are used to obtain data of value in a variety of areas of science and technology.

The angular width of a diffracted X-ray beam from a thick crystal is proportional to the structure amplitude and is typically of the order of a few seconds of arc or 10^{-5} radians. Hence very small deviations in orientation of the lattice planes may cause large changes of intensity. It has been known for many years that a slight strain applied to a thick crystal, such as caused by the thermal gradient due to a finger placed near the crystal will change the diffracted intensity by some orders of magnitude. The methods of X-ray topography, developed by Lang [1959] have been used to record the strain fields due to isolated dislocations or impurity aggregates in near-perfect crystals. The apparatus used is as sketched in Fig. 9.8. A review of recent results obtained with this technique has been given by Lang [1970], and Authier [1970] has reviewed the theory of contrast of the images.

15.2. X-ray interferometry

Measurements of high precision have been made possible by the development of X-ray interferometers by Bonse and Hart [1965, 1970]. A number of variants of these instruments have been developed employing both transmission and reflection of the X-ray beam from near-perfect crystals of silicon or germanium. The first and simplest is that sketched in Fig. 15.1. Three near-perfect thick crystals are arranged to be exactly parallel. The transmitted and diffracted beams given by the first crystal are diffracted again by the second crystal to give beams which come together at the third crystal. Interference between these beams produces a wave field modulated with the periodicity of the lattice planes. If the lattice of the third crystal is an exact continuation of the lattice of the first two the intensity maxima of the wave field will coincide with the gaps between the planes of atoms and there will be maximum transmission. A lateral shift of either the third crystal, or the modulations of the wave field, by half the interplanar spacing will give the condition for maximum absorption in the third crystal and hence minimum transmission. Hence the intensity transmitted by this device is extremely sensitive to any variation in orientation or lattice constant of the third crystal or any change of phase of either X-ray beam, resulting for example from the insertion of a thin object in the position indicated in Fig. 15.1(a).

The problem of maintaining three crystals parallel and in register with a lateral precision of a few Å and an angular precision of better than about

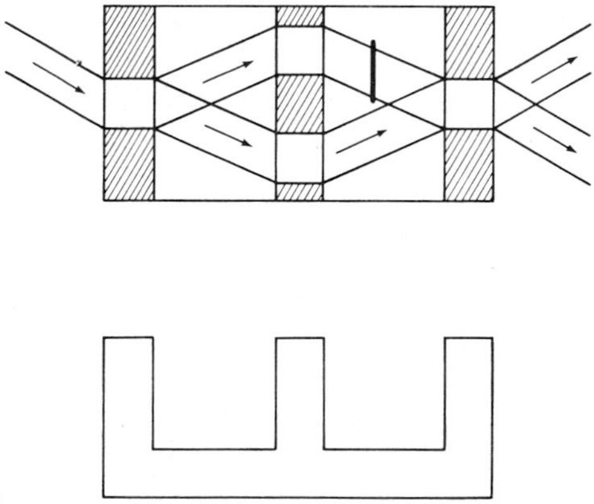

Fig. 15.1. The configuration of an X-ray interferometer formed by cutting three parallel connected slabs from a large near-perfect crystal of silicon.

10^{-6} radians has been overcome by cutting the three crystals and a massive connecting base from one large-near-perfect single crystal as suggested by Fig. 15.1(b). The lateral dimensions of the crystals and the width of the X-ray beams used may be several cm and mm, respectively.

Instruments of this sort may be used to measure the thickness or refractive index for X-rays of any object inserted in one ray path of the interferometer. The sensitivity of the transmission to the alignment of the third crystal means that dilations of the lattice, $\Delta d/d$, as small as 10^{-8} and angular rotations as small as 10^{-8} radians can be detected. Intrinsic strains due to variations of impurity content of a crystal by 1 part in 10^7 or due to individual dislocations or other defects may be seen; or the device may be used to detect or measure strains due to external causes such as thermal gradients or elastic deformations.

Methods have been devised for combining an X-ray interferometer with an optical interferometer so that both radiations are used to measure the same displacement of a crystal. By this means the X-ray wavelengths and crystal lattice parameters may be referred to the fundamental standards of length, defined in terms of the wavelengths of particular optical spectral lines. This promises an improvement in the absolute accuracy of these quantities by several orders of magnitude.

15.3. *n*-beam and 2-beam dynamical diffraction

Before proceeding to discuss the methods used for deriving information from dynamical electron diffraction effects it may be as well to discuss in more detail the applicability of the two-beam approximation. For X ray or neutron diffraction, as we have seen this approximation is good for almost all cases and it is only necessary to ensure that the geometry of the experiment is not such that a third beam of appreciable amplitude is introduced. The assumption of 2-beam conditions was the only practical basis for the initial attempts to derive information from electron diffraction dynamical effects and persisted until it became abundantly clear that even in the most favorable of cases the neglect of *n*-beam interactions was leading to errors which were serious for the levels of accuracy being sought.

With the advent of detailed *n*-beam computations of intensities it was shown that the variation of intensities with crystal thickness and orientation could be as complicated as suggested, for example by Fig. 10.5 and 10.6. On the other hand it was confirmed that a relatively simple variation of intensity with thickness could be achieved in two types of special circumstance: one when a two-beam situation is carefully selected for a strong inner reflection;

the other when the incident beam is very close to being parallel to a principal axis of the crystal.

Even for a two-beam orientation, with the Bragg condition for a strong inner reflextion satisfied and large excitation errors for all other strong reflections, it is recognized that for a very thin crystal a multitude of diffracted beams will appear. Then the manner of the convergence to the two-beam situation with increasing thickness is of interest. Initial *n*-beam calculations assuming no absorption suggested that the relative intensities of all beams were established with the first "extinction distance" i.e. the principal periodicity shown by the incident beam. Beyond that, all beams maintained their relative intensities, averaged over a few extinction distances, for all thicknesses.

This conclusion must be modified slightly in the light of the results of calculations including absorption, such as those of Fisher [1969, 1972]. In order to avoid scaling difficulties and to make the result clearer, Fisher eliminated

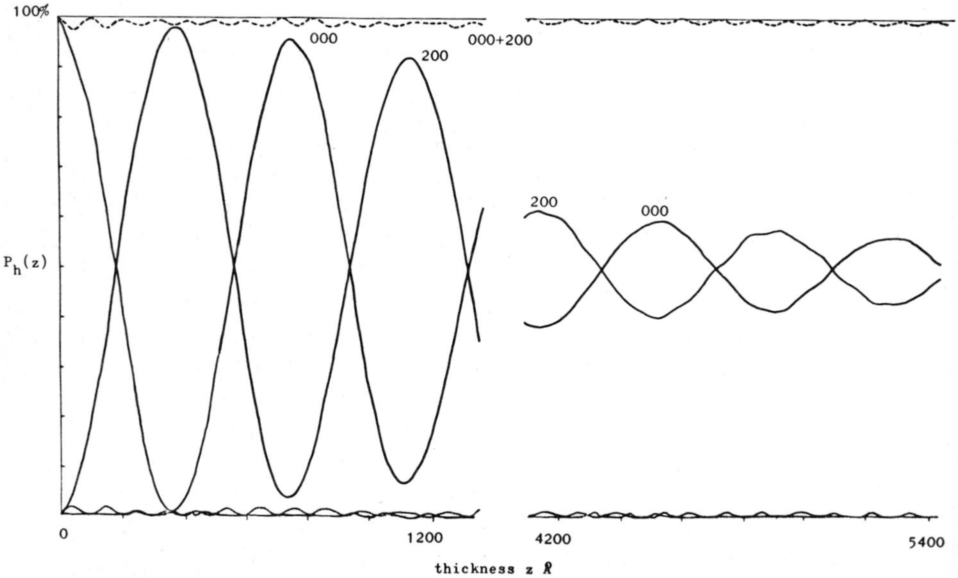

Fig. 15.2. Variation of the relative intensities of the diffracted beams with thickness, calculated for an aluminum crystal set for Bragg reflection of the (200) reflection with systematic interactions only, with 100 keV electrons. The intensities are given as percentages of the sum of all beam intensities for each thickness. (Fischer [1969].)

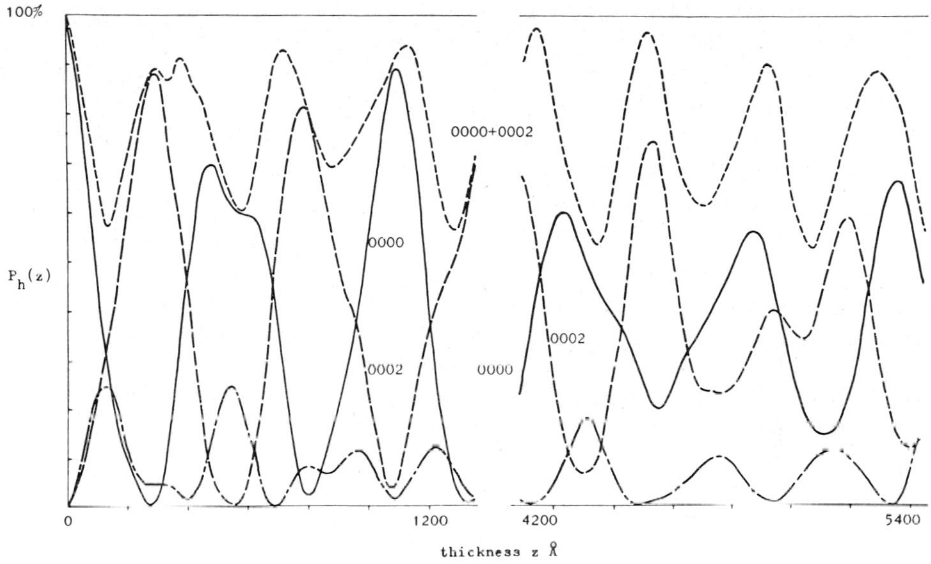

Fig. 15.3. Plot similar to 15.2 but for a crystal of CdS in ⟨120⟩ orientation with Bragg condition satisfied for (0002) reflection and systematic interactions only. (Fisher [1969].)

the effect of the over-all decrease in intensity due to the average absorption effects and plotted the intensities of each beam as a percentage of the sum of the intensities in all beams for each thickness. Typical results are shown in Figs. 15.2 and 15.3. In the case of the 200 reflection from aluminum for 100 keV electrons with a tilt away from the ⟨001⟩ orientation which ensures that very little effect of non-systematic interactions will be involved, the percentage of the energy contained in the two main beams is about 99 at a thickness of one extinction distance. With increasing thickness there is a gradual increase in the percentage energy in the two main beams, to 99.67 per cent for very large thicknesses. Thus the weak beams lose energy slowly. In the thickness range of a few thousand Å which represents the normal limit of observability, their contribution decreases by a factor of about one half.

With increasing thickness it is seen that the two strong beams share the energy almost equally but oscillate less and less. The absorption coefficient related to the intensity oscillations is greater than the overall absorption coefficient. This behavior is characteristic of the Borrmann effect, although for electrons this effect is by no means as striking as for X-rays.

For heavier atoms or larger unit cells or for orientations giving more non-systematic interactions the percentage of energy contained in the weak beams is greater. For example in the case of gold with the Bragg condition satisfied for the (200) reflection and systematic interactions only the percentage energy in the weak beams decreases from about 13 per cent at 200 Å to 4.2. per cent for thicknesses greater than about 5000 Å. For the non-centrosymmetric CdS structure with the (002) reflection satisfied, the two-beam condition is again less well defined, with considerable deviations from a sinusoidal variation of intensities of the main beams and about 25 per cent of the energy in weak beams for large thicknesses (Fig. 15.3).

We may conclude that within a few extinction distances something approaching a two beam situation can be set up. For favorable orientations of simple crystal structures the two main beams may have all but a few per cent of the energy. The average combined strength of the weak beams then decreases slowly with thickness until, for a few thousand Å, which is beyond the usual range of observations for 100 keV electrons, it may be as small as one per cent of the total energy. The weak beams continue to oscillate while the oscillations of the main beams become relatively less. This limiting situation may be upset by crystal imperfections. Calculations by Goodman [1968] show that a single fault may renew strong oscillations of the main beams and increase the weak beams considerably.

Apart from the decrease in energy of the two main beams, the presence of weak beams also has the effect of modifying the apparent extinction distance as measured by the periodicity of the oscillations. For the strong inner reflections from simple materials the effect is to decrease the extinction distance by from 5 to 10 per cent (Hirsch et al. [1965]; Goodman and Lehpfuhl [1967]).

It is now fully appreciated that the importance of n-beam interactions increases with increasing accelerating voltage. Observations on the periodicities of thickness fringes have been used to illustrate this dependence. For example, Mazel and Ayroles [1968] have shown theoretically and experimentally that the extinction distance for the 111 reflection of aluminum differs from the two-beam value by about 10 per cent at 50 keV, 16 per cent at 500 keV and 22 per cent at 1200 keV.

For reasons which are not so well understood as in the two-beam case, a relatively simple variation of diffracted beam intensity with thickness may occur in the case of an incident beam very nearly parallel to a principal axis of the crystal. The calculations of Fisher [1969] for a beam parallel to the [001] axis of a copper-gold crystal, reproduced in Fig. 10.4, show that for the case of no absorption, all beams oscillate sinusoidally with the same periodicity. All diffracted beam intensities oscillate in phase with each other but out of phase with the incident beam.

In the presence of absorption and for other structures this relationship of the periodicities and phases is not so exact. This is shown for example by the calculations made by P.L. Fejes in relation to studies on the repetition with thickness of the lattice images obtained with the beam parallel to the *b*-axis of a crystal of $TiNb_{24}O_{62}$ (Fejes, Iijima and Cowley [1973]) (see Chapter 13).

15.4. The determination of structure amplitudes from thickness fringes

For strong reflections obtained under two-beam conditions from a near-perfect crystal which has the form of a wedge, the intensities of transmitted and diffracted beams will show damped sinusoidal oscillations with thickness, giving thickness fringes parallel to the edge of the wedge, as described in Chapter 9. From the relationship, equation (9.1) and (9.2), the structure amplitude can be derived from the periodicity of the fringes if the wedge angle and the deviation from the Bragg angle are known accurately.

For X-rays the periodicities of the oscillations of intensity with thickness are of the order of hundreds of microns. Measurements may be made on small-angle wedges, cut from near-perfect crystals, by use of the X-ray topography technique illustrated in Fig. 9.8 (Kato and Lang [1959]). However for X-rays greater accuracy, reliability and convenience can be obtained by use of the section topographs (Chapter 9). A summary of experiments of this type is given by Kato [1969]. It is estimated that accuracies of determination of absolute values of structure amplitudes obtained in this way may be about 1 per cent but relative values may be obtained to about 0.1 per cent. Still greater accuracy is claimed for measurements made by use of an X-ray interferometer. Hart and Milne [1970] estimate an accuracy of about 0.2 per cent for their measurement of the structure factor for the (220) reflection of silicon made by observing the interference fringes arising when a small non-diffracting gap separates two thick perfect parallel regions of single crystal.

For electrons, the observations of thickness fringes are made in bright and dark-field electron microscope images. In principle measurements could be made on wedge-shaped crystals only a few thousand Å in size such as the small cubic crystals of MgO smoke. However there are practical difficulties of determining the orientation of such crystals with high accuracy and of maintaining the orientation constant. The most accurate measurements made so far have utilized wedge-shaped edges of large crystals which can be held rigidly. Then orientations can be determined accurately from Kikuchi line patterns obtained from a relatively large area of the crystal.

Many-beam dynamical effects are inevitably present and important for the electron diffraction case, although for relative simplicity of observation and computation the orientation is usually chosen to give systematic interactions only. Within this limitation the orientation may be selected to give maximum sensitivity for each parameter to be refined.

It may be noted that when strong *n*-beam dynamical interactions take place the thickness fringes no longer have the simple sinusoidal variation of the two-beam case. The combination of structure amplitudes and the excitation errors involved provide a unique form for the intensity variation with thickness which can be recognized independently of the scale of the image. Hence there is no need for an accurate determination of the crystal wedge angle or the magnification of the electron microscope.

This was found to be the case in the studies of the silicon (*hhh*) reflections by Pollard [1970, 1971] who used crystal wedges formed by fracture of large crystals. The wedges usually had good flat faces but the wedge angles varied greatly.

Typical results are shown in Fig. 15.4 in which part (a) is the measured intensity of elastically scattered electrons in the (222) reflection from a crystal set at the Bragg angle for (222). The measurement was made by electron-counting using the computerized EMMIE system (Holmes et al. [1970]) with energy filtering to remove inelastic scattering and an accelerating voltage of

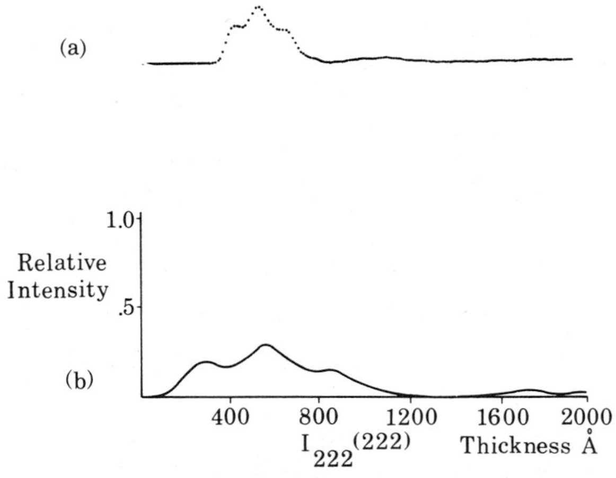

Fig. 15.4. Variation of intensity with thickness for the (222) reflection from a silicon crystal set at the (222) Bragg condition. (a) Measurement of intensity of a dark-field electron microscope image of a crystal wedge. (b) Calculated curve.

65 keV. Part (b) is the theoretical intensity curve computed for refined values of the structure amplitudes. The calculations were made using 11 beams for the systematic reflections with careful investigation of the possible errors due to the effects of non-systematic interactions, inelastic scattering and finite beam convergence.

From these, and other curves obtained for other orientations, the structure amplitudes found were 5.02 volts and –0.105 volts for the (111) and (222) reflections respectively, with a conservatively estimated limit of error of about 5 per cent. For the (222) reflection the structure amplitude would be zero for spherically symmetrical atoms. The value obtained is a measure of the covalent tetrahedral bonding present. It is in excellent agreement with the value of 0.107 volts deduced by Dawson [1967] from an analysis of X-ray diffraction results.

15.5. Structure amplitudes from rocking curves

A further set of experimental methods for deducing structure amplitudes depends on the measurement of diffraction intensities as a function of the angle of incidence of the incident beam. For X-ray diffraction the experiment giving the most direct and accurate data on structure amplitudes seems to be that of measuring the width of the Bragg-case reflection from the surface of a large perfect crystal under strict two-beam conditions (Kikuta et al. [1970], Kikuta [1971]). The theoretical curve (see Chapter 9) has a width proportional to $|F_h|$ This is broadened in practice by the angular width of the incident beam but if this incident beam is provided by a double-crystal spectrometer using asymmetric reflections from perfect crystals (Kohra and Kikuta [1968]) the angular width may be as small as 0.10″ and accurate corrections for this can be made.

In this way the width of the 422 reflection from silicon was measured to be 2.95″ from which the value of the atomic scattering factor was found with an estimated error of about 0.3 per cent which could be improved upon by slight modifications of the equipment. Improvements of accuracy and versatility of the method have resulted from the use of a triple-crystal arrangement (Nakayama et al. [1971]).

In electron diffraction experiments the situation is complicated by the presence of n-beam dynamical diffraction effects. In addition, for the Bragg case of reflection from the surface of a large crystal, difficulties arise for 20 to 100 keV electrons because, with small Bragg angles and small angles of incidence (1 or 2 degrees with the surface), the intensities and widths of the reflections are extremely sensitive to deviations from planarity of the surface and to modifications of the crystal structure in the surface layers of atoms (Colella and Menadue [1972]). However for transmission through thin perfect crystals the correlation of experimental observations with theoretical calculations can be made with much greater confidence. Rocking curves for crystals of silicon 1200 and 2700 Å thick have been made by Kreutle and Meyer-Ehmsen [1969] using 71 kV electrons. By comparison with theoretical curves obtained by including the interactions of 6 or 18 systematic reflections, they deduced structure amplitude values for the (111) and (222) reflections (5.16 and 0.10 volts) in reasonable agreement with results from other methods.

The most extensive and accurate determinations of structure amplitudes have been made by use of the convergent beam diffraction method (Chapter 9) which is equivalent to a rocking curve method except that instead of measuring the diffracted intensities as a function of angle of rotation of a crystal for a parallel incident beam, one has a range of incident angles in the incident beam and the corresponding variation of diffracted beam intensity appears in the profile of the broadened diffraction spots. This method was developed and applied to the study of the structure amplitudes of MgO by Goodman and Lehmpfuhl [1967]. A further study of the method and an evaluation of several possible sources of error was made by McMahon [1969]. With this technique an area of perfect parallel-sided crystal of diameter as small as 200 Å has been used. It should be possible to find suitable regions for almost any stable crystal. However MgO offers the advantage that perfect, parallel-sided unbent crystals having no defects visible in the electron microscope and areas of several square microns are available.

The results of the work on MgO by these authors may be summarized briefly as follows.

(a) The thickness of a crystal may be found approximately from the near-kinematical rocking curves (equation (9.5)) of weak outer beams. Then the thickness value becomes a parameter in the calculations and is refined, along with the structure amplitudes, until it is determined with an accuracy of perhaps 5 Å for a 1000 Å crystal.

(b) The extent of the agreement between experimental and theoretical intensities for the $h00$ systematics of MgO is suggested in Fig. 15.5.

(c) For orientations carefully chosen (by use of K-line patterns obtained with a wide angle cone of incident radiation) the effect of non-systematic interactions could be minimized. Inclusion of non-systematic interactions in a two-dimensional calculation for $h00$ reflections with these orientations required a change in the structure amplitudes by about 1 per cent for best agreement with observations. Fig. 15.6 shows that such calculations using

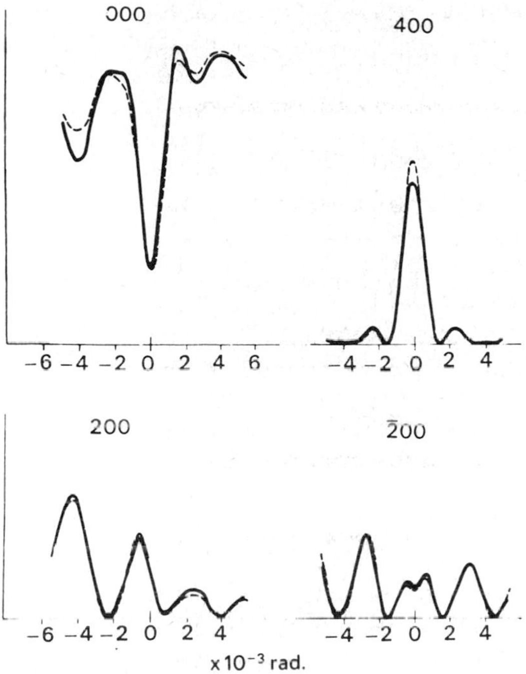

Fig. 15.5. Experimental (broken lines) and calculated (full lines) intensity distributions across spots in the convergent beam diffraction pattern from a thin MgO crystal, set at the Bragg angle for the 400 reflection with systematic interactions only (Goodman and Lehmpfuhl [1967]).

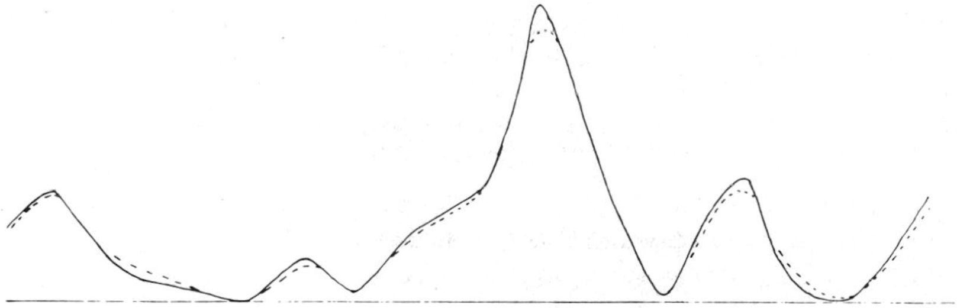

Fig. 15.6. The effect of non-systematic interactions on the spot profile for the 400 reflection in a convergent beam diffraction pattern of MgO, near the Bragg angle for 200. Good agreement is seen for the calculation with systematics-only using V_{200} = 7.00 volts and the two-dimensional calculation including non-systematics but with V_{200} = 6.93 volts (McMahon [1969]).

V_{200} = 6.93 volts agree with systematics-only calculations using V_{200} = 7.00 volts. Hence the neglect of the non-systematic interactions for the most favorable orientations would give an error of about 1 per cent in a determination of structure amplitudes. For other apparently favorable orientations the error may be 2 per cent or higher while for arbitrarily chosen orientations the error would be higher still. For the *hhh* systematics the effect of non-systematics seems to be less important.

(d) The intensity distributions of diffracted beams other than the central 000 beam are not very sensitive to the form assumed for the absorption function or the corresponding imaginary parts of the structure amplitudes.

(e) Taking all known sources of possible error into account the values of the structure amplitude, with conservative estimates of error, were given as $V_{111} = 1.78 \pm 0.03$, $V_{200} = 6.87 \pm 0.04$ and $V_{222} = 3.99 \pm 0.06$ volts.

Convergent beam studies made for other substances have not been used to the same extent for refinement of structure amplitudes Smith and Lehmpfuhl [1975] have used the method to obtain structure amplitudes for silicon and have been able to demonstrate the changes in these values associated with the doping of the silicon with small quantities of arsenic and other elements. The work by Goodman and Lehmpfuhl [1968] on CdS was concerned more with the symmetry in the diffraction patterns and demonstrated the breakdown of Friedel's law in the absence of a center of symmetry. The study of gold by Lynch [1971] served to demonstrate that appreciable diffraction intensities may be produced by reciprocal lattice points outside of the reciprocal lattice plane giving the main reflection of a two-dimensional pattern.

Various other experimental situations have been explored as possible means for deriving accurate structure amplitudes. Steeds [1970] has made use of bend contours in electrons micrographs of bent wedge-shaped crystals which display the variation of transmitted and diffracted intensities with both thickness and orientation of the crystal. Comparison is made with computer-generated pictures utilising the same type of limited grey-scale as in the dislocation-image technique of Head [1967] (see Chapter 18).

Lehmpfuhl and Reissland [1968] have explored the possibilities for deriving information from the refraction fine structure of diffraction spots given by wedge-shaped crystals. Gjønnes and Høier [1971] suggested that structure amplitudes could be derived from the geometry of intersecting Kikuchi or Kossel lines for three-beam cases in which the lines interact as suggested in Fig. 14.8 or show zeroes of intensity at particular excitation errors. Terasaki et al. [1979] have investigated the possibility of obtaining higher accuracy by applying this method using high-energy (1 MeV) electrons. They obtained accuracies for silicon comparable with those for other methods.

15.6. Relativistic effects

With increasing electron accelerating voltage the n-beam dynamical effects in electron diffraction become increasingly important as a result of the relativistic increase of mass. This is seen from the wave equation (8.1) where the potential $\phi(\boldsymbol{r})$ is multiplied by $2me/h^2$ to give $v(r)$ and the Fourier coefficients which enter into the dynamical equations. Then for a voltage E,

$$v_h(E) = (1 + eE(m_0c^2)\, v_h(0) \equiv R\, v_h(0)\,. \tag{1}$$

The relativistic factor R is approximately 1.2 for 100 keV and 2 for 500 keV since $m_0c^2 = 510.9$ keV.

From the Born series type of development, equations (1.16) or (1.17), it is seen that R enters as second, third and higher powers for the double, triple and multiple scattering terms which therefore become increasingly important with increasing voltage. If these terms are not all of the same sign there may be some voltage for which the intensity of a particular reflection goes to zero. This has been observed to happen in the relatively simple cases involving only the interactions of a systematic set of reflections for some uncomplicated crystal structures and has been used to obtain relationship between structure amplitudes and so accurate values for particular structure amplitudes.

Nagata and Fukuhara [1967] and more recently, Fisher et al. [1970] observed the vanishing of the extinction contour corresponding to a second order reflection at a certain critical voltage E_c. The corresponding disappearance of the second or higher-order Kikuchi lines at the critical voltage in diffraction patterns from thick crystals was reported by Uyeda [1968] and by Watanabe et al. [1968] and similar effects in convergent beam diffraction patterns were observed by Bell [1971] and more recently by Imeson et al. [1979].

The original explanation of the effect by Uyeda [1968] and Watanabe et al. [1968] was made in terms of the Bethe potentials (see Chapter 8). The second Bethe approximation gives

$$v_h'(E) = R\left[v_h(0) - R \sum_{g \neq 0,h} \frac{v_g(0)\, v_{h-g}(0)}{\kappa^2 - k_g^2}\right] \tag{2}$$

where κ is the wave number for the average crystal potential and k_g is that for the g reflection. The denominator is thus proportional to the excitation error ζ_g. If the Bragg condition for the h reflection is satisfied, this becomes simply $g(h-g)/d_1^2$ where d_1 is the spacing for the first order reflection.

For a second order reflection $h = 2$ and the main contribution to the sum is for $g = 1, h - g = 1$. This is a relatively large term for simple structures so that for many metals and compounds having small unit cells $v_h^1(E)$ becomes zero for a value of R corresponding to a voltage in the range 0 to 1000 keV.

Putting the bracket of (2) equal to zero for an experimentally determined value of R gives a relationship between structure amplitudes. The interest is usually in the determination of the structure amplitudes of the first order reflections since these are most affected by the ionization and bonding of atoms in crystals. It can be assumed that the structure amplitudes for outer reflections are given to a good approximation by the theoretical values for isolated atoms. The relationships derived from (2) have been used to find values for the first order structure amplitudes for a number of metals by Watanabe et al. [1969].

The use of the Bethe second approximation for the interpretation of the critical voltages in this way has been compared with the results of complete n-beam calculations (Watanabe et al. [1969]; Fejes [1970]) and shown to give an accuracy of better than 0.5 per cent in the critical voltage (and hence an even smaller error in v_1) for many substances.

A further description of the extinction phenomenon may be made in terms of the Bloch wave formalism by recognizing that at the disappearance voltage two branches of the dispersion surface touch i.e. that two eigenvalues of the matrix in equation (10.8) coincide (Lally et al. [1972]).

The values of the structure amplitudes entering equation (2) or the corresponding full n-beam formulation are dependent on such factors as the Debye–Waller factor, the long-range order parameter for a solid solution, the concentration of impurity atoms and so on. Preliminary experiments have demonstrated the dependence of the critical voltage on such quantities and indicate the possible application of critical voltage measurements for their determination.

15.7. Absorptions and temperature effects

The first attempts to derive experimental values for the Fourier coefficients of the absorption function, μ_0 and μ_h, for electrons were based on the 2-beam formulas such as (9.6) and (9.7). Careful measurements showed, however, that these formulas are inadequate (e.g. Uyeda and Nonoyama [1965]). The absorption coefficients derived on this basis were in error because in a first approximation n-beam diffraction effects could be disguised as absorption effects. It was subsequently shown by Goodman and Lehmpfuhl [1967] and others that the absorption coefficients are even more sensitive to the adequacy of the n-

beam treatment than the Fourier coefficients of the potential. Thus Goodman and Lehmpfuhl found a value of the imaginary part of the structure amplitude for the 200 reflection from MgO, $V^i_{200}(= \mu_h/\sigma)$ to be 0.14 volts, whereas if they had interpreted their measurements according to a two-beam theory the result would have been 0.3 volts.

As we have previously noted (Chapter 4), the absorption coefficients for electrons are not intrinsic properties of the crystal but will have values depending on the experimental conditions used for a particular investigation. Hence, while agreement with theoretical treatments is usually found within a factor of two or three there may not be the same fundamental significance for the results as in the case of the determination of the structure amplitudes for elastic scattering.

Except in the case of the μ_0 term, which gives just an overall reduction of intensity of all diffracted beams or images, the main contribution to the absorption coefficients μ_h usually comes from the thermal motion of the atoms. Hence the values for these coefficients will vary strongly with temperature. This has been demonstrated, for example, by the observations of Meyer [1966].

Diffraction or image intensities will also show variation with temperature through the Debye–Waller factors applied to the structure amplitudes. For a two-beam case this gives a simple smooth variation of the extinction distance. For more complicated n-beam cases the temperature variation may be a complicated function of crystal orientation, as shown by the experiments and calculations of Goodman [1971]. Correspondingly the absorption coefficients depend on the number and strength of the diffracted beams occuring.

15.8. The determination of crystal symmetries

As mentioned in Chapters 5 and 6, diffraction experiments for which the kinematical approximation applies can give information regarding some of the symmetry elements of crystal structure but are severely limited in that they cannot reveal symmetry elements whose effect is to be found in the relative phases, rather than the amplitudes of reflections. This limitation is expressed by Friedel's law and the most obvious example is that the presence or absence of a center of symmetry cannot be determined.

For dynamical scattering involving the interaction of two or more diffracted beams, plus the incident beam, this limitation no longer applies. The observed intensities depend on the relative phases of the participating reflections. This has been known for many years for X-ray diffraction work and has been investigated theoretically and experimentally, for example by Collela [1974] who investigated the relative intensities of various "umweg" reflections from a germanium crystal. Post [1979] showed that by investigating three-beam intersections of Kossel lines it is possible in principle, and to a limited extent in practice, to overcome the phase problem of X-ray diffraction by determining the relative phases of reflections.

Observations of the failure of Friedel's law in reflection electron diffraction patterns from large crystals were reported and discussed by Miyake and Uyeda [1950] and treated theoretically by Kohra [1954] and Miyake and Uyeda [1955] who applied a three-beam dynamical treatment.

In transmission, as well as reflection, electron diffraction patterns it has been common experience that the information on symmetry may appear to be less than that provided by kinematical scattering in that the systematic absences to be expected in the presence of screw axes or glide planes were not apparent. The "forbidden" reflections could, on occasion, be as strong as the "allowed" ones.

It was suggested by Cowley and Moodie [1959] that for the special case of a perfect crystal with the incident beam in a principal orientation, the transmission electron diffraction should show evidence of the symmetry of the projection of the structure of the slices of the structure perpendicular to the beam, including systematic absences. This concept was refined by Cowley et al. [1961] and a definitive statement on the nature of the information on symmetries given by electron diffraction in the presence of strong dynamical scattering effects was given by Gjønnes and Moodie [1965].

The conclusions may be summarized loosely as follows. If every plane perpendicular to the beam shows the same symmetry elements, and the unique axes related to these symmetry elements for the various slices are exactly superimposed in the beam direction, then the systematic absences related to these symmetry elements will hold strictly for any thickness of crystal. For example, if there is a glide plane parallel to the beam with the relevant axis perpendicular to the beam, the odd order reflections in the direction of this axis will be absolutely forbidden.

On the other hand if there is a screw axis or glide plane for which this is not strictly true but for which the projection of the unit cell in the beam direction has a symmetry which would give forbidden reflections, then for a thick crystal these reflections would not be absolutely forbidden but would probably be very weak. This would apply, for example, to the case of a 3-fold screw axis perpendicular to the beam. Three atoms related by this screw axis would not be exactly equivalent with respect to the electron wave field in the crystal because this wave field will be a function of the z-coordinate (the beam directions) and the three atoms will not have identical z coordinates.

The more precise arguments presented by Gjønnes and Moodie [1965] were in terms of the reciprocal space quantities. From such arguments it follows immediately that for a non-centrosymmetric crystal the intensities for $\boldsymbol{h}$ and $\bar{\boldsymbol{h}}$ reflections are not necessarily the same.

Strong deviations from Friedel's law and the consequent possibility of determining the absence of a center of symmetry and the absolute polarity of a non-centrosymmetric structure have been clearly demonstrated by the convergent-beam diffraction study of CdS by Goodman and Lehmpfuhl [1968] and by the various dark-field electron micrographs of 180° twin boundaries in crystals of $BaTiO_3$ (Tanaka and Lehmpfuhl [1972]).

The convergent beam patterns of CdS also demonstrated clearly that for an incident beam exactly parallel to a principal axis the systematic absences corresponding to symmetry of the projection of the unit cell structure were observed, although the reflections appeared for incident beam orientations only a small fraction of one degree away from the principal axis. (See also Fig. 13.4(c).)

A systematic study and tabulation has been made of the symmetries to be expected in diffraction patterns under dynamical diffraction conditions by

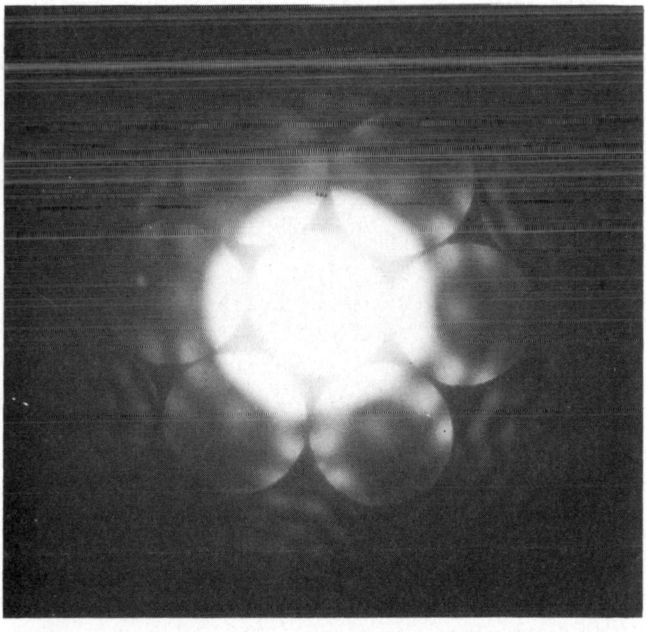

Fig. 15.7(a)

Buxton et al. [1976]. On the basis of similar considerations Goodman [1975, 1978] has developed a practical method whereby the observations of dynamical diffraction symmetry in convergent beam patterns can be used to determine the three dimensional space group symmetry of crystals without the ambiguities which arise in kinematical X-ray diffraction work due to the difficulty of determining the presence of a center of symmetry. The most difficult, but feasible, determination is that of "handedness" (Goodman and Johnson [1977]). A number of space group determinations giving serious on X-ray studies, have been resolved in this way.

An important point to be made in this connection is that the symmetry and intensities of a transmission electron diffraction pattern reflect the symmetries of the crystal and not the symmetries of the unit cell. Thus for graphite with an incident beam in the *c*-axis direction the projection of the structure is hex-

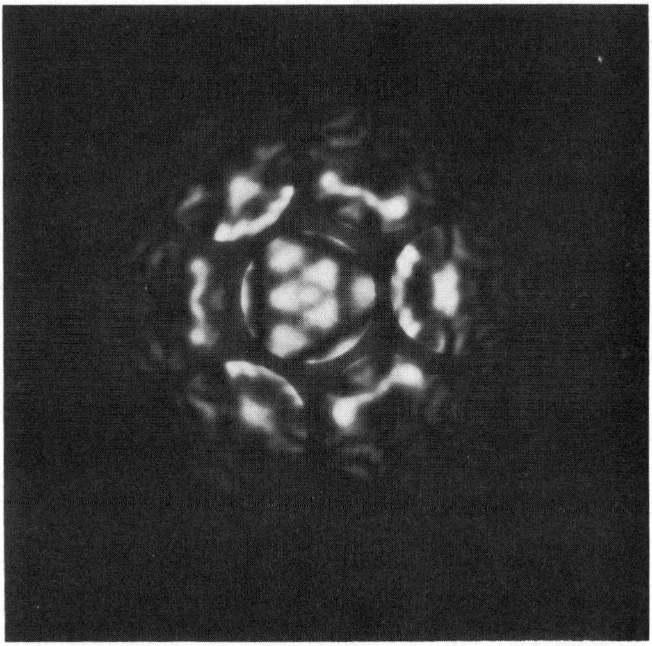

Fig. 15.7(b)

Fig. 15.7. Convergent beam diffraction patterns from graphite crystals with the center of the incident beam parallel to the *c*-axis, showing (a) the hexagonal symmetry for a perfect crystal and (b) the trigonal symmetry corresponding to a crystal having one or more stacking faults.

agonal and most diffraction patterns (Fig. 15.7(a)) show a hexagonal symmetry. But if there is a stacking fault in the graphite crystal (but not in the middle of it) such that a rhombohedral stacking sequence occurs (... ABABCACAC ...), the symmetry of the projection of the crystal will be trigonal and a diffraction pattern having trigonal symmetry can result (Fig. 15.7(b)) (Johnson [1972]).

A further example is provided if the periodicity, or symmetry, of individual layers of atoms perpendicular to the beam is not that of the projection of the unit cell. Then, if the crystal contains a non-integral number of unit cells, the symmetry or periodicity of the projection of the crystal will be that of the individual layers. For example, gold crystals, viewed in the [111] direction, show a strong periodicity of intensities of the fractional order reflections (such as 4/3, 2/3, −2/3), as individual layers of gold atoms are added beyond an integral number of unit cells (Lynch [1971]). Also intensities for MoO_3 crystals show a strongly varying contribution of the "forbidden" odd order reflections with the additions of fractions of the unit cell contents on the surface of crystals (Goodman and Moodie [1974]). This suggests important possibilities for the study of surface steps and defects on crystals by the use of dark-field electron microscopy from these reflections (Cherns [1974]).

CHAPTER 16

Mosaic crystals and polycrystalline materials

16.1. General

For the application of kinematical diffraction theory it is necessary to make the assumption that crystals are either very small or else are "ideally imperfect". On the other hand dynamical theory has been developed for ideally perfect crystals, with extensions to the treatment of crystal imperfections which become increasingly complicated and intractable as the departure from idealized situations increases. In between the limiting cases which may be approached by these relatively simple theories lies the vast majority of situations encountered in practice. The materials normally available for study by diffraction methods are often far from ideal with respect to either set of approximation. They may have a very complicated array of both extended and localized defects which are neither random nor isotropic in their distribution. The range of orientations of the crystal lattice may be very small or very large and the changes in orientation may take place discontinuously on well-defined planes (grain boundaries) or may be gradual (involving distortions of the lattice).

A general diffraction theory capable of taking all these factors into account is scarcely feasible and the detailed data on form and distribution of the crystal defects and discontinuities, which would be needed for such a theory, is rarely available.

To deal with the variety of experimentally important situations a number of convenient idealized models have been introduced and useful theoretical approximations have been made. The acceptance of these models and approximations have depended more on considerations of practical utility than on their accuracy in representing the physical situation or on theoretical rigor.

In many cases the possibility of producing a relatively simple model depends on the averaging of intensities which is appropriate when the coherence of the incident radiation and the resolution of the diffraction pattern are limited. This is the basis, for example, of the assumption that the intensities given by crystallites having different orientations may be added incoherently.

16.2. Mosaic crystals

16.2.1. The mosaic crystal model

The earliest model introduced to explain the kinematical nature of the diffraction from macroscopic crystals was the "mosaic crystal" model of Darwin [1914]. It was assumed that a crystal is made up of small perfect-crystal blocks, small enough so that the diffraction from each individually is given by a kinematical approximation and having a random distribution in orientation over a range of angles which is much greater than the angular width of reflection from an individual crystal.

This model has been retained as a useful one even after detailed studies of the structure of crystals by electron microscopy and other methods have shown that its relationship to the actual state of affairs is often quite remote. In some materials it is seen that most dislocations have aggregated to form dislocation networks, forming small-angle grain boundaries separating regions of crystal which are only slightly distorted. For many metals there are tangles of dislocations and other defects giving statistical accumulations of distortions and a variation of orientation of the lattice which is more or less random but continuous. In other materials stacking disorders, micro-twins, impurities and other defects provide much the same effect. Frequently there is a strong anisotropy since the defects occur preferentially on particular crystallographic planes or in particular directions.

The diffuse scattering due to the defects and the modification of the sharp Bragg reflection intensities by pseudo-temperature effects have been treated in other chapters. Here we consider only the Bragg intensities in relationship to the kinematical and dynamical approximations.

16.2.2. Kinematical integrated intensities.

For a mosaic crystal we can consider the distribution of scattering power around all reciprocal lattice points to be broadened equally by the average of the shape transforms for the individual crystallites. In addition the distribution of orientations of the crystallites will spread the scattering power maximum in two dimensions over the surface of a sphere centered at the reciprocal space origin.

For most X-ray diffraction work the measurement is made of the scattering power integrated in three dimensions over the whole of the scattering power maximum, giving the sum of the intensities from all individual crystallites. For a crystal rotated with angular velocity ω through the Bragg position, this integrated intensity is then (Warren [1969]):

$$E = \frac{I_0}{\omega}\left(\frac{e^4}{m^2c^4}\right)\frac{\lambda^3 V|F_T|^2}{\Omega^2}\left(\frac{1+\cos^2 2\theta}{2\sin 2\theta}\right), \tag{1}$$

where I_0 is the incident intensity, V is the volume of the sample, F_T is the structure amplitude modified by the temperature factor, $(1 + \cos^2 2\theta)/2$ is the polarization factor and $\sin^{-1} 2\theta$ is the Lorentz factor.

In the case of electron diffraction, intensities approaching those for kinematical scattering from single crystals are given only by very thin crystal sheets which are normally very much greater in lateral dimension than in thickness. The most common origin of misorientation between different parts of the crystal is bending of the crystal by rotation about axes lying approximately parallel to the sheet. The scattering power around reciprocal lattice points is greatly elongated in the direction nearly parallel to the incident beam by the small crystal dimension. This extension may be represented by convolution of the scattering power distribution by $\pi^{1/2} C \exp\{-\pi^2 C^2 w^2\}$, where C is the average crystal thickness and w is the appropriate reciprocal space coordinate.

If the mean angular spread of orientations is α, there is an additional extension of the scattering power spikes represented approximately by convolution with

$$(\pi^{1/2} d_h/\alpha) \exp\{-\pi^2 d_h^2 w^2/\alpha^2\},$$

where d_h is the lattice spacing for the $\boldsymbol{h}$ reflection, giving the distribution suggested in Fig. 16.1. The fact that the misorientation gives the spikes a finite curvature and width may be ignored as a first approximation.

The intensity of the diffraction spot integrated over the photographic plate is then given by the intersection of the Ewald sphere with the corresponding spike as

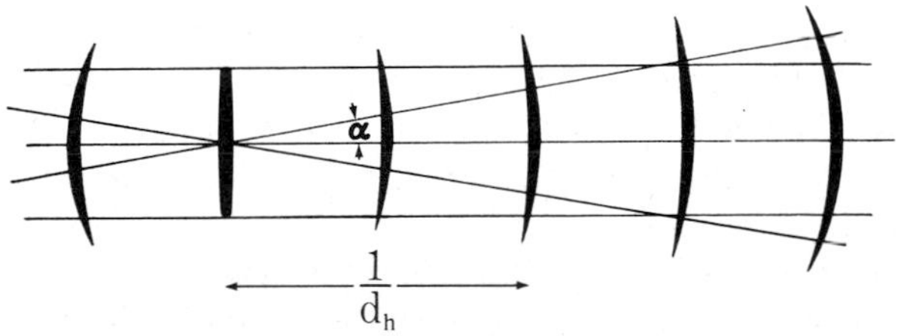

Fig. 16.1. The elongation and bending of the "shape-transform" spikes around reciprocal lattice points for a thin crystal, bent to give an angular spread of orientations, α.

$$I_h = I_0 \frac{\sigma^2 |\Phi_h|^2}{\Omega^2} \frac{\pi^{1/2} C d_h V}{(C^2\alpha^2 + d_h^2)^{1/2}} \exp\left\{-\frac{\pi^2 C^2 d_h^2 w^2}{C^2\alpha^2 + d_h^2}\right\} \tag{2}$$

with w equal to the excitation error ζ_h for the $\boldsymbol{h}$ reflection. Here Φ_h is the Fourier coefficient of the potential distribution. For high voltage electrons and a large angular range of orientations it may be assumed as a first approximation that the exponent is zero and the expressions becomes

$$I_h = I_0 \frac{\sigma^2 |\Phi_h|^2}{\Omega^2} \frac{\pi^{1/2} d_h V}{\alpha} . \tag{3}$$

If the exponent in (2) is not zero, the effect will be similar to that of an additional temperature factor on the intensities.

16.2.3. Extinction effects

It was recognized by Darwin [1914] that the approximation of kinematical scattering from an ideal mosaic crystal could give appreciable errors for a variety of reasons. He distinguished two types of "extinction" effect which could act to reduce the intensity below the kinematical value.

Primary extinction is said to be present if individual perfect crystal blocks are too large to allow the assumption of kinematical scattering. Under the two-beam diffraction assumption this approximation fails for appreciable values of $F_h H$ for X-rays or $\sigma\Phi_h H$ for electrons where H is a measure of the crystal dimensions. Thus for a given crystal size one would expect the stronger reflections to be affected most and the weaker reflections to be most nearly kinematical. Under the n-beam diffraction conditions of electron diffraction, of course, this does not apply.

Even in the absence of primary extinction, there may be an effect known as secondary extinction associated with the repeated diffraction of an incident beam by several separate mosaic blocks. The intensity incident upon one mosaic block may be attenuated by diffraction in several previous blocks having almost the same orientation. This effect will be most pronounced for inner reflections, for which the angular range of reflection from a small volume is greater, for reflections having large structure amplitudes and for samples having small mosaic spread or large dimensions.

Secondary extinction has been found particularly important for neutron diffraction for which the mosiac blocks are usually much too small to give primary extinction effects, but relatively large samples are normally used in order to get useful diffraction intensities from relatively weak sources.

Since one or the other type of extinction is common for the crystals used for X-ray diffraction structure analysis, a practical method of making corrections for extinction is of great value. This has been provided by Zachariasen [1968, 1969] and refined by Cooper and Rouse [1970] and Becker and Coppens [1974]. The difficulty of treating the coherent interactions of diffracted beams in a highly imperfect and inhomogeneous crystal is avoided by the assumption that the averaging out of phase-sensitive terms can be done before, rather than after, the consideration of the interactions of the beams. Hence a set of differential equations similar to (10.32) is used, but for intensities rather than amplitudes, thus;

$$\left.\begin{aligned} \frac{\partial I_0}{\partial x_1} &= -\sigma(I_0 - I) \\ \frac{\partial I}{\partial x_2} &= -\sigma(I - I_0) \\ \frac{\partial I_0}{\partial x_1} + \frac{\partial I}{\partial x_2} &= 0 . \end{aligned}\right\} \tag{4}$$

Here I_0 and I are the intensities of the incident and diffracted beams at the point in the crystal defined by coordinates x_1, x_2 measured along the incident and diffracted beam directions (Becker and Coppens [1974]), and σ is diffracting cross-section per unit volume. These equations are then integrated over all paths of the radiation through the crystal, giving expressions for the extinction correction factor in terms of the kinematical scattered intensity per unit volume, the crystal dimensions and the mosaic spread.

In a series of papers, Kato [1976, 1979] has provided a more elaborate and complete treatment of the problem by introducing parameters to characterize correlation functions which express the ranges over which the lattice phase factors are correlated or the distances over which the crystal is essentially perfect.

16.2.4. Dynamical electron diffraction intensities

A treatment of the dynamical scattering of electrons by imperfect crystals on the same basis as this "extinction" treatment for X-ray or neutron diffraction is scarcely feasible because strong dynamical effects take place within crystal regions much smaller than the usual "mosaic block" size and because the number of mosaic blocks along the path of an electron beam through a "single crystal" sample is rarely greater than one or two.

For the most part the spot patterns given by imperfect, bent or mosaic crystals must be considered as the sum of dynamical diffraction patterns from individual near-perfect crystals. In some cases where non-parallel crystals overlap, there will be the effects of double diffraction as each diffracted beam of the first crystal acts as a primary beam for the second crystal, giving diffraction spots whose positions are defined by the sums of the operative diffraction vectors for the individual crystals. When there is no simple relationship between the orientations of the two crystals the directions of the doubly diffracted beams prevent any coherent interactions with the singly diffracted beams. This is the case termed "secondary elastic scattering" (Cowley, Rees and Spink [1951]) to distinguish it from coherent multiple (dynamical) scattering.

In general electron diffraction spot patterns from large areas of thin crystal films may be considered as representing the averaging of dynamical diffraction intensities over a range of thickness and orientation.

Attempts which have been made to find analytical expressions for the electron diffraction intensities from a crystal averaged over orientation of thickness have not produced any very useful result (Kogiso and Kainuma [1968], Cowley [1969]). The averaging of intensities, computed using n-beam diffraction, over thickness is readily performed but averaging over angle of incidence is laborious. The indications are that when the intensities in a diffraction pattern are averaged in either or both of these ways, they may show a pattern of spots which is distinctive and characteristic of the crystal structure but not readily related to the kinematical diffraction pattern. In averaging over thickness for a perfect crystal in a principal orientation, the characteristic absences corresponding to some symmetry properties of the crystal are maintained (Chapter 14) but this is not necessarily the case for imperfect crystals or when there is an averaging over orientation.

16.3. Polycrystalline material

16.3.1. Idealized models

The ideal polycrystalline material contains a very large number of small, independently-diffracting crystallites having complete randomness in orientation about at least one axis. It may consist of fine powder, either loose or compacted, or a continuous solid having closely-spaced large-angle grain boundaries, as in a fine-grained piece of metal.

If the orientations of the crystallites are completely random, with no preferred direction for any crystal axis, the diffracted beams form continuous cones of radiation with the incident beam as axis and a half-angle equal to

$2\theta_h$ where θ_h is the Bragg angle for the $\boldsymbol{h}$ reflection. The intersection of these cones of radiation with cylindrical films in the case of X-ray diffraction, or with flat plates in the case of electron diffraction, give the well-known powder patterns of continuous lines or rings.

Preferred orientations of crystals may result from particular conditions of specimen preparation. If the crystallites have a thin, plate-like habit, they may tend to lie flat on a supporting film when the specimen is formed by settling from suspension, by growth on a flat substrate or by a mechanical spreading action. Then, ideally, the one crystallographic axis perpendicular to the plane of the plates has a strongly preferred orientation, but the orientations about this axis are completely random. This occurs frequently for electron diffraction specimens consisting of very small thin crystals supported on a thin supporting film of carbon or other light-element material.

To illustrate the form of the diffraction patterns produced by such a sample we may consider the case of crystals for which the a and b axis lie in a plane which is preferentially oriented parallel to a supporting surface, so that there is preferred orientation of the c^* axis perpendicular to the support. In reciprocal space, the randomness of orientation will spread the scattering power maximum around each reciprocal lattice point into a continuous circle about the c^* axis, as suggested in Fig. 16.2. This circle will be a sharp line if the preferred orientation of the c^* axis is exact but will be spread into a ring of finite width by any spread of orientations.

For convenience we consider an orthorhombic cell for which c^* and c are parallel. Then if an incident beam of fast electrons is parallel to the c axis, the Ewald sphere will be tangential and close to the $a^* - b^*$ plane and parallel to the rings of scattering power for the $hk0$ reflections. The diffraction pattern will therefore consist of continuous circles but will include those for $hk0$ reflections only.

If the incident beam is tilted at an angle ϕ to the c axis, as suggested in Fig. 16.2(a), the near-planar Ewald sphere will intersect the rings of scattering power in a set of short arcs. The $hk0$ arcs will be on a line through the origin. Other parallel lines of arcs will be given for the hkl arcs, with all reflections having the same l value giving arcs lying on the same line as suggest in Fig. 16.2(b). These patterns, which resemble the rotating-crystal patterns familiar in X-ray diffraction, have been used extensively for purposes of structure analysis by the Soviet school (Vainshtein [1964]) who have named them "oblique texture patterns".

A similar type of pattern may be given by fine needle-shaped crystals for which the needle axis is given a preferred orientation, for example by drawing or extruding of a sample into a thin rod, or also by some special growth process.

(a)

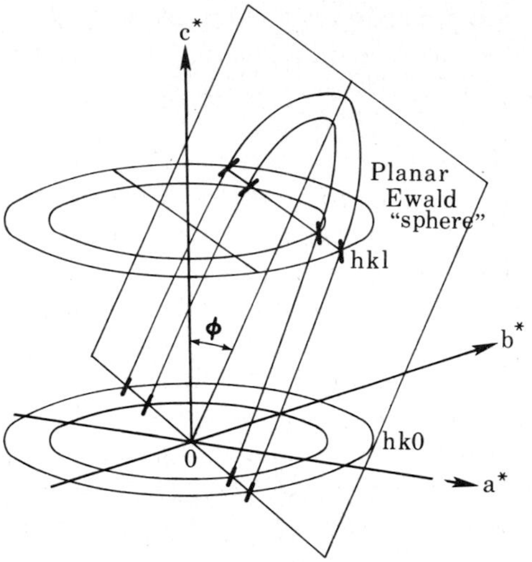

(b)

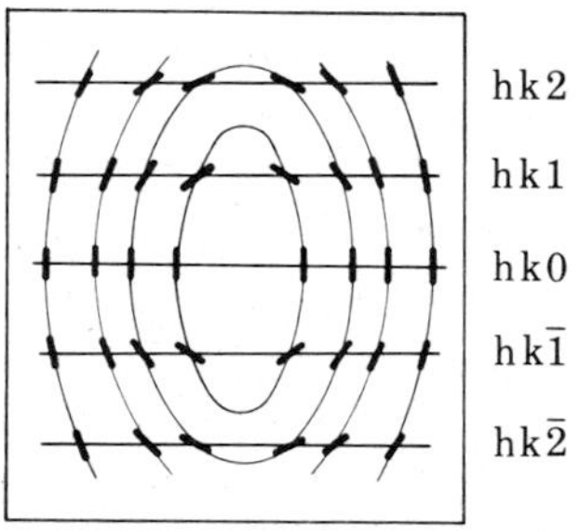

Fig. 16.2. The production of an "oblique texture" arc pattern when the Ewald sphere intersects the set of rings generated in reciprocal space by the reciprocal lattice points of crystals randomly oriented except that their *c*-axes are all parallel.

In this case it is normally a real-space axis (the *c* axis rather than $\boldsymbol{c}^*$) which has the preferred orientation. The incident beam is then usually almost perpendicular to the fiber axis and gives a diffraction pattern which is the same as a conventional rotating-crystal pattern.

16.3.2. Kinematical diffraction intensities

From the expression (1) for the integrated intensity from a single small crystal we may derive the expression for the total intensity in a diffraction cone given by a random powder as (Warren [1964])

$$P = I_0 \left(\frac{e^4}{m^2 c^4}\right) \frac{V\lambda^3 m |F_T|^2}{4\Omega^2} \left(\frac{1 + \cos^2 2\theta}{2 \sin \theta}\right). \tag{5}$$

Hence m is the multiplicity factor, giving the number of equivalent reflections, having the same Bragg angle, which contribute to the ring intensity.

The measurement usually made is that of the intensity per unit length of a diffraction ring, given by dividing P of (5) by the circumference of the ring,

$$P' = P/2\pi R \sin 2\theta \ , \tag{6}$$

when R is the distance from specimen to film.

For oblique texture patterns we may similarly distinguish between the total intensity of an arc and the intensity per unit length at the center of the arc, which is often easier to measure. For the total intensity within the arc we may write (Vainshtein [1964])

$$I_h = I_0 \frac{\sigma^2 |\Phi_h|^2}{\Omega^2} \frac{VR\lambda m}{2\pi r' \sin \phi} \tag{7}$$

where r' is the radius of the ring of scattering power for the h reflection.

If the angular spread of the arc is α and r is the radial distance of the arc from the center of the pattern, the local intensity per unit length will be,

$$I'_h = I_h / r\alpha \ . \tag{8}$$

16.3.3. Line profile analysis

For a small near-perfect single crystal the distribution of scattering power in reciprocal space around each reciprocal lattice point is given by the square of the Fourier transform of the shape function of the crystal. If the crystal is bent or distorted or if there are many such crystal which are closely parallel with some distribution of orientation or of lattice constants, the reciprocal space distribution will be modified in a characteristic way, as illustrated for a particular case by Fig. 16.1. A detailed investigation of the distribution of scattering power in reciprocal space should therefore allow a great deal of information to be derived about the crystal sizes, the spread of orientations and also the spread of unit cell dimensions.

For many materials of industrial or scientific significance it is not possible to work with single crystals. This is the case particularly for X-ray diffraction studies of microcrystalline materials, such as metals which have been subjected to various degrees of cold working. Then only powder patterns can be obtained and the only information on the form of the reciprocal space distributions around the reciprocal lattice points of the small crystallites is the statistically averaged data contained in the intensity profiles of the diffraction rings.

The use of powder patterns suffers, on the one hand, because on averaging over all orientations of the Ewald sphere, the scattering function in three dimensions is reduced to a one-dimensional function. On the other hand, the intensities and dimensions of the diffraction features can be measured with considerable accuracy in powder patterns and a detailed analysis is possible in terms of a limited number of well-chosen parameters.

An elegant method introduced by Warren and Averbach [1950, 1952] allows the contributions of crystal size and strain to be distinguished through analysis of the Fourier transforms of sets of powder pattern lines. Since that time, the development of the line-profile analysis techniques has been extensive and a considerable literature on the subject has accumulated (see Young, Gerdes and Wilson [1967]).

16.3.4. Dynamical diffraction intensities

For X-ray and neutron diffraction, extinction effects are present in somewhat modified form for powder samples. Primary extinction becomes important if the crystals are too large in comparison with the structure amplitudes. Secondary extinction effects may occur if the sample size is too large although, for random orientations of the crystals, the incident beam will be weakened by all possible diffraction processes at once. The effect will be the same for all reflections and will resemble a uniform absorption effect.

For electron diffraction it is necessary to average the dynamic diffraction intensities over angle of incidence. This was done initially for the pure two-beam case without absorption by Blackman [1939].

If we start from the expression for the intensity of the diffracted beam I_h in the Laue case, equation (8.29), we can replace integration over angle by integration over the variable w and use the relationship

$$\int_{-\infty}^{\infty} \frac{\sin^2[A(1+w^2)^{1/2}]}{1+w^2}\,\mathrm{d}w = \pi A \int_0^A J_0(2x)\,\mathrm{d}x\,. \tag{9}$$

In this case

$$A \equiv A_h = v_h \lambda H/4\pi = \sigma H \Phi_h\,. \tag{10}$$

Hence we deduce that the ratio of observed dynamical to kinematical intensities should be

$$I_{\rm dyn}/I_{\rm kin} = A_h^{-1} \int_0^{A_h} J_0(2x)\,{\rm d}x\,. \tag{11}$$

Then for small value of A_h, the integral is proportional to A_h and the kinematical result is the limiting case. For large values of A_h, the integral tends to its limiting value of 0.5 and since $I_{\rm kin}$ is proportional to $|A_h|^2$, we see that the dynamical intensity is proportional to $|A_h|$ and the ratio (11) tends to zero through slight oscillations.

The variation of the ratio (11) with A_h is the "Blackman curve", plotted in Fig. 16.3 together with experimental measurements of the relative intensities measured from ring patterns given by thin aluminum films (Horstmann and Meyer [1962, 1965]). The experimental measurements were made for a number of different reflections (different Φ_h), for various values of the average

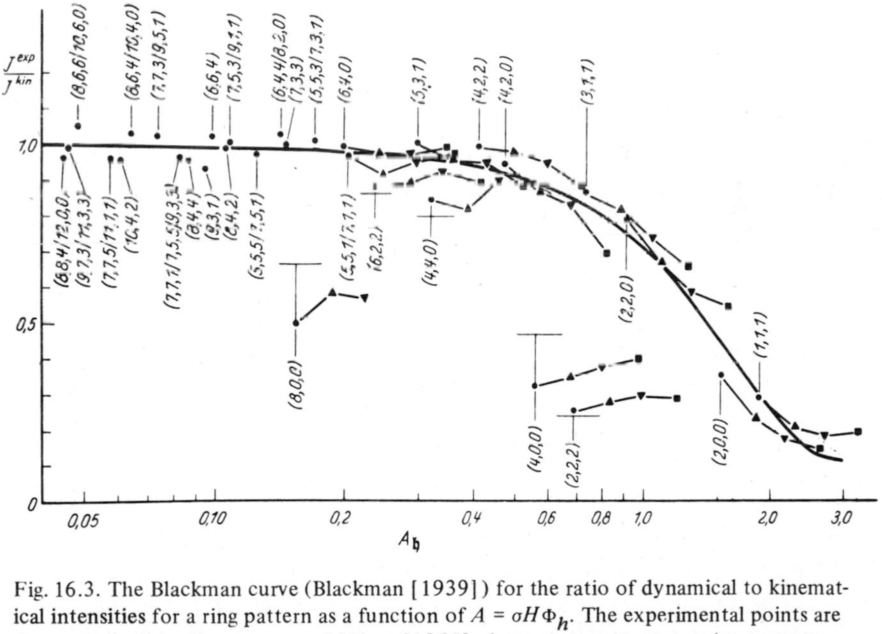

Fig. 16.3. The Blackman curve (Blackman [1939]) for the ratio of dynamical to kinematical intensities for a ring pattern as a function of $A = \sigma H \Phi_h$. The experimental points are those obtained by Horstmann and Meyer [1965], from measurements on ring patterns from aluminum films at various voltages. The short horizontal lines indicate values calculated using the Bethe potentials, equation (12). (After Horstmann and Meyer [1965].)

thickness, H, and for various electron wavelengths (and hence σ values). It is seen that in general the agreement with the Blackman curve is quite good. Comparable results have been obtained for ring patterns of other metals (Wedel [1963]; Kuwabara [1961, 1962]) and various inorganic salts (Kuwabara [1967]).

In each case a few of the reflections showed large deviations from the Blackman curve, as in the case of the (400) and (222) reflections of Fig. 16.3. These are the higher orders of strong inner reflections and for simple structures they show uniformly low intensity values.

For the most part it appears that the strong n-beam dynamical interactions which occur for many orientations of small crystals are averaged out and have little effect on the powder pattern intensities but, as might be anticipated, the "systematic" interactions remain since for the higher and lower orders of a given reflection the excitation error will be the same for all orientations when the given reflection is excited.

Some account of this effect is given by use of the Bethe potentials (8.34). As Bethe [1928] suggested, the effect of weak beams may be included into a two-beam theory by replacing the potential coefficients by, for example,

$$U_{0,h} = v_h - \sum_g{}'' \frac{v_g v_{h-g}}{\kappa^2 - k_g^2}, \tag{12}$$

and it is these modified potential coefficients which enter into the expressions for the diffracted intensities.

For a reflection $2h$, which is the second order of a strong inner reflection $\boldsymbol{h}$, the most important contribution to the second term of (12) is for $g = h$ so that

$$U_{0,2h} \approx v_{2h} - \{v_h^2/(\kappa^2 - k_h^2)\}. \tag{13}$$

Since the denominator will be positive, the effective scattering factor is reduced.

Applying these corrections gives reasonable agreement with the experimentally observed intensities for the case of Fig. 16.3. However, as mentioned in 8.6 the Bethe potentials are, in principle, not valid for very thin crystals. In the phase-object approximation, which is valid for very thin crystals, the ratio of the first and second order terms (from (11.44)) in the expansion of the structure amplitude is proportional to the thickness. Hence some modified form of the Bethe potentials must be used (Horstmann and Meyer [1964]; Gjønnes [1962]).

The expression (11) has been used as the basis for a sort of "extinction correction" in a great deal of structure analysis work based on ring or arc patterns from polycrystalline materials (Vainshtein [1964]). It has been assumed that the relationship of the integrated intensity of a reflection to the structure amplitude $|\Phi_h|$ varies from a second order to a first-order dependence with increasing strength of the dynamical effects and that for a particular diffraction pattern the appropriate fractional order dependence can be determined and used as a basis for the interpretation of the intensities. The use of Bethe potentials to apply corrections for some particular intensities offers some improvement on this somewhat oversimplified concept for crystals which are not too thin.

16.3.5. n-beam diffraction effects

It has been argued with some justification that the structures of crystals which are of interest for structural investigations are not usually simple and may have many atoms in a relatively large unit cell. Then the absolute values of the structure amplitudes will be less and the dynamical effects correspondingly weaker than for the metals or other crystals of simple structure normally used to test the dynamical theory expressions. On the other hand, for substances having relatively large unit cells the density of reciprocal lattice points is high, the number of simultaneous reflections for any orientation of a crystal will be large and the n-beam dynamical effects may be important even when the individual reflections are relatively weak. It is therefore important for structure analysis applications that the effects of n-beam diffraction effects on ring and arc patterns should be evaluated.

In the absence of any convenient analytical method of integrating n-beam dynamical intensities over orientation and thickness, the only feasible procedure appears to be the very laborious one of calculating intensities for a sufficiently finely spaced set of orientations. An attempt to do this was made by Turner and Cowley [1969] who made n-beam calculations of intensities of arc patterns for thin BiOCl crystals and for the substance $AgTlSe_2$ for which the structure analysis had been performed by Imamov and Pinsker [1965]. The experimental measurements had been analysed on the basis of the two-beam dynamical approximation and it was concluded from this that the intensities were purely kinematical. However, the n-beam dynamical calculations showed that there was probably sufficient dynamical modification of the intensities to have the effect of introducing appreciable errors in the details of the structure deduced.

Later Imamov et al. [1976] made systematic tests to reveal the presence of n-beam dynamical interactions. The zero layer-line intensities of oblique texture patterns of PbSe and Bi_2Se_3 were measured for various angles of tilt of the axis of preferred orientation and thus for various strengths of n-beam interaction with upper layer line reflections. For some particular tilts the zero-layer line intensities were modified by 40 percent or more. A more detailed study with close coordination of the theoretical calculations and the experimental measurements, performed for a variety of materials, would be of considerable value.

CHAPTER 17

Ordering of atoms in crystals

17.1. The nature and description of disordered states

The disordered arrangement of different kinds of atoms or molecules on the lattice sites of a crystal is of interest to a wide variety of scientists for a very varied set of reasons. For the theoretical physicist it represents one example of the three-dimensional lattice ordering problem, similar to the ordering of spins in a ferromagnet. The analysis of the statistical mechanics of ordering starts with the idealized Ising model and gets no further except for approximate or asymptotic solutions (Brout [1965]).

From this point of view, the interest is in the simplest possible ordering systems, such as that of β-brass, in which Cu and Zn atoms occupy the two sites of a body-centered cubic (B.C.C.) lattice in a disordered way above the critical temperature and with partial ordering over large distances (many unit cells) below T_c (see Fig. 17.1). From the completely ordered state, the equilibrium state at 0 K, the disordering increases more and more rapidly with temperature until it becomes catastrophic at T_c, giving ideally, a second-order phase transformation.

For the metallurgist ordering is an important parameter in determining the physical properties of materials such as the electrical and thermal resistivity, the hardness, ductility and so on. From the high-temperature disordered state the ordering may proceed in various ways. At one extreme, unlike atoms may tend to alternate, with each atom tending to surround itself by unlike neighbors and the system moves towards an ordered super-lattice. At the other extreme atoms of the same kind may tend to clump together, leading to a segregation of phases at lower temperature. Various intermediate situations provide materials of great commercial significance (Cohen [1968]).

High temperature chemists are concerned usually with the ordering of only a fraction of the ions in a crystal. For example the anions may have a disordered distribution among the possible sites defined by an ordered cation lattice, as in the case of many complex oxide or sulfide phases. The interest

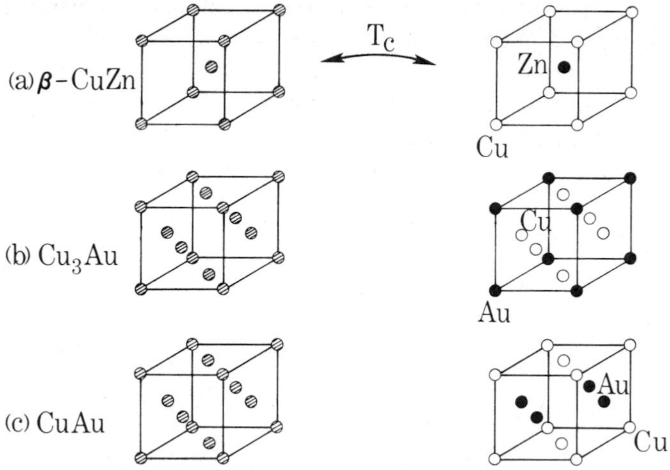

Fig. 17.1. The disordered and ordered structures of (a) β-brass, (b) Cu_3Au and (c) CuAu.

in the nature and amount of ordering derives from its relationship to the non-stoichiometry and phase transitions of such structures (Anderson [1964]).

Crystallographers involved with structure analysis of crystals have, in the past, tended to regard disorder chiefly as a hindrance to their purpose and have avoided dealing with disordered systems. Recently the increasing evidence for the widespread occurrence of partially or imperfectly ordered materials and the increasing importance of such materials in many areas of science has forced them to consider the implications for structure analysis very seriously. The disorder may be of many types: one or more types of atom may be disordered, or replaced by vacancies, on the lattice sites. There may be static or dynamic displacements of atoms sometimes correlated over large distances to form partially or fully ordered superstructures, commensurate or incommensurate with the basic sublattice. Molecules or parts of molecules may show disordered or partially or fully correlated rotations or distortions. Accounts of many such phenomena are included in the conference report edited by Cowley et al. [1979].

For diffraction physicists, disorder represents an interesting example of diffraction from an imperfect crystal coming within the first of the two main classes discussed in Chapter 7. We treat it as such here, adding some comments on the additional considerations involved with the practical complications of dynamical scattering and the possible implications of the combination of scattering theory with statistical mechanics. For simplicity we confine our attention to simple binary alloy solid solutions, composed of A and B atoms in fractional proportions m_A and m_B. The disordered alloys are assumed to have simple structures such as B.C.C. as for β-CuZn or F.C.C. as for Cu-Au alloys (Fig. 17.1a, b). The extension to systems with more than two types of atoms and of more complicated structure follows with elaboration of the algebra but little conceptual innovation.

17.2. Order parameters

17.2.1 Short range order

The state of order of the crystal is defined by order parameters which define correlations between the occupancy of sites and so specify, for example, the probability of finding a B atom at a given distance (vectorial) away from an A atom. This probability is written P_{ij}^{AB} where i and j signify the sites specified by the vectors $\boldsymbol{r}_i$ and $\boldsymbol{r}_j$ from some origin.

We start by defining occupation parameters (Flinn [1956])

$$\sigma_i^A = \begin{cases} 1 & \text{for an A at site } i, \\ 0 & \text{for a B at site } i, \end{cases}$$
$$\sigma_i^B = \begin{cases} 1 & \text{for a B at site } i, \\ 0 & \text{for an A at site } i. \end{cases} \tag{1}$$

Since these parameters are interconnected we may replace both by a single parameter

$$\sigma_i = \begin{cases} m_B & \text{for an A at } i \\ -m_A & \text{for a B at } i, \end{cases} \tag{2}$$

$$\sigma_i = \sigma_i^A - m_A = m_B - \sigma_i^B, \qquad \langle\sigma_i\rangle = 0, \tag{3}$$

where the ⟨ ⟩ brackets are used to devote an average over all sites.

The probability P_{ij}^{AB} is readily seen to be given by

$$\begin{aligned} P_{ij}^{AB} &= \langle \sigma_i^A \sigma_j^B \rangle \\ &= m_A m_B - \langle \sigma_i \sigma_j \rangle . \end{aligned} \tag{4}$$

Similarly

$$P_{ij}^{AA} = \langle \sigma_i^A \sigma_j^A \rangle = m_A^2 + \langle \sigma_i \sigma_j \rangle . \tag{5}$$

Then the average $\langle \sigma_i \sigma_j \rangle$ is the pair-correlation parameter, giving the difference between the probability for a particular pair of atoms occurring at a distance $\boldsymbol{r}_i - \boldsymbol{r}_j$ apart, and the probability for a completely random array.

In the same way we may define a three-atom correlation parameter $\langle \sigma_i \sigma_j \sigma_k \rangle$ derived from the probability of occupation of three sites by particular types of atom. For example,

$$\begin{aligned} P_{ijk}^{ABB} &= \langle \sigma_i^A \sigma_j^B \sigma_k^B \rangle \\ &= m_A m_B^2 + m_A \langle \sigma_j \sigma_k \rangle - m_B \langle \sigma_i \sigma_k \rangle - m_B \langle \sigma_i \sigma_j \rangle - \langle \sigma_i \sigma_j \sigma_k \rangle . \end{aligned} \tag{6}$$

Manipulation of the σ_i operators is aided by use of the relationship (Shirley [1970]; Shirley and Wilkins [1972])

$$(\sigma_i)^n = A_n + B_n \sigma_i ,$$

where

$$\begin{aligned} A_n &= \{ m_A m_B^n + (-1)^n m_B^n m_B \} , \\ B_n &= m_B^n - (-m_A)^n . \end{aligned} \tag{7}$$

This expression allows us to assign values to correlation parameters when two or more of the sites coincide: for example,

$$\begin{aligned} \langle \sigma_i \sigma_i \rangle &= \langle \sigma_i^2 \rangle = m_A m_B , \\ \langle \sigma_i \sigma_j \sigma_j \rangle &= (m_B - m_A) \langle \sigma_i \sigma_j \rangle , \end{aligned} \tag{8}$$

and so on.

The two-atom correlation parameters, or "Warren short-range order-parameters", commonly used are α_{ij}, given by

$$\langle \sigma_i \sigma_j \rangle = m_A m_B \alpha_{ij} . \tag{9}$$

The whole set of correlation parameters, for all numbers of atoms, can be considered to specify the state of order of the system. Usually, however, only two-atom correlations are considered. The values of $\langle\sigma_i\sigma_j\rangle$ specify the degree to which the neighbors of one sort of atom tend to be preferably of the same sort or of the opposite sort. If $\langle\sigma_i\sigma_j\rangle$ is positive for all short vector lengths $\boldsymbol{r}_i - \boldsymbol{r}_j (= \boldsymbol{r}_{ij})$, it is seen from (5) that like atoms tend to clump together. If $\langle\sigma_i\sigma_j\rangle$ alternates positive and negative with increasing $|\boldsymbol{r}_{ij}|$ values, being negative for nearest neighbors, there is a tendency for the two sorts of atom to form an ordered superlattice.

Above the critical temperature for ordering, the values of $\langle\sigma_i\sigma_j\rangle$ decrease rapidly with r_{ij}, typically falling off by an order of magnitude in a few unit cell lengths. This indicates a state of "short-range order" (s.r.o.). The transition to long-range order (l.r.o.) as the crystal is cooled through T_c implies that the range of the correlation between the occupation of lattice sites extends effectively to infinity. The values of the two-atom correlation parameters decrease over short distances but then tend to limiting, finite values as a regular superlattice is formed with particular atoms predominating on particular sites of the superlattice unit cell.

17.2.2. Long-range order

The limiting values of the correlation parameters are not the same for vectors from a given lattice site to the different sites of the superlattice unit cell. For example for the F.C.C. lattice of a Cu_3Au type structure, the sites given by the vectors $h_1\boldsymbol{a}_1 + h_2\boldsymbol{a}_2 + h_3\boldsymbol{a}_3$ have one limiting value, say s_1, if $\boldsymbol{a}_1, \boldsymbol{a}_2, \boldsymbol{a}_3$ are the unit cell axial vectors and h_1, h_2, h_3 are integers, while for vectors $h_1\boldsymbol{a}_1 + h_2\boldsymbol{a}_2 + h_3\boldsymbol{a}_3$ plus $(\boldsymbol{a}_1+\boldsymbol{a}_2)/2$ or $(\boldsymbol{a}_2+\boldsymbol{a}_3)/2$ or $(\boldsymbol{a}_3+\boldsymbol{a}_1)/2$, the limiting value will be equal to $s_2 \neq s_1$.

It is customary to assume a fixed ratio of s_1 to s_2, namely $s_1 = -3s_2$ for this case, although this involves an additional assumption of uniform composition (Cowley [1965]). With this assumption, the s_1 and s_2 values are simply related to the square of the traditional Bragg-Williams l.r.o. parameter S, defined as a measure of the fraction, r_α, of atoms sitting on their "correct" lattice sites. For the case of Cu_3Au structure (Cowley [1950])

$$S = \frac{3}{4}\left(\frac{r_\alpha - m_A}{1 - m_A}\right) + \frac{1}{4}\left(\frac{r_\beta - m_B}{1 - m_B}\right)$$

and

$$s_1 = \lim_{r_{ij}\to\infty}\left\{\frac{\langle\sigma_i\sigma_j\rangle}{m_A\, m_B}\right\} = \tfrac{16}{3} m_A\, m_B\, S^2\,. \qquad (11)$$

It is usually assumed for convenience that a state of l.r.o. involves correlations over infinite distances. In practice the correlations will not extend beyond the limits set by grain boundaries, crystal defects, or the anti-phase boundaries to be described later. Hence the correlation functions will actually decrease with a half-width of the average size of the perfect single crystal regions.

17.3. Patterson function

If, instead of site occupancies, we consider correlations in terms of specific continuous functions which may imply the site occupancy, such as the electron density function $\rho(\boldsymbol{r})$ (or nuclear density function or potential) we may recall that correlations may be described by the Patterson function

$$P(\boldsymbol{r}) = \rho(\boldsymbol{r}) * \rho(-\boldsymbol{r}) \, . \tag{12}$$

If $\rho(\boldsymbol{r})$ is considered to consist of symmetrical contributions from each atom, $\rho_{\mathrm{A}}(\boldsymbol{r})$ or $\rho_{\mathrm{B}}(\boldsymbol{r})$, the Patterson becomes

$$\begin{aligned} \rho(\boldsymbol{r}) * \rho(-\boldsymbol{r}) &= \sum_i (\sigma_i^{\mathrm{A}} \rho_{\mathrm{A}} + \sigma_i^{\mathrm{B}} \rho_{\mathrm{B}}) * \delta(r - r_i) * \sum_j (\sigma_j^{\mathrm{A}} \rho_{\mathrm{A}} + \sigma_j^{\mathrm{B}} \rho_{\mathrm{B}}) * \delta(\boldsymbol{r} + \boldsymbol{r}_j) \\ &= \sum_i \sum_j [\sigma_i^{\mathrm{A}} \sigma_j^{\mathrm{A}} (\rho_{\mathrm{A}} * \rho_{\mathrm{A}}) + \sigma_i^{\mathrm{B}} \sigma_j^{\mathrm{B}} (\rho_{\mathrm{B}} * \rho_{\mathrm{B}}) + 2\sigma_i^{\mathrm{A}} \sigma_j^{\mathrm{B}} (\rho_{\mathrm{A}} * \rho_{\mathrm{B}})] \\ &\quad * \delta(\boldsymbol{r} - \boldsymbol{r}_i + \boldsymbol{r}_j) \, . \end{aligned} \tag{13}$$

This is expressed in terms of σ_i and σ_j by use of the equation (3). The double summation may be replaced by the number of atoms, N, times the average surroundings of a particular site. Putting $\boldsymbol{r}_j = 0$ in (13) then gives, using (3),

$$\begin{aligned} \rho(\boldsymbol{r}) * \rho(-\boldsymbol{r}) &= N \sum_i [(\rho_{\mathrm{A}} * \rho_{\mathrm{A}}) (m_{\mathrm{A}}^2 + \langle \sigma_0 \, \sigma_i \rangle) + (\rho_{\mathrm{B}} * \rho_{\mathrm{B}}) (m_{\mathrm{B}}^2 + \langle \sigma_0 \, \sigma_i \rangle) \\ &\quad + 2(\rho_{\mathrm{A}} * \rho_{\mathrm{B}}) (m_{\mathrm{A}} \, m_{\mathrm{B}} - \langle \sigma_0 \, \sigma_i \rangle)] * \delta(\boldsymbol{r} - \boldsymbol{r}_i) \\ &= N \sum_i [(m_{\mathrm{A}} \, \rho_{\mathrm{A}} + m_{\mathrm{B}} \, \rho_{\mathrm{B}}) * (m_{\mathrm{A}} \, \rho_{\mathrm{A}} + m_{\mathrm{B}} \, \rho_{\mathrm{B}}) \\ &\quad + \langle \sigma_0 \, \sigma_i \rangle \{(\rho_{\mathrm{A}} - \rho_{\mathrm{B}}) * (\rho_{\mathrm{A}} - \rho_{\mathrm{B}})\}] * \delta(\boldsymbol{r} - \boldsymbol{r}_i) \, . \end{aligned} \tag{14}$$

This is in the form of the Patterson of the average disordered lattice plus the Patterson of the deviations from the average lattice since, from (4), $\langle \sigma_0 \sigma_i \rangle$ represents the deviation from the average probability for site-pair occupancies.

17.4. Size effects

In making the transition from (13) to (14) however, we have made the important assumption that the set of $\boldsymbol{r}_i$ vectors is the same when referred to any member of the set i.e. that the vectors are the vectors $\boldsymbol{R}_i$ between lattice points of a perfect crystal lattice. This is so only in the limiting case that the two types of atom may be interchanged without any perturbation of atom positions, i.e. are of exactly the same size. Only in rare cases is this true, even to within one per cent. For example for Cu-Au alloys although the effective radius of atoms in alloys is not necessarily that in the monatomic crystals, the difference of about 10 per cent between the lattice constants for Cu and Au (3.65 and 4.07 Å resp.) suggests a considerable difference in effective atom size in the alloys.

If a single large atom is inserted into a lattice of smaller atoms one would expect it to displace the neighboring atoms or, more precisely, the atom will have a displacement field displacing all atoms around it by amounts and in directions depending on their vector separations from it. A smaller atom than the average may be expected to have a displacement field of opposite sign.

The extension of this concept to the case of binary alloys in which every atom is either larger or smaller than the average is not obvious. As a first approximation it may be assumed that each atom produces its own displacement field, acting on all other atoms and depending only on their vectoral separations from it, not on their natures. With a further stage of approximation we can assume that the displacement of one atom by another depends not on the actual vector separation, $\boldsymbol{r}_i - \boldsymbol{r}_j$, but on the separation of the nearest lattice points, which we may write as $\boldsymbol{R}_i - \boldsymbol{R}_j$. Then we assume that the total displacement of any one atom from the lattice point of the average lattice for the crystal is given by the sum of the displacements due to all other atoms. The vector displacement of the atom from the i lattice point is thus:

$$\boldsymbol{\Delta}_i = \sum_k (\sigma_k^{\mathrm{A}} \boldsymbol{\Delta}_{ik}^{\mathrm{A}} + \sigma_k^{\mathrm{B}} \boldsymbol{\Delta}_{ik}^{\mathrm{B}}) \,. \tag{15}$$

The displacement $\boldsymbol{\Delta}_{ik}^{\mathrm{A}}$ is for an A atom at k and $\boldsymbol{\Delta}_{ik}^{\mathrm{B}}$ is for a B atom at k. It may further be assumed that for an average atom at k the displacement would be zero i.e.

$$m_A \boldsymbol{\Delta}_{ik}^{A} + m_B \Delta_{ik}^{B} = 0\,. \tag{16}$$

Then the displacements due to A and B atoms are collinear and

$$\boldsymbol{\Delta}_{ik}^{B} = -(m_A/m_B)\,\boldsymbol{\Delta}_{ik}^{A}$$

and (15) becomes

$$\boldsymbol{\Delta}_i = \sum_k \sigma_k \boldsymbol{\Delta}_{ik} \tag{17}$$

where

$$\boldsymbol{\Delta}_{ik} = \boldsymbol{\Delta}_{ik}^{A}/m_B\,.$$

Hence the delta function in (13) is replaced by

$$\delta\left(\boldsymbol{r} - \boldsymbol{R}_i + \boldsymbol{R}_j - \sum_k \sigma_k(\boldsymbol{\Delta}_{ik} - \boldsymbol{\Delta}_{jk})\right),$$

and when the averaging process is performed, instead of (14) we obtain

$$\begin{aligned} P(\boldsymbol{r}) = N \sum_i \Big[&(m_A\rho_A + m_B\rho_B) * (m_A\rho_A + m_B\rho_B) \\ &* \left\langle \delta\left\{\boldsymbol{r} - \boldsymbol{R}_i - \sum_k \sigma_k(\boldsymbol{\Delta}_{ik} - \boldsymbol{\Delta}_{jk})\right\}\right\rangle \\ &+ \left\langle (m_A\rho_A + m_B\rho_B) * (\rho_A - \rho_B)(\sigma_i + \sigma_0) * \delta\left\{\boldsymbol{r} - \boldsymbol{R}_i - \sum_k \sigma_k(\boldsymbol{\Delta}_{ik} - \boldsymbol{\Delta}_{jk})\right\}\right\rangle \\ &+ \left\langle (\rho_A - \rho_B) * (\rho_A - \rho_B)(\sigma_0\sigma_i) * \delta\left\{\boldsymbol{r} - \boldsymbol{R}_i - \sum_k \sigma_k(\boldsymbol{\Delta}_{ik} - \boldsymbol{\Delta}_{jk})\right\}\right\rangle\Big]. \end{aligned} \tag{18}$$

Thus for the Patterson of the average lattice, the first term of (18), there is a spread of the peaks around the average lattice vector positions, $\boldsymbol{R}_i$. For the contributions from the deviations from the average lattice, the Patterson peaks will likewise be spread but the displacements given by individual atom-pair vectors from the $\boldsymbol{R}_i$ positions are correlated with the site occupancies so that there can be displacements of the mean positions of the peaks as well as an order-dependent broadening of the peaks.

As indicated earlier, this simplified formulation of the Patterson function rests on a number of assumptions which are not necessarily justified, but it will serve to illustrate the type of diffraction effects to be observed.

17.5. Kinematical diffraction

17.5.1. Diffraction with ordering only

As a basis for obtaining the intensities in diffraction patterns under the assumptions of kinematical diffraction conditions we obtain the scattering power distributions in reciprocal space by Fourier transform of the Patterson function.

In the absence of size effect, Fourier transform of (14) gives simply,

$$I(\boldsymbol{u}) = N(m_A f_A + m_B f_B)^2 \sum_i \exp\{2\pi i \boldsymbol{u} \cdot \boldsymbol{R}_i\}$$

$$+ N(f_A - f_B)^2 \sum_i \langle \sigma_0 \sigma_i \rangle \exp\{2\pi i \boldsymbol{u} \cdot \boldsymbol{R}_i\}, \qquad (19)$$

where f_A and f_B are the atomic scattering factors for the radiation used and may be assumed to include the temperature factors such as $\exp\{-M_A\}$. The first term of (19) gives the delta-function peaks at the reciprocal lattice points for the average lattice which produce the so called "fundamental" diffraction peaks which are independent of ordering. The second term, due to the deviations from the average lattice, is order dependent, giving rise to diffuse scattering and is written separately as I_d.

In the limiting, high temperature case of complete randomness all $\langle \sigma_0 \sigma_i \rangle$ values will be zero except that, from (8), $\langle \sigma_0 \sigma_0 \rangle = m_A m_B$ and the second term of (19) becomes

$$I_d = N m_A m_B (f_A - f_B)^2, \qquad (20)$$

which gives rise to a uniform background scattering, falling off smoothly with scattering angle roughly in proportion to f^2.

Correlation of occupancy of neighboring sites gives a modulation of this background. For a tendency for clustering of like atoms, all $\langle \sigma_0 \sigma_i \rangle$ values will tend to be of the same sign so that maxima in I_d appear around the reciprocal lattice points for the average structure. If there is a tendency for ordering into a superlattice, with an alternation of the two kinds of atom on lattice sites,

Table 1
Measured values of i (for the ith neighbor shell) given by Chen et al. [1979] for Cu_3Au

Shell number	Typical coordinates	Perfect order	α_i $T = 396$ (°C)	$T = 420$ (°C)	$T = 685$ (°C)
i	lmn	$\alpha_i(T=0)$			
1	110	−1/3	−0.176	−0.125	−0.130
2	200	1	+0.214	0.154	0.106
3	211	−1/3	+0.005	0.023	0.032
4	220	1	+0.062	0.049	0.019
5	310	−1/3	−0.079	−0.078	−0.066
6	222	1	+0.022	0.010	−0.009
7	321	−1/3	−0.010	−0.012	−0.002
8	400	1	+0.073	0.071	0.029
9	330	−1/3	−0.030	−0.022	−0.000
10	411	−1/3	+0.026	0.018	0.007
11	420	1	+0.034	0.024	0.000

$\langle\sigma_0\,\sigma_i\rangle$ will be negative for nearest neighbors, positive for second nearest neighbors and so on. Table 1 gives the values of the order parameters, α_{0i} $(= (16/3)\langle\sigma_0\sigma_i\rangle)$, for Cu_3Au at three temperatures above T_c (394°C), given by Chen et al. [1979] and Fig. 17.2 shows the corresponding reciprocal space distribution $I_d(\boldsymbol{u})$.

Broad maxima of scattering power appear at the 100, 110 and similar reciprocal lattice points which are the positions of delta function maxima for the ordered Cu_3Au lattice, Fig. 17.1(b) but not for the average lattice i.e. at the positions of the sharp superlattice reflections of the ordered state.

Below the critical temperature, T_c, the values of $\langle\sigma_0\,\sigma_i\rangle$ tend to constant limiting values $m_A\,m_B\,s_i$ as R_i becomes large. Then we may separate out the contributions from these limiting values and write

$$I_d(\boldsymbol{u}) = Nm_A\,m_B(f_A - f_B)^2 \sum_i s_i \exp\{2\pi i\boldsymbol{u}\cdot\boldsymbol{R}_i\}$$

$$+ N(f_A - f_B)^2 \sum_i \{\langle\sigma_0\,\sigma_i\rangle - m_A\,m_B\,s_i\} \exp\{2\pi i\boldsymbol{u}\cdot\boldsymbol{R}_i\}\,. \tag{21}$$

The first part of this expression comes from a periodic structure and gives the sharp superlattice peaks. The second part includes the differences of the

Fig. 17.2 Contour map of diffuse scattering intensity due to short-range order in the $hk0$ reciprocal lattice plane for a disordered crystal of Cu_3Au (after Cowley [1950b]).

$\langle\sigma_0\,\sigma_i\rangle$ from their limiting values, which will fall off rapidly with distance and so give rise to some residual diffuse scattering. In the limiting case of zero temperature the equilibrium structure would be perfectly ordered. Both $\langle\sigma_0\,\sigma_i\rangle$ and $m_A\,m_B\,s_i$ would tend to the values for the perfect superlattice structure and the second term of (21) would vanish.

The integrated intensity of the superlattice reflections will be proportional to the s_i values and so, from (11) to S^2. Hence the Bragg-Williams l.r.o. parameter S may be found (apart from an ambiguity of sign) from measurements of intensity of superlattice reflections. In this way accurate values have been obtained for the l.r.o. parameter for β-brass and compared with the prediction of the Ising model and various approximate, although more realistic theories (Chipman and Walker [1972]).

From the general expression (19) it is seen that the values of the correlation coefficients may be derived from the diffuse scattering by Fourier transform:

$$\langle\sigma_0\,\sigma_i\rangle = \int \frac{I_\mathrm{d}(\boldsymbol{u})}{N(f_\mathrm{A}-f_\mathrm{B})^2}\,\exp\{-2\pi\mathrm{i}\boldsymbol{u}\cdot\boldsymbol{R}_i\}\,\mathrm{d}\boldsymbol{u}\,, \tag{22}$$

where the integral is taken over one unit cell in reciprocal space. In this way the values of α_i in Table 1 and similar results were obtained from single-crystal X-ray diffraction measurements, after correction of the data for Compton scattering, thermal diffuse scattering, instrumental background and the size effects discussed below.

17.5.2. Diffraction with ordering and size effects

When size effects are present, the Patterson function, to a first approximation, is given by (18). The reciprocal space scattering power given by Fourier transform, is then written,

$$
\begin{aligned}
I(\boldsymbol{u}) &= N(m_\mathrm{A} f_\mathrm{A} + m_\mathrm{B} f_\mathrm{B})^2 \sum_i \exp\{2\pi \mathrm{i} \boldsymbol{u}\cdot \boldsymbol{R}_i\} \\
&\times \left\langle \exp\left\{2\pi \mathrm{i}\boldsymbol{u}\cdot \sum_k \sigma_k(\boldsymbol{\Delta}_{ik} - \boldsymbol{\Delta}_{jk})\right\}\right\rangle + N(m_\mathrm{A} f_\mathrm{A} + m_\mathrm{B} f_\mathrm{B})(f_\mathrm{A} - f_\mathrm{B}) \\
&\times \sum_i \exp\{2\pi \mathrm{i}\boldsymbol{u}\cdot \boldsymbol{R}_i\} \left\langle (\sigma_i + \sigma_0)\exp\left\{2\pi \mathrm{i}\boldsymbol{u}\cdot \sum_k \sigma_k(\boldsymbol{\Delta}_{ik} - \boldsymbol{\Delta}_{jk})\right\}\right\rangle \\
&+ N(f_\mathrm{A} - f_\mathrm{B})^2 \sum_i \exp\{2\pi \mathrm{i}\boldsymbol{u}\cdot \boldsymbol{R}_i\} \left\langle \sigma_0\, \sigma_i \exp\left\{2\pi \mathrm{i}\boldsymbol{u}\cdot \sum_k \sigma_k(\boldsymbol{\Delta}_{ik} - \boldsymbol{\Delta}_{jk})\right\}\right\rangle .
\end{aligned}
\tag{23}
$$

The averages in this expression are not easy to evaluate. One method for finding the scattering would be to do the averaging in the real space of the Patterson function, and for each Patterson function peak find the displacement from the lattice site and the spread function which broadens it. Then the Fourier transform is evaluated in detail.

The alternative, which has been most widely used, is to assume that all displacement terms $\boldsymbol{\Delta}_{ik}$ are small and to expand the exponentials in (23) in power series. It is necessary to take into account at least the second order terms in order to include qualitatively all significant diffraction effects.

From the first term in (23) we obtain for the fundamental reflections,

$$
\begin{aligned}
I(\boldsymbol{u}) &= N(m_\mathrm{A} f_\mathrm{A} + m_\mathrm{B} f_\mathrm{B})^2 \sum_i \exp\{2\pi \mathrm{i}\boldsymbol{u}\cdot \boldsymbol{R}_i\} \\
&\times \left[1 - 2\pi^2 \sum_k \sum_l \langle \sigma_k \sigma_l\rangle \{\boldsymbol{u}\cdot(\boldsymbol{\Delta}_{ik} - \boldsymbol{\Delta}_{0k})\}\{\boldsymbol{u}\cdot(\boldsymbol{\Delta}_{il} - \boldsymbol{\Delta}_{0l})\} + \ldots\right]
\end{aligned}
\tag{24}
$$

From the last bracket we may separate out the terms independent of the vector $\boldsymbol{R}_i$. Since by symmetry the summations over k and l of $(\boldsymbol{u}\cdot\boldsymbol{\Delta}_{ik})(\boldsymbol{u}\cdot\boldsymbol{\Delta}_{il})$ must be the same as the summations of $(\boldsymbol{u}\cdot\boldsymbol{\Delta}_{0k})(\boldsymbol{u}\cdot\boldsymbol{\Delta}_{0l})$, these are

$$\left[1 - 4\pi^2 \sum_k \sum_l \langle\sigma_k\, \sigma_l\rangle (\boldsymbol{u}\cdot\boldsymbol{\Delta}_{0k})(\boldsymbol{u}\cdot\boldsymbol{\Delta}_{0l}) + \ldots\right]. \tag{25}$$

This has the form of an order-dependent effective Debye-Waller factor, $\exp\{-M'\}$. In the absence of any s.r.o. it becomes

$$1 - 4\pi^2 \sum_k (\boldsymbol{u}\cdot\boldsymbol{\Delta}_{0k})^2 + \ldots, \tag{26}$$

and the second part is just the sum of the squares of all displacements of atoms from the lattice sites. With s.r.o. present this effective Debye-Waller factor will in general come closer to unity, which must be its value for a completely ordered lattice.

The terms of the last bracket of (24) which are dependent on the vectors R_i, such as

$$2\pi^2 \sum_k \sum_l \langle\sigma_k\, \sigma_l\rangle (\boldsymbol{u}\cdot\boldsymbol{\Delta}_{ik})(\boldsymbol{u}\cdot\boldsymbol{\Delta}_{0l}) \tag{27}$$

will be centrosymmetric and decreasing more-or-less uniformly with $|\boldsymbol{R}_i|$ and so will give rise to a diffuse scattering around the fundamental reciprocal lattice points, the Huang scattering (Huang [1947]).

For the second term of (23), the average is rewritten

$$\begin{aligned}\langle\ldots\rangle = 0 &+ 2\pi\mathrm{i} \sum_k \{\langle\sigma_i\sigma_k\rangle + \langle\sigma_0\,\sigma_k\rangle\}\{\boldsymbol{u}\cdot(\boldsymbol{\Delta}_{ik}-\boldsymbol{\Delta}_{0k})\}\\ &- 2\pi^2 \sum_k \sum_l \{\langle\sigma_i\,\sigma_k\,\sigma_l\rangle + \langle\sigma_0\,\sigma_k\,\sigma_l\rangle\}\{\boldsymbol{u}\cdot(\boldsymbol{\Delta}_{ik}-\boldsymbol{\Delta}_{0k})\}\{\boldsymbol{u}\cdot(\boldsymbol{\Delta}_{il}-\boldsymbol{\Delta}_{0l})\}\\ &- \mathrm{i}\ldots\end{aligned} \tag{28}$$

and from the third term we obtain

$$\begin{aligned}\langle\ldots\rangle = \langle\sigma_0\,\sigma_i\rangle &+ 2\pi\mathrm{i} \sum_k \langle\sigma_0\,\sigma_i\sigma_k\rangle\{\boldsymbol{u}\cdot(\boldsymbol{\Delta}_{ik}-\boldsymbol{\Delta}_{0k})\}\\ &- 2\pi^2 \sum_k \sum_l \langle\sigma_0\,\sigma_i\sigma_k\,\sigma_l\rangle\{\boldsymbol{u}\cdot(\boldsymbol{\Delta}_{ik}-\boldsymbol{\Delta}_{0k})\}\{\boldsymbol{u}\cdot(\boldsymbol{\Delta}_{il}-\boldsymbol{\Delta}_{0l})\} + \ldots\end{aligned} \tag{29}$$

Apart from the s.r.o. diffuse scattering of equation (19) given by the first term in (29), we now have other terms dependent on both correlation parameters and pair-wise displacement vectors. The anti-symmetric terms containing $(2\pi i)$ will give rise to anti-symmetric diffuse scattering around each fundamental reciprocal lattice peak except the origin, giving a displacement of the apparent center of the Huang scattering. Also they will provide anti-symmetric contributions to the diffuse scattering around the s.r.o. diffuse maxima and this will produce an apparent displacement of the s.r.o. diffuse peaks by an amount increasing with $|\boldsymbol{u}|$. This effect was first noticed and interpreted as due to size-effect displacements by Warren et al. [1951].

The next terms in (28) and (29) with double summations over k and l may be resolved in much the same way as the similar, simpler term in (24) into the equivalent of a Debye-Waller factor on the s.r.o. diffuse scattering and further symmetrical diffuse scattering terms.

It may be noted that three-atom correlation parameters enter into both the symmetrical and anti-symmetrical contributions to the diffuse scattering at the first stage beyond the elementary s.r.o. and size-effect contributions. With expansion of the exponentials to higher powers of $(\boldsymbol{u}\cdot\boldsymbol{\Delta})$, terms containing all multiple-atom correlation parameters are introduced (Cowley [1968]). The complications introduced with consideration of size effect prevent the straightforward derivation of correlation parameters as in (22).

The size effect perturbation of the s.r.o. scattering will be small when the displacements $\boldsymbol{\Delta}_{ik}$ are small and the range of $\boldsymbol{u}$ values is restricted to the first one or two reciprocal lattice unit cells about the origin. Under these conditions it has been shown by Borie and Sparks [1971] and Gragg [1970] that the size effect contributions can be separated by making use of the differences in their symmetry characteristic from those of the s.r.o. scattering. In this way diffraction data containing size effects can be used in the determination of both the $\langle\sigma_0\sigma_i\rangle$ parameters characterizing the atom displacements. The methods used have been summarized by Bardhan and Cohen [1976] and Chen et al. [1979].

As a function of the reciprocal space coordinates u, v, w, the diffuse scattering intensity including up to second order terms in interatomic displacements, can be written

$$I_{\mathrm{D}}(uvw) = Nm_{\mathrm{A}}m_{\mathrm{B}}|f_{\mathrm{A}} - f_{\mathrm{B}}|^2 [I_{\mathrm{SRO}} + uQ_x + vQ_y + wQ_z + u^2R_x + v^2R_y + w^2R_z + uvS_{xy} + vwS_{yz} + wuS_{zx}] .$$

Here $m_{\mathrm{A}}m_{\mathrm{B}}I_{\mathrm{SRO}}$ is the last part of (19), the scattering due to short range order in the absence of size effects. The quantities Q, R, S are introduced by the size effects and can be expressed in terms of parameters such as $\gamma_i^x, \delta_i^x, \epsilon_i^x$ which depend on combinations of the short range order parameters α_i or $\langle\sigma_0\sigma_i\rangle$, the displacement parameters x_{0i}^{AA}, y_{0i}^{AB}, etc. which are the x, y components of the displacements of an atom at r_i due to an atom at the origin, and also the parameters of the type $\langle x_0^{\mathrm{A}} y_i^{\mathrm{B}}\rangle$ which measures the correlation of the y displacement of a B atom at r_i and the x displacement of an A atom at r_0. The parameters γ, δ, ϵ also depend on the ratios of scattering amplitudes $\eta = f_{\mathrm{A}}/|f_{\mathrm{A}} - f_{\mathrm{B}}|$ and $\zeta = f_{\mathrm{B}}/|f_{\mathrm{A}} - f_{\mathrm{B}}|$.

The quantity Q_x is written

$$Q_x = \sum_i \gamma_i^x \sin(2\pi u x_i)\cos(2\pi v y_i)\cos(2\pi w z_i)$$

where x_i, y_i, z_i are the components of r_i in terms of unit cell dimensions, and so represents contributions to the diffuse scattering which are antisymmetric about the reciprocal lattice points in the x-direction, contained in our expressions (28) and (29). The function R_x gives scattering centrosymmetric about the reciprocal lattice points and so describes the Huang scattering, while S_{xy} gives terms antisymmetric in both x and y directions.

If the statistical symmetry of the disordered system is used, so that for a cubic system the x, y, z and u, v, w dependencies can be assumed to be equivalent, it is possible to derive the functions $I_{\mathrm{SRO}}, Q_x, R_x$ and S_{xy} from linear combinations of measurements of the observed diffuse intensity from symmetry related points in reciprocal space. Then if it can be assumed that the scattering amplitude ratios η and ζ are constants, independent of the reciprocal space coordinates, the values of the parameters α (or $\langle\sigma_0\sigma_i\rangle$), γ, δ and ϵ may be derived by Fourier transform, as described by Cohen and Schwarz [1978].

The assumption that η and ζ are constants is obviously good for neutron diffraction. For X-ray diffraction, the approximation may be poor and may lead to serious errors in some cases, especially for clustering of atoms in the presence of appreciable size effects.

An alternative method of analysis has been suggested by Tibballs [1975] and developed into a practical procedure by Georgopolous and Cohen [1977]. In this, η and ζ are separated from the Q, R, S functions and inserted explicitly. The short range order and size effect parameters may then be obtained directly by solving a large array of linear relationships by least squares methods. This technique works very well for X-ray measurements for which varying values of η and ζ provide sufficiently large numbers of independent linear relationships. It does not work for neutron diffraction.

These methods of analysis have not been extended beyond the terms of second order in displacement parameters or to include higher order correlation parameters. To do so would complicate the analysis enormously. No systematic method has been devised to evaluate correlation parameters such as $\langle \sigma_0 \sigma_i \sigma_j \rangle$ or parameters involving correlations of greater numbers of atoms. It is not clear to what extent terms containing these parameters and occurring in equations (24) to (29) do affect the intensities in any practical cases or to what extent omission of these terms will modify the values of the parameters derived from treatments limited to second order terms. Clearly such terms will become more important for larger values of $|\boldsymbol{u}|$.

The values of the size effect parameters derived experimentally from the diffraction methods are as yet insufficient to provide very clear pictures of the displacement fields around atoms. Some earlier calculations of the size effects to be expected were made using a simple model for the displacements, namely that the displacements are radial, falling off with the inverse square of the distance from the origin as in the macroscopic case of the perturbation of an isotropic solid by a center of dilation (Borie [1957]. All available experimental evidence and approximate theoretical models suggest that the displacement fields of point defects are not isotropic and vary in a complicated way with distance and direction from the defect, and the effective displacement fields to be used for a disordered alloy should no doubt be similar.

The amount of experimental data which could be gathered should, in principle, be sufficient to allow derivation of considerably more than the s.r.o. parameters. To assist this process the scattering power $I(\boldsymbol{u})$ could be measured for the whole range of $\boldsymbol{u}$ and also other factors could be varied. Variations of composition of alloys could be useful since, for the 50:50 composition for example, the three-atom correlation parameters $\langle \sigma_0 \sigma_i \sigma_k \rangle$ are all zero. The relative values of the structure factors f_A and f_B are different for different radiations. Thus for X-rays f_{Cu} and f_{Au} are widely different but f_{Cu} and f_{Zn} are almost the same. For neutron diffraction the situation is reversed. The terms of (23) containing $(f_A - f_B)$ almost disappear for neutron diffraction from Cu-Au alloys or for X-ray diffraction from β-CuZn.

On the other hand for some alloys containing atoms with negative scattering lengths it is possible to adjust the composition or isotopic abundances in such a way that for neutron diffraction the average scattering factor $(m_A f_A + m_B f_B)$ is zero, as in the case of Cu-Ni alloys studied by Moser et al. [1968]. For such "null-matrix" alloys, the fundamental reflections and their attendant thermal diffuse and Huang scattering are zero and only the third part of (23) remains.

17.6. Relationship with ordering energies

The tendency for either like or unlike atoms to come together, giving either clustering or ordering towards a superlattice is a consequence of the balance of the interaction energies between atoms. It is customary to make the approximation, at least for alloys, that the energy for a configuration of atoms is the sum of interactions between pairs of atoms. The configurational energy is then written

$$U = N\{P_{ij}^{\mathrm{AA}} V_{ij}^{\mathrm{AA}} + P_{ij}^{\mathrm{BB}} V_{ij}^{\mathrm{BB}} + P_{ij}^{\mathrm{AB}} V_{ij}^{\mathrm{AB}} + P_{ij}^{\mathrm{BA}} V_{ij}^{\mathrm{BA}}\}\,, \tag{30}$$

where V_{ij}^{AB} is the interaction energy for an A atom at the position i and a B atom at the position j.

By use of the expressions (4) and (5) we obtain

$$U = \tfrac{1}{2} \sum_i \sum_j (m_{\mathrm{A}}^2 V_{ij}^{\mathrm{AA}} + 2m_{\mathrm{A}} m_{\mathrm{B}} V_{ij}^{\mathrm{AB}} + m_{\mathrm{B}}^2 V_{ij}^{\mathrm{BB}}) + \sum_i \sum_j \langle \sigma_i \sigma_j \rangle V_{ij} \tag{31}$$

where

$$V_{ij} = \tfrac{1}{2}\{(V_{ij}^{\mathrm{AA}} + V_{ij}^{\mathrm{BB}} - 2V_{ij}^{\mathrm{AB}})\}\,,$$

which is the increase in energy when like atom pairs replace unlike atom pairs. The first part of (31) is order-independent and can be ignored.

The problem of finding the correlation parameters as a function of temperature for given values of V_{ij} is an unsolved problem of statistical mechanics, equivalent in its simplest form to the three-dimensional Ising model problem. However for temperatures above the critical ordering temperature T_{c}, several approximate solutions are available (Brout [1965]) and from these can be obtained the relationship (Clapp and Moss [1968])

$$\alpha(\boldsymbol{k}) = \frac{G_2(T)}{1 + G_1(T)\, V(\boldsymbol{k})}\,, \tag{32}$$

where we have used the reciprocal space functions

$$\alpha(\boldsymbol{k}) = \sum_i \alpha_{0i} \exp\{\mathrm{i}\boldsymbol{k} \cdot \boldsymbol{r}_{0i}\}\,, \tag{33}$$

$$V(\boldsymbol{k}) = \sum_i V_{0i} \exp\{\mathrm{i}\boldsymbol{k} \cdot \boldsymbol{r}_{0i}\}\,. \tag{34}$$

The function $G_1(T)$ is approximately proportional to $(T/T_c)^{-1}$ and $G_2(T)$ is almost a constant, being only weakly dependent on T. From (19) it is seen that $\alpha(\boldsymbol{k})$ is directly proportional to the component of the diffuse scattering intensity due to short-range ordering, and so is an observable quantity.

From (32) $V(\boldsymbol{k})$ and hence the interaction energies may be obtained directly from the diffuse scattering measurements. For reasons of practical convenience the determinations of interaction energies to date have relied on the real-space relations between α_{ij} and V_{ij}, equivalent to (32).

One immediate consequence of (32) is that the maxima of the diffuse scattering will occur at the points where $V(\boldsymbol{k})$ has minima. Hence for various assumptions of the nature of the interatomic energy functions and so the relative values for the V_{ij}, the positions of the diffuse scattering maxima may be predicted and hence the type of ordered structure which will tend to form in the alloy may be deduced. This aspect of the situation has been explored by Clapp and Moss [1968] who found interesting correlations with the ordered structures of real alloy systems.

Determinations of V_{ij} values from observed diffraction intensities have been used, in particular, to investigate the contributions of conduction electrons to the configuration energy of alloys. It has been shown that minima of $V(\boldsymbol{k})$ and hence maxima of $\alpha(\boldsymbol{k})$ may occur for k values corresponding to $\boldsymbol{k}$ vectors between flat areas of the Fermi surface for the alloy. Hence the form of the Fermi surface may strongly influence the form of the diffuse scattering and so the type of superlattice which tends to be formed. An account of the derivation of the V_{ij} values is given by Wilkins [1970]. The relationship with Fermi surfaces is discussed by Cowley and Wilkins [1972] and a more general discussion including an account of the formation of long-period (10 to 40 Å) out-of phase domain superlattices in relation to long-range oscillatory potentials and speculations on the situation for non-metals is given by Cowley [1971].

17.7. Dynamical scattering from disordered crystals

17.7.1. Dynamical effects in diffuse scattering

In passing through the crystal both the sharp (fundamental or superlattice) reflections and the diffuse scattering will be subject to further scattering. For the sharp reflections this will be coherent dynamical scattering of the usual sort but involving an absorption coefficient because energy will be lost from the sharp reflections to the diffuse background.

For the diffuse scattering the strongest dynamical effect will be dynamical interactions of strength F_h between diffuse amplitudes separated by vectors $\boldsymbol{h}$, where $\boldsymbol{h}$ and F_h refer to the fundamental reflections and their reciprocal lattice points. Such interactions should give rise to Kossel lines or related effects in X-ray diffraction patterns but because the crystals used commonly are imperfect with a relatively large mosaic spread, these lines have been observed only very weakly and the dynamical effects are usually ignored.

For electron diffraction, however, dynamical diffraction effects are inevitably strong and can not be ignored. Since electron diffraction patterns are being increasingly used in studies of disordered alloys and superlattice formation because of the relative ease by which observations can be made, it is important to gain at least some approximate indication of the extent to which dynamical effects might modify the configurations and relative intensities of the kinematical diffuse scattering. This has been done by Fisher [1965, 1970] and Cowley and Murray [1968].

17.7.2. Calculations of diffuse scattering

Fisher made detailed calculations of diffuse scattering intensities for thin crystals of disordered copper-gold alloys using a method formulated by Gjønnes [1962] and developed into a general n-beam treatment by Cowley and Pogany [1968]. Considering first-order diffuse scattering only, the total diffuse scattering is taken as the sum of the diffuse intensities produced by each thin slice of the crystal separately. The range of correlation of atomic positions may be considered to be small so that diffuse scattering from separate slices is incoherent and intensities, not amplitudes, are added.

For a slice of thickness Δz at a depth z in the crystal of thickness H, as in Fig. 12.3, it is considered that an incident beam is first diffracted by the average lattice in the region 0 to z giving a set of fundamental beam amplitudes $\Psi_z(\boldsymbol{h})$. Each one of these beams is then scattered in the region of thickness Δz, giving fundamental reflections plus diffuse scattering. The diffuse scattering is given by the planar section, $w = 0$, of the distribution $I_d(u)$ of diffuse scattering power given as in (19). Then in the final section of the crystal, from z to H, all parts of the diffuse scattering undergo dynamical interaction through the fundamental Bragg reflection. A diffuse scattering beam in the direction u', v' for example interacts with all beams $h + u'$, $k + v'$ where h, k are reflection indices for the average lattice. This gives the diffuse scattering intensity from the slice at position z and this is integrated over z from 0 to H to give the total diffuse scattering.

The results of these calculations for two-dimensional diffraction patterns are of considerable interest. It appears that, in general, the ratio of dynamical to kinematical intensities for pure s.r.o. diffuse scattering may be represented by a smoothly varying function increasing with distance from the origin. Hence it is to be expected that the positions and shapes of the diffuse scattering maxima will not be affected by dynamical effects (unless for a near-perfect crystal, a strong Kikuchi line is generated, passing through a maximum) but there will be an over-all modification of relative intensities from one region of the pattern to the next. The more complete treatment by Fields and Cowley [1978] leads to the same conclusion.

On the other hand the calculations showed that if the size-effect displacement of s.r.o. peaks is included in the scattering from the individual slices, this displacement may be eliminated almost completely when strong two-dimensional dynamical scattering takes place. This result is in accord with the speculation by Cowley [1965], based on crude arguments, that strong dynamical scattering could eliminate contributions to diffuse intensities due to the displacements of atoms, but would not affect contributions due to the interchange, or variation of scattering power, of atoms. It is also in agreement with experimental observations. For example Fig. 17.3 shows the intensity distribution along the *h*00 line in reciprocal space for disordered $CuAu_3$ obtained by X-ray diffraction (Batterman [1957]) and by electron diffraction (Watanabe and Fisher [1965]). The size-effect displacement of the diffuse s.r.o. peak is considerable for X-ray diffraction but absent for the electron diffraction case. However if electron diffraction observations are made on a crystal tilted in such a way as to avoid strong dynamical interactions for particular diffuse peaks, then the size effect displacement of these peaks is visible.

17.7.3. Strong scattering and multi-atom correlations

One defect of these considerations which could, in principle, be serious is the approximation made that the scattering from a slice of the crystal should be kinematical even when the slice thickness is sufficiently large to allow the assumption that the correlation of atom positions does not extend from one slice to the next. For a heavy-atom alloy such as those of Cu and Au, even the difference term $(f_A - f_B)$ is equivalent to scattering by a medium weight atom for which the kinematical approximation may be expected to fail for a few atoms thickness.

A re-formulation of the problem in terms of a phase-grating approximation for a single slice has been made by Cowley and Murray [1968]. When the potential distribution in a slice is projected, the maxima of projected potential vary with the numbers of atoms of either kind in the rows of atoms in the incident beam direction. Putting these maxima into the complex exponential of

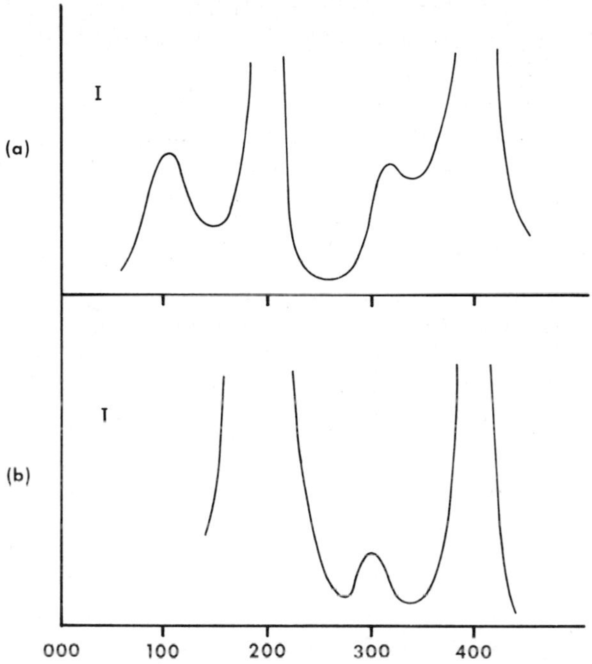

Fig. 17.3. Diffuse scattering intensity measured along the $h00$ line in reciprocal space for a disordered crystal of $CuAu_3$. (a) X-ray measurement (Batterman [1957]) showing size-effect displacement of the 300 diffuse scattering peaks. (b) Electron diffraction measurement (Watanabe and Fischer [1965]) showing no displacement of the 300 peak.

the phase grating transmission function, the scattering is no longer a linear function of the numbers and kinds of atoms. The scattered amplitudes will depend on the probabilities of occurrence of, say, lines of three or four gold atoms. The sharp fundamental reflections will be modified by a pseudo-temperature factor (see Chapter 12) which, like the diffuse scattering intensity, will depend on the values of particular many-atom correlation parameters.

Preliminary calculations indicate however that, apart from an over-all decrease of diffuse scattering intensities, the influence on the diffraction patterns will not be great. For the possible range of values of the many-atom correlation parameters the differences in diffuse scattering intensities are scarcely measurable with the present experimental systems, especially when complicated by dynamical scattering effects. Hence, pending the improvement of measuring techniques and more detailed calculations for representative cases, it appears that the simple kinematical approximation for slice scattering may be trusted to give qualitatively correct results.

17.8. Out-of-phase domains

17.8.1. Ordered out-of-phase superlattices

In many binary-alloy and other systems, there are, in addition to the simple ordered structures we have been considering, other superlattices having much larger unit cells, generated by the periodic occurrence of shift faults in the ordered structure. For the Cu_3Au structure, Fig. 17.1(b), for example, the Au atom could equally well be located preferentially on any one of the four equivalent sites of the F.C.C. unit cell. Thus there are four "variants" of the structure related by vector shifts having coordinates (1/2, 1/2, 0), (0, 1/2, 1/2) etc. A superlattice may be formed by a periodic alternation of two (or more) of these variants in one, two or three dimensions. (See Sato and Toth [1963]; Cowley [1971].)

The best known and most thoroughly investigated of these structures is the CuAu II structure, formed by alternation in one dimension of five unit cells of each of the two variants of the CuAu I structure illustrated in Fig. 17.1(c). One variant is as shown, with a plane of Cu atoms through the origin of the unit cell. In the other variant the Cu and Au atoms are interchanged. The resultant superlattice is orthorhombic with $a = 3.96$, $b = M \times 3.97$, $c = 3.68$ Å, as illustrated by Fig.'17.4(a). Here M is the number of unit cells in half the long-period repeat unit, taken as $M = 5.0$. As a first approximation the structure can be described as the ordered CuAu I unit cell contents convoluted by a distribution function which has one point per unit cell and a shift of (1/2, 0, 1/2) every five unit cells, as suggested in Fig. 17.4(b).

The Fourier transform of this distribution function can be written

$$\sum_k e^{2\pi i 10k} [1 + 2\cos 2\pi k + 2\cos 2\pi 2k]\ 2\cos\pi(U/2+5k) \tag{35}$$

where $U = h + l$ and h, k, l refer to the CuAu I unit cell. The first term gives sharp peaks at intervals of $k/10$. For $h + l$ even, there are maxima only for k integral. For $h + l$ odd there are maxima for k an odd multiple of 1/10. Hence each non-fundamental reciprocal lattice point of the CuAu I structure is

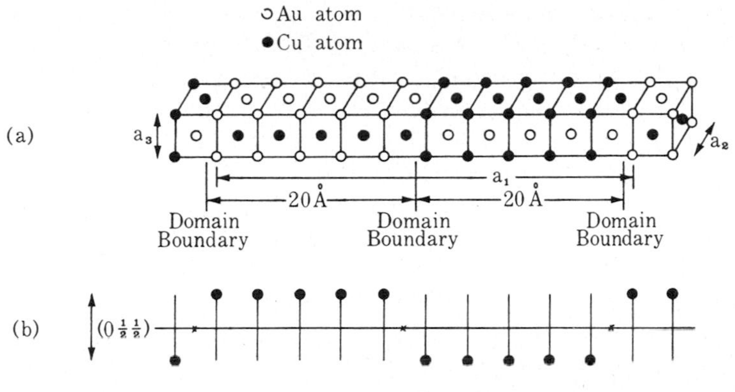

Fig. 17.4. (a) Diagram of the periodic out-of-phase domain structure of the Cu-Au-II superlattice. (b) The distribution function of one point per unit cell used in the derivation of equation (17.35).

replaced by two strong maxima, separated by 1/5 of $\boldsymbol{b}^*$ plus weaker subsidiary maxima. Since the axis of the out-of-phase superlattice may occur in any of the cube axis directions, the total diffraction pattern produced is the sum of a number of equivalent distributions and the electron diffraction pattern for a beam in a (100) direction is as illustrated in Fig. 17.5, with a characteristic

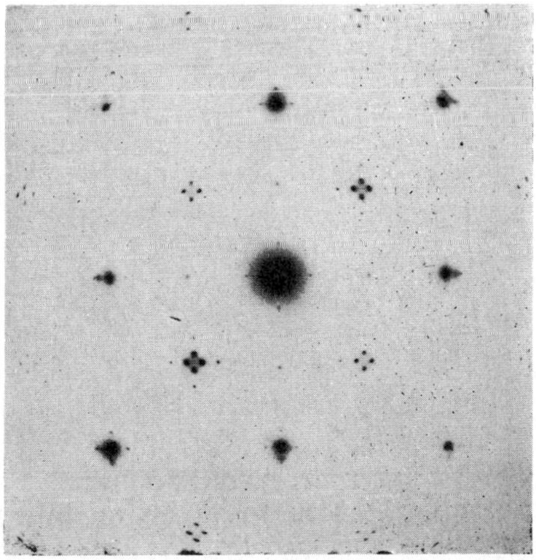

Fig. 17.5. Electron diffraction pattern from CuAu II in [100] orientation showing the characteristic groups of superlattice spots (after Ogawa [1962]).

group of spots replacing each CuAu I superlattice spot.

The periodicities of the out-of-phase superlattices for alloys have been attributed to the occurrence of long-range oscillatory pseudo-potentials, associated with the restriction of the conduction electron energies and momenta at the Fermi surface (Sato and Toth [1963]; Tachiki and Teramoto [1966]). It has been demonstrated very clearly (Hashimoto and Ogawa [1970]) that, if the electron-atom ratio in an alloy is varied by the addition of substitutional impurity atoms of different valency, the periodicity of the long-period superlattice varies in such a way that the energy of the electrons at the Fermi surface is minimized.

17.8.2. Out-of-phase domains in disordered alloys

When an alloy is heated so that long-range ordering disappears, it is to be expected that the long-range oscillatory potential describing the electron-energy terms will remain virtually unchanged. Hence there will be a tendency in the short-range ordered state for a correlation of atom positions, related to this long-range oscillatory potential through the relationship (17.32), to persist.

In fact minima of $V(k)$ and hence maxima of the diffuse intensity, proportional to $\alpha(k)$, do exist around the superlattice reciprocal lattice positions. The diffuse maxima given by many alloys having only short-range order do appear to be split in much the same way as the sharp maxima for ordered alloys.

This splitting was first observed by Raether [1952] using electron diffraction, investigated by Watanabe and Fisher [1965] and finally observed using X-ray diffraction by Moss [1965] when the X-ray experimental methods were refined to provide sufficient resolution.

Since the presence of this splitting in the diffuse diffraction maxima implies that correlations between atom site occupancies exist over distances of, perhaps, 20 to 40 Å, it is inconvenient to describe the state of short-range order in terms of correlation parameters. The number of order coefficients required would be very large. As an alternative it has become a common practise to describe the state of short-range order in terms of an assembly of microdomains, separated by out-of-phase boundaries and similar in many cases, but not always, to the type of anti-phase domains which are present in the alloys having long-range order. Computer simulation methods (Gehlen and Cohen [1965]) have allowed the correlation between microdomain structures and short-range order coefficients to be investigated and visualized in a very illiminating manner.

This model for short-range order is extremely useful but has, perhaps, been interpreted rather too literally by investigators who attempt to image the microdomains in short range ordered alloys by obtaining dark-field electron micrographs, using only the diffuse maxima of the diffraction patterns to obtain the image. Some of the difficulties and pitfalls of this technique have been pointed out by Cowley [1973].

17.8.3. Modulated structures

The results that we have described for the relatively simple binary alloy systems in the previous two Subsections are paralleled by observations made recently on many types of materials. Accounts of many experimental observations and theoretical descriptions of ordered or disordered superstructure formation in which the superstructure periodicity may or may not be commensurate with the sublattice unit cell, have been given in the report of the conference on Modulated Structures (Cowley et al. [1979]). The examples are drawn from the fields of solid-state structural chemistry, mineralogy, metallurgy and solid-state physics and the theoretical treatments are correspondingly diverse.

In the case of the out-of-phase domain superstructures of Subsection 17.8.1, the long period repeat distance is usually not an integral multiple of the basic unit cell size: i.e. M is not an integer. Correspondingly the superstructure reflections in patterns such as Fig. 17.5 do not form a regular subdivision of the sublattice spot separations.

It is not to be concluded that the domain boundaries such as are illustrated in Fig. 17.4 occur regularly at intervals of, say, 4.7 rather than 0.5 unit cells. Rather, it seems that for $M = 4.7$ the domain boundaries will maintain the same form and occur at intervals of either 4 or 5 unit cells, with a random distribution of the 4 and 5 cell spacings occurring with relative frequencies such that the average spacing is 4.7 unit cells. It has been shown by Fujiwara [1957] that this arrangement can give sharp spots at intervals $u = 1/4.7a$ in reciprocal space, plus some weak diffuse scattering.

Thus the structure may be regarded as a locally disordered sequence of commensarute superstructure units with a statistically long-range ordered, incommensurate superstructure. The long-range order with the $M = 4.7$ periodicity is presumably induced by a corresponding periodicity in the long-range oscillatory pseudopotential mentioned in Subsection 17.8.1.

There are many similar cases among the nonstoichiometry oxides and minerals in which the formation of a superstructure involves a distortion of the unit cell or a displacement or replacement of atoms which can occur only at definite sites within the unit cell so that a long period incommensurate structure is made up of locally commensurate units. On the other hand, many examples can be found for which there is apparently a progressive distortion of atom groups within a structure, forming an incommensurate long range periodicity which appears to be quite independent of the subcell periodicity. These include materials such as K_2SeO_4 in which it is shown by neutron diffraction (Iizumi et al. [1977]) that the incommensurate superstructure appears as the limiting, zero frequency case of a soft-mode optical phonon which can be related to the elastic properties of the material. Also a wide range of both organic and inorganic materials show incommensurate and commensurate long range periodicities which are related to their striking one-dimensional or two-dimensional electrical conduction properties. In some cases incommensurate-to-commensurate (lock-in) transitions occur with changes of temperature. The concept of charge-density waves, with associated lattice distortions or atomic ordering, has been invoked to account for many observations of diffuse maxima or sharp spots seen clearly in electron diffraction patterns or investigated by X-ray diffraction.

Changes of sign or relative magnitudes of the short-range order parameters can lead to a segregation of the individual types of atom rather than an ordering towards superlattice formation. The operation of a long-range oscillatory pseudo-potential in these cases can give rise to metastable periodic variations of composition as in the spinodal decomposition of alloys. The additional closely spaced sharp or diffuse spots in diffraction patterns then appear around the strong fundamental lattice reflections and so are often difficult to detect. Similar phenomena, often with periodic changes of unit cell dimensions and symmetry, occur in minerals such as the felspars giving rise to superstructure spots in diffraction patterns and a mottling of the contrast in electron micrographs.

Problems

1. Find the limiting values of the order parameters α_{0i} and $\langle \sigma_0 \sigma_i \rangle$ for the perfectly ordered lattices of Cu_3Au and CuAu. How do the values of these order parameters vary with temperature and with the vector length, $|\boldsymbol{r}_{0i}|$?

2. Given that the energy term for nearest neighbors is positive and the ratios $V_2/V_1 = -0.5$, $V_3/V_1 = 0.2$, where V_2 and V_3 refer to second- and third-nearest neighbor atomic pairs, find the minima in the values of $V(k)$ and hence the positions of diffuse scattering maxima for the Cu_3Au structure and the type of ordered lattice which will tend to form. Do this also for $V_2/V_1 = +0.75$, $V_3/V_1 = +0.2$.

CHAPTER 18

Extended defects

18.1. Introduction

A vast literature exists concerning the extended defects which occur in most types of crystals. They are of fundamental importance for the consideration of the physical and chemical properties of solids and of great technological significance. The simplest and best-known are the stacking faults, twins and the various forms of dislocation. To these one can add defects clusters, impurity aggregates, segregated concentrations of particular atoms as in G-P zones, coherent and incoherent precipitates, vacancy clusters, voids, ordering nuclei, and so on.

Here we plan to do little more than illustrate the methods by which such defects can be studied by diffraction methods, and will confine our considerations to stacking faults and dislocations.

There have been two main avenues of approach to the study of these faults. Historically the first, and still important, is the observation of the streaking or diffuse scattering in diffraction patterns. The classical example is that in close-packed structures the sequence of stacking of the hexagonal close-packed planes of atoms may not follow the regular two-plane periodicity of the hexagonal close-packed structure or the regular three-plane periodicity of the face-centered cubic structure but may show faults in either type of sequence, or there may even be an almost complete randomness in switching from one type of sequence to another. The effect in reciprocal space is to produce continuous lines of scattering power, perpendicular to the close-packed planes and passing through some reciprocal lattice points. The initial analysis was on stacking faults in hexagonal cobalt (Wilson [1942]) but other examples followed and analogous effects were found to exist for a very wide range of metallic and non-metallic materials (Guinier [1963]; Warren [1969]).

The analysis of the nature of the faults and their distributions from the diffraction patterns, assuming kinematical scattering, is necessarily in terms of statistical averages over very large numbers of faults. Starting from the diffraction intensities one can derive, and then attempt to interpret, the generalized Patterson function (Chapter 7). The means for doing this have included the construction of a "probability tree" (see Warren [1969]) or the more formally elegant matrix methods developed for example by Hendricks and Teller [1942] and Kakinoki and Komura [1951, 1952]. We have chosen to present here a slightly different method developed in terms of the probabilities of occurrence of various types of fault in the structure.

With the extensive development of electron microscopy in the 1950s, the study of individual defects in thin crystalline films became possible. Fault planes gave the appearance of bands of fringes. Dislocations appeared as dark or dark-and-light lines. The basis for interpretation of the observations was essentially the dynamical theory of electron diffraction and an extensive body of experience on both the configurations of the defects and the rules for the interpretation of the images was rapidly assembled (see Hirsch et al. [1965]). Equivalent observations of defects in near-perfect crystals by X-ray diffraction under dynamical scattering conditions followed a few years later (Lang [1958, 1959]; Kato and Lang [1959]), and the appropriate X-ray diffraction theory was developed on the initial work of Kato [1960, 1961]. More recently, more exact treatments in terms of n-beam dynamical theories have been developed for electron diffraction and for all radiations the difficult task has been tackled of providing an adequate dynamical theory for imperfect crystals (e.g. Kato [1973]; Kuriyama [1973]). We will follow these developments in outline only.

18.2. Stacking faults – statistical, kinematical theory

18.2.1. Patterson method for a simple case

There are many substances for which the ideal crystal structure can be considered to be built up by the regular superposition of identical layers, each layer being one unit cell thick. In practise the regular superposition is occasionally interrupted by a fault which gives a displacement of one layer, and all subsequent layers, relative to the previous layers. If these faults are not too numerous, we may assume as a first approximation that they occur at random. We suppose that there is a probability, α, that a fault will occur in which the displacement is defined by the vector $\boldsymbol{s}$.

If the vector $\boldsymbol{s}$ is not parallel to the plane of the layers, there will in general be a subtraction or addition of atoms to a layer at the fault to maintain approximately the same density of material. We defer consideration of this point until later.

Since the structure can be described in terms of an electron density distribution

$$\rho(\boldsymbol{r}) = \rho_0(\boldsymbol{r}) * d(\boldsymbol{r}) \,,$$

where $\rho_0(\boldsymbol{r})$ is the electron density of one layer and $d(\boldsymbol{r})$ is some distribution function, the generalized Patterson function can be written, as in (7.12) as

$$P(\boldsymbol{r}) = \rho_0(\boldsymbol{r}) * \rho_0(-\boldsymbol{r}) * \mathcal{D}(\boldsymbol{r}) \,,$$

where $\mathcal{D}(\boldsymbol{r})$ is a distribution function Patterson describing the probability that if one layer is centered at the origin, another layer will be centered at $\boldsymbol{r}$. After n layers the probability that a number, m, of faults should have occurred is given by the Poisson distribution function as follows:

$$\text{Probability of 0 faults} = \exp\{-\alpha n\} \,,$$

$$\text{Probability of 1 fault} = \alpha n \exp\{-\alpha n\} \,,$$

$$\text{Probability of } m \text{ faults} = \frac{(\alpha n)^m}{m!} \exp\{-\alpha n\} \,.$$

The vectors between the origins of two layers separated by n normal translations plus m faults will be $n\boldsymbol{a} + m\boldsymbol{s}$. The generalized Patterson function is then

$$P(\boldsymbol{r}) = N\rho_0(\boldsymbol{r}) * \rho_0(-\boldsymbol{r}) * \left[\sum_{n=1}^{\infty} \sum_{m=0}^{\infty} \delta(\boldsymbol{r} - n\boldsymbol{a} - m\boldsymbol{s}) \frac{(\alpha n)^m}{m!} \exp\{-\alpha n\} \right.$$

$$\left. + \sum_{n=1}^{\infty} \sum_{m=0}^{\infty} \delta(\boldsymbol{r} + n\boldsymbol{a} + m\boldsymbol{s}) \frac{(\alpha n)^m}{m!} \exp\{-\alpha n\} + \delta(\boldsymbol{r}) \right] . \qquad (1)$$

Fourier transforming gives the intensity distribution

$$I(\boldsymbol{u}) = |F(\boldsymbol{u})|^2 \left[\sum_{n=1}^{\infty} \sum_{m=0}^{\infty} \exp\{2\pi \mathrm{i} \boldsymbol{u} \cdot \boldsymbol{a} n\} \exp\{2\pi \mathrm{i} m \boldsymbol{u} \cdot \boldsymbol{s}\} \frac{(\alpha n)^m}{m!} \exp\{-\alpha n\} \right.$$

$$\left. + \sum_{n=1}^{\infty} \sum_{m=0}^{\infty} \exp\{-2\pi \mathrm{i} \boldsymbol{u} \cdot \boldsymbol{a} n\} \exp\{-2\pi \mathrm{i} m \boldsymbol{u} \cdot \boldsymbol{s}\} \frac{(\alpha n)^m}{m!} \exp\{-\alpha n\} + 1 \right] . \qquad (2)$$

The summation over m gives $\exp\{n\alpha \exp[2\pi \mathrm{i} \boldsymbol{u} \cdot \boldsymbol{s}]\}$ and the summation over n is made using the relation $\sum_{n=0}^{\infty} x^n = (1-x)^{-1}$ so that

$$I(\boldsymbol{u}) = |F(\boldsymbol{u})|^2 [(1 - \exp\{2\pi i \boldsymbol{u}\cdot\boldsymbol{a} - \alpha + \alpha \exp(2\pi i \boldsymbol{u}\cdot\boldsymbol{s})\})^{-1}$$

$$+ (1 - \exp\{-2\pi i \boldsymbol{u}\cdot\boldsymbol{a} - \alpha + \alpha \exp(-2\pi i \boldsymbol{u}\cdot\boldsymbol{s})\})^{-1} - 1]$$

$$= |F(\boldsymbol{u})|^2$$

$$\times \left[\frac{1 - \exp\{2\alpha(\cos 2\pi\boldsymbol{u}\cdot\boldsymbol{s}) - 1\}}{1 + \exp\{2\alpha(\cos 2\pi\boldsymbol{u}\cdot\boldsymbol{s} - 1)\} - 2\exp\{\alpha(\cos 2\pi\boldsymbol{u}\cdot\boldsymbol{s} - 1\}\cos(2\pi\boldsymbol{u}\cdot\boldsymbol{a} + \alpha\sin 2\pi\boldsymbol{u}\cdot\boldsymbol{s})}\right]. \tag{3}$$

If α is small it can be seen readily that this function has fairly sharp maxima which are displaced from the reciprocal lattice points for an unfaulted crystal if $\boldsymbol{s}$ is not parallel to the layers. The positions of the maxima are given by

$$\boldsymbol{u}\cdot\boldsymbol{a} - h/a = -(2\pi)^{-1}\alpha \sin 2\pi\boldsymbol{u}\cdot\boldsymbol{s}\,. \tag{4}$$

These maxima are of height $[\alpha(\cos 2\pi\boldsymbol{u}\cdot\boldsymbol{s} - 1)]^{-1} |F(\boldsymbol{u})|^2$ and of half width $(\alpha/\pi)(\cos 2\pi\boldsymbol{u}\cdot\boldsymbol{s} - 1)$.

As an example we may quote the case of magnesium fluoro-germanate (Bless et al. [1972]) in which the structure can be considered as made up of four layers of metal atoms per unit cell within the close-packed oxygen-fluorine structure. The layers are perpendicular to the $\boldsymbol{c}$-axis. The presence of fluorine is associated with faults where one of the four layers is omitted. The component of the $\boldsymbol{s}$ vector in the $\boldsymbol{c}$-axis direction is then $-c/4$. For the $00l$ spots, the 001 spot is shifted by an amount $+\alpha/2\pi$ away from the origin and has a half-width α/π, the 002 spot is unshifted but has a half width $2\alpha/\pi$, the 003 spot is shifted $-\alpha/2\pi$ towards the origin and has width α/π while the 004 spot is sharp and unshifted. This can be seen to be the case in the electron diffraction pattern, Fig. 18.1, from which it can be deduced readily that $\alpha = 0.2$.

In all cases such as this when $\boldsymbol{s}$ is not parallel to the layers, the presence of a fault implies the addition or subtraction of part of a layer. In order to treat such cases adequately it is necessary to use a somewhat different, more general approach (Cowley [1976a]) although, as we will see, it is often possible to reduce the more general result to a relatively simple form in many cases of practical significance, especially if the experimental data is not strictly quantitative.

18.2.2. A general treatment

We consider the crystal to be made up of an arbitrary number of different

Fig. 18.1. Electron diffraction pattern from a crystal of magnesium fluoro-germanate showing streaking of spots along the c^*-direction due to planar faults in the crystal (courtesy of P. Kunzmann).

types of layers. The ith type of layer has an electron density distribution $\rho_i(\boldsymbol{r})$ and if such layers were stacked regularly the translation vector would be $\boldsymbol{R}_i$. If a fault occurs to change the layer from $\rho_i(\boldsymbol{r})$ to $\rho_j(\boldsymbol{r})$ which has a translation vector $\boldsymbol{R}_j$ there will be the addition of an electron density $\Delta_{ij}(\boldsymbol{r})$ and the vector $\boldsymbol{R}_i$ is modified to $\boldsymbol{R}_i + \boldsymbol{S}_{ij}$. The probability of such a fault is α_{ij}.

The assumption that such faults occur at random with this probability is not as restrictive as it may at first appear. If particular sequences of planes occur commonly, each of these sequences may be regarded as a separate type of layer. Hence, the preference for clumps of layers of particular types may be included in the description.

The probability, g_i, that the ith type of layer will occur at any particular position is given by equating the number of transitions to and from the i-type:

$$\sum_j g_j \alpha_{ji} = g_i \sum_j \alpha_{ij} . \tag{5}$$

The generalized Patterson function can then be written as a series of terms corresponding to 0, 1, 2... interlayer vectors $\boldsymbol{R}_n$. For the interlayer vector of zero length,

$$P_0(\boldsymbol{r}) = N \sum_i g_i [(1 - A_i)\{\rho_i(\boldsymbol{r}) * \rho_i(-\boldsymbol{r})\}$$

$$+ \sum_j \alpha_{ij}(\rho_i(\boldsymbol{r}) + \Delta_{ij}(\boldsymbol{r})) * (\rho_i(-\boldsymbol{r}) + \Delta_{ij}(-\boldsymbol{r}))] . \tag{6}$$

Here we have put $A_i = \Sigma_j \alpha_{ij}$ so that $(1 - A_i)$ is the probability that an i layer will not be modified by a fault. The second term in the square bracket comes from layers modified by faults.

The contribution from (6) to the observed intensity is given by Fourier transforming as

$$I_0/N = \sum_i g_i \ (1 - A_i)|F_i|^2 + \sum_j \alpha_{ij}|F_i + G_{ij}|^2 \tag{7}$$

where F_i and G_{ij} are the Fourier transforms of ρ_i and Δ_{ij}.

The contribution to the intensity due to terms with a single interlayer vector $\boldsymbol{R}_i$ to which there may be added $\boldsymbol{S}_{ij}$ if a fault occurs, is given by the same sort of reasoning as

$$I_1/N = \sum_i g_i \left[(1 - A_i)F_i^* \left\{F_i + \sum_j \alpha_{ij} G_{ij}\right\}\right] \exp\{2\pi i \boldsymbol{u}\boldsymbol{R}_i\}$$

$$+ \sum_i g_i \sum_j \alpha_{ij} \left[(F_i^* + G_{ij}^*) \cdot \left\{F_j + \sum_k \alpha_{jk} G_{ji}\right\}\right] \exp\{2\pi i \boldsymbol{u}(\boldsymbol{R}_i + \boldsymbol{S}_{ij})\} , \tag{8}$$

which represents the sum of terms for which there is no fault plus the sum of times for which a fault occurs between neighboring layers.

To simplify this and subsequent expressions we define $F_i' \equiv F_i + \Sigma_j \alpha_{ij} G_{ij}$ which represents the average structure amplitude for a layer, as modified by the possibility of faults.

For subsequent terms we can simplify the expressions further by writing

$$B_i \equiv (1 - A_i)F_i' + \sum_j \alpha_{ij} F_j' \exp\{2\pi i\boldsymbol{u} \cdot \boldsymbol{S}_{ij}\}$$

which represents the average neighbor of a given layer, including the possibility of a fault or no faults. For vectors between second nearest neighbors, the contribution to the intensity is then

$$I_2/N = \sum_i g_i F_i^* B_i \exp\{2\pi i\boldsymbol{u} \cdot 2\boldsymbol{R}_i\}$$

$$+ \sum_i g_i \sum_j \alpha_{ij}(F_i^* + G_{ij}^*) B_j \exp\{2\pi i\boldsymbol{u} \cdot (\boldsymbol{R}_i + \boldsymbol{R}_j + \boldsymbol{S}_{ij})\}\,. \tag{9}$$

If we take, from each of these terms, the contribution from the cases where no fault occurs we obtain the series

$$1 + (1 - A_i)\exp\{2\pi i\boldsymbol{u}\boldsymbol{R}_i\} + (1 - A_i)^2 \exp\{2\pi i\boldsymbol{u} \cdot 2\boldsymbol{R}_i\} + \dots$$

which sums to give

$$[1 - (1 - A_i)\exp\{2\pi i\boldsymbol{u} \cdot \boldsymbol{R}_i\}]^{-1}\,.$$

For vectors $-\boldsymbol{R}_i$ we obtain terms $I_{-n} = I_n^*$. Then summing for all interplanar vectors we obtain the general expression:

$$I/N = \sum_i g_i (1 - A_i)|F_i|^2 + \sum_i g_i \sum_j \alpha_{ij} |F_i + G_{ij}|^2$$

$$+ \sum_i g_i (1 - A_i) F_i^* \exp\{2\pi i\boldsymbol{u} \cdot \boldsymbol{R}_i\}$$

$$\times \left[F_i' + \frac{\exp\{2\pi i\boldsymbol{u} \cdot \boldsymbol{R}_i\}}{1 - (1 - A_i)\exp\{2\pi i\boldsymbol{u} \cdot \boldsymbol{R}_i\}} \right.$$

$$\times \left[B_i + \sum_j \alpha_{ij} \frac{\exp\{2\pi i\boldsymbol{u} \cdot (\boldsymbol{R}_j + \boldsymbol{S}_{ij})\}}{1 - (1 - A_j)\exp\{2\pi i\boldsymbol{u} \cdot \boldsymbol{R}_j\}} \left[B_j \right.\right.$$

$$\left.\left.\left. + \sum_k \alpha_{jk} \frac{\exp\{2\pi i\boldsymbol{u} \cdot (\boldsymbol{R}_k + \boldsymbol{S}_{jk})\}}{1 - (1 - A_k)\exp\{2\pi i\boldsymbol{u} \cdot \boldsymbol{R}_k\}} [B_k + \dots] \right] \right] \right]$$

$$+ \sum_i g_i \sum_j \alpha_{ij}(F_i^* + G_{ij}^*) \exp\{2\pi i\boldsymbol{u} \cdot (\boldsymbol{R}_i + \boldsymbol{S}_{ij})\}$$

$$\times \left[F_j' + \frac{\exp\{2\pi i\boldsymbol{u}\boldsymbol{R}_j\}}{1 - (1 - A_j)\exp\{2\pi i\boldsymbol{u} \cdot \boldsymbol{R}_j\}} [B_j + \dots] \right] + \text{c.c.} \tag{10}$$

The first two terms represent intralayer contributions. The next two terms represent the contributions from interlayer terms starting on nonfault layers and layers with faults respectively for positive vectors. The initial parts of these two terms are different but from the bracket starting with B_j they are identical. The complex conjugate gives the terms for negative vectors.

If the probabilities for faults are low only a few of the brackets need be considered. From the nature of the denominators it is seen that maxima of intensity will tend to occur whenever $\boldsymbol{u} \cdot \boldsymbol{R}_i$ is close to an integer, i.e. around the reciprocal lattice points for an ordered stacking of each of the individual types of layers.

The general expression (10) can be used to treat cases of much greater complexity than are normally considered in practice. To date quantitative comparisons of experimental and theoretical intensities have been rare and intensities are usually calculated for only relatively simple models of faulted structures. The virtue of (10) is that it may readily be simplified and often leads to series which can be summed to give algebraic expressions from which the form of the intensity distribution may readily be deduced.

A particularly simple case is that when there is only one type of layer and a probability α that a fault occurs with a displacement vector $\boldsymbol{S}$ but no modification of the layer structure at the fault. Then (10) becomes

$$I/N = \frac{F^*}{1-(1-\alpha)\exp\{2\pi i\boldsymbol{u}\cdot\boldsymbol{R}\}}\left[F + \frac{\alpha\exp\{2\pi i\boldsymbol{u}\cdot(\boldsymbol{R}+\boldsymbol{S})\}}{1-(1-\alpha)\exp\{2\pi i\boldsymbol{u}\cdot\boldsymbol{R}\}}\right.$$

$$\left.\cdot\left[F + \frac{\alpha\exp\{2\pi i\boldsymbol{u}\cdot(\boldsymbol{R}+\boldsymbol{S})\}}{1-(1-\alpha)\exp\{2\pi i\boldsymbol{u}\cdot\boldsymbol{R}\}}\,[F+\ldots]\right]\right] + \text{c.c.} - |F|^2\,.$$

The expression in the square brackets may be summed as a geometric series to give

$$I/N = \frac{|F|^2\alpha(1-\alpha)(1-\cos 2\pi\boldsymbol{u}\cdot\boldsymbol{S})}{1-\alpha+\alpha^2-(1-\alpha)\cos 2\pi\boldsymbol{u}\cdot\boldsymbol{R}-\alpha\cos 2\pi\boldsymbol{u}\cdot(\boldsymbol{R}+\boldsymbol{S})+\alpha(1-\alpha)\cos 2\pi\boldsymbol{u}\cdot\boldsymbol{S}}. \tag{11}$$

This expression appears different from (3) but gives an intensity distribution which is almost identical for small α. The intensity maxima are close to the reciprocal lattice points for which $\boldsymbol{u} \cdot \boldsymbol{R}$ is an integer, unless $\boldsymbol{u} \cdot \boldsymbol{S}$ is close to an integer. The maxima are actually displaced from the reciprocal lattice points by an amount ϵ given by

$$\epsilon = -\alpha \sin(2\pi\boldsymbol{u}\cdot\boldsymbol{S})/\{1-\alpha(1-\cos 2\pi\boldsymbol{u}\cdot\boldsymbol{S})\}$$

and the maximum intensity is

$$I/N \approx 2|F|^2(1+\alpha)/[\alpha(1-\cos 2\pi \boldsymbol{u}\cdot\boldsymbol{S})] .$$

For the case of the magnesium fluoro-germanate discussed above and illustrated by Fig. 18.1, roughly one quarter of the layer content is omitted when there is a fault so that it is necessary to consider the slightly greater complication that $C_{ij} \neq 0$. This modifies the treatment only slightly. The result is an expression for which the denominator is exactly the same as for (11) but the numerator becomes

$$\begin{aligned}\alpha(1-\alpha)[&|F|^2(1-\cos 2\pi\boldsymbol{u}\cdot\boldsymbol{S})\\ &+(\mathrm{Re}F*G)\{1+\cos 2\pi\boldsymbol{u}\cdot(\boldsymbol{R}+\boldsymbol{S})-\cos 2\pi\boldsymbol{u}\cdot\boldsymbol{R}-\cos 2\pi\boldsymbol{u}\cdot\boldsymbol{S}\}\\ &-(\mathrm{Im}F*G)\{\sin 2\pi\boldsymbol{u}\cdot\boldsymbol{R}+(1-2\alpha)\sin 2\pi\boldsymbol{u}\cdot\boldsymbol{S}\\ &-\sin 2\pi\boldsymbol{u}\cdot(\boldsymbol{R}+\boldsymbol{S})\}\\ &+|G|^2(1-\cos 2\pi\boldsymbol{u}\cdot\boldsymbol{R})] ,\end{aligned} \tag{12}$$

Here Re and Im indicate the real and imaginary parts of the product $F*G$.

In eq. (12), G appears to occur with much the same weighting as F even though the faults are much less frequent than the unfaulted layers. However it is seen that for x small the terms involving G are very small in the neighborhood of the intensity maxima which are close to the reciprocal lattice points given by integer values of $\boldsymbol{u}\cdot\boldsymbol{R}$. The nature of the faults then influences the intensity distribution very little except in the low intensity regions well away from the intensity maxima (see Cowley [1976]). The assumption that $G = 0$ may often be used to simplifly the calculations for α small, especially if the aim of the investigation can be satisfied by a determination of α and $\boldsymbol{S}$ from the position and width of the intensity maxima.

This treatment of stacking faults is most useful if the assumption can be made of random faults. In many cases however, faults do tend to cluster together or else to avoid each other. If the clustering is well defined these cases can often be treated by considering particular groupings of layers, possibly with particular sequences of faults, as the layers of different types in the above equations. For other cases, particularly with $G = 0$, it is possible to consider combonations of two fault probabilities α and γ giving an intensity distribution

$$I(u) = aI_\alpha(u) + cI_\gamma(u) \tag{13}$$

with $a + c = 1$ and $a\alpha + c\gamma > 0$. The probability of no fault in n layers then takes the form

$$a(1-\alpha)^n + c(1-\gamma)^n .$$

If $\gamma > \alpha$ and c is positive, the faults tend to cluster together. If c is negative with $\gamma > \alpha$, the faults will tend to avoid each other.

18.2.3. Faults in close-packed structures

The close-packing of atoms in simple structures is conventionally described in terms of the sequence of positions of the origins of hexagonal two-dimensional layers of atoms as illustrated in Fig. 18.2. Relative to an origin layer labelled A, the neighboring layers may take the positions B or C. The hexagonal close-packed (HCP) sequence is the ABABAB... or ACACAC... or equivalent, and the face-centered cubic sequence is ABCABC... or ACBACB... or

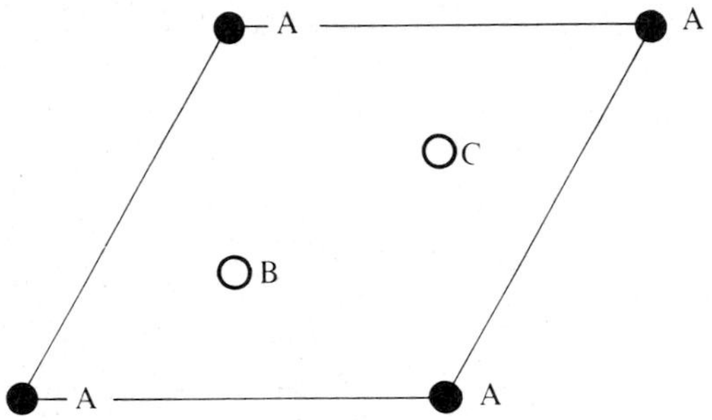

Fig. 18.2a

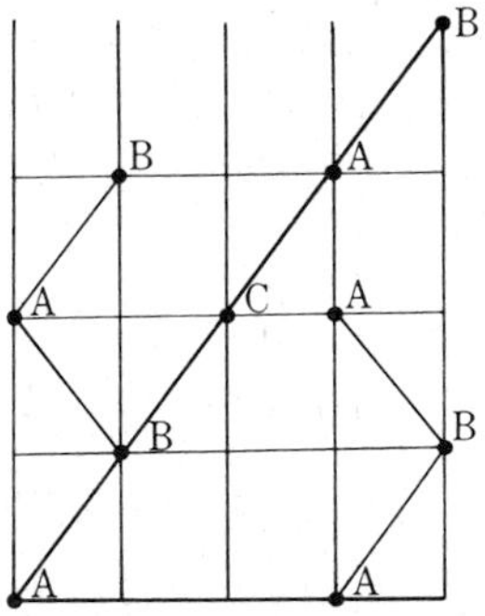

Fig. 18.2b

Fig. 18.2. The stacking sequences of close-packed structures. (a) The labelling of alternate stacking positions for two-dimensional close-packed layers. (b) Stacking sequences illustrated by the atom positions in (110) planes.

equivalent. Convenient diagrams for the sequences and their faults are obtained by considering sections of the structure on (110) planes as in Fig. 18.2b; and axes $\boldsymbol{a}$ and $\boldsymbol{c}$ can be drawn within this plane. Then $\boldsymbol{a} = 3^{1/2}\boldsymbol{a}_0$, where $\boldsymbol{a}_0$ is the hexagonal layer periodicity and $\boldsymbol{c}$ may be chosen equal to the H.C.P. cell axis, the F.C.C. $\boldsymbol{c}$-axis or, for convenience, equal to the thickness of one layer.

In order to describe the faults in HCP structures according to the scheme given above, we take a $\boldsymbol{c}$-axis equal to the HCP $\boldsymbol{c}$-axis so that $\boldsymbol{R} = \boldsymbol{c}$. Then growth faults, giving sequences such as ABABCBCBCB... can be described in terms of two types of fault occurring with equal probability, one being a fault at the $z = 0$ layer and the other a fault occurring at the $z = 1/2$ layer; the first gives a shift $\boldsymbol{s} = \boldsymbol{c}/2 + \boldsymbol{a}/3$ and one plane of atoms is added; the second gives a shift $-\boldsymbol{s}$ and one plane of atoms is subtracted.

Then, neglecting initially the added and subtracted planes,

$$I(\boldsymbol{u}) = \sum_N \exp\{2\pi i \boldsymbol{u}\cdot\boldsymbol{R}N\}\, |F_0(\boldsymbol{u})|^2$$
$$\times\, [(1-2\alpha)^N + \alpha(\exp\{2\pi i\boldsymbol{u}\cdot\boldsymbol{s}\} + \exp\{-2\pi i\boldsymbol{u}\cdot\boldsymbol{s}\})\, C_1^N (1-2\alpha)^{N-1} + \ldots$$
$$+\, \alpha^n (2\cos 2\pi\boldsymbol{u}\cdot\boldsymbol{s})^n C_n^N (1-2\alpha)^{N-n} + \ldots] \,. \tag{14}$$

Summing over N then gives:

$$\frac{I(\boldsymbol{u})}{|F_0(\boldsymbol{u})|^2} = \sum_{n=0}^{\infty} \left[\frac{\exp\{2\pi i\boldsymbol{u}\cdot\boldsymbol{R}n\}\alpha^n (2\cos 2\pi\boldsymbol{u}\cdot\boldsymbol{s})^n}{[1-(1-2\alpha)\exp\{2\pi i\boldsymbol{u}\cdot\boldsymbol{R}\}]^{n+1}} + \text{c.c.} \right] - 1$$
$$= \frac{1-[1-2\alpha(1-\cos 2\pi\boldsymbol{u}\cdot\boldsymbol{s})]^2}{1+[1-2\alpha(1-\cos 2\pi\boldsymbol{u}\cdot\boldsymbol{s})]^2 - 2[1-2\alpha(1-\cos 2\pi\boldsymbol{u}\cdot\boldsymbol{s})]\cos 2\pi\boldsymbol{u}\cdot\boldsymbol{R}} \,. \tag{15}$$

It is readily confirmed that this result gives the same intensity distribution as derived by other methods (Warren [1969]). The hl spots of our two dimensional unit cell are sharp for $h = 3n$. For $h \neq 3n$ there are continuous lines of intensity in the $\boldsymbol{c}^*$ direction with maxima for l integral, the maxima for l odd being 9 times the height and one third of the width of those for l even.

For the F.C.C. structure we take an oblique two dimensional unit cell within the (110) plane, one axis being the a axis and the other being the vector from an A atom in one layer to a B or C atom position in the next layer. Thus we have two "layer" types ρ_1 and ρ_2 differentiated by repetition vectors $\boldsymbol{R}_1$ and $\boldsymbol{R}_2$. A growth fault in an F.C.C. structure is then one for which $\boldsymbol{R}_1$ changes to $\boldsymbol{R}_2$ or vice versa, with $\boldsymbol{s} = 0$. This case has been treated by Cowley [1976a].

For "displacement" or "deformation" faults of the F.C.C. lattice, if the structure is characterized by a repetition vector $\boldsymbol{R}_1$, a fault gives a shift $\boldsymbol{s} = \boldsymbol{a}/3$ and for $\boldsymbol{R}_2$ the shift is $-\boldsymbol{s} = -\boldsymbol{a}/3$, but the type of structure is not changed from one variant to the other. The various other types of fault can be described similarly and treated by the appropriate simplification of the general formulation given above. For example, the case of frequent twinning which sometimes occurs for minerals such as the felspars, has been treated in this way by Cowley and Au [1978].

18.3. Dynamical diffraction by stacking faults

The possibility of interpreting the images of individual stacking faults, appearing in electron micrographs, has provided a great expansion in the knowledge of the form and variety of stacking faults plus, more importantly, the interactions of faults with other planar faults, dislocations, grain boundaries and so on. It is this latter type of information which is almost completely inaccessible if one is limited to the diffraction studies of statistical distributions which we have discussed above.

The interpretation of fault images normally involves the use of the column approximation, Fig. 10.7. For a fault plane intersecting a thin foil sample, we may consider diffraction in the perfect crystal region above the fault, a translation of the lattice by a shift vector $\boldsymbol{s}$ at the fault, approximated by a shift perpendicular to the column, and then diffraction in the perfect crystal region following the fault. The calculation of amplitudes for the perfect crystal regions may be made by any of the available n-beam dynamical treatments and the shift modulates the structure amplitudes for the subsequent part of the crystal by a factor $\exp\{2\pi i \boldsymbol{h}\cdot\boldsymbol{s}\}$.

The most familiar features of fault images may be derived simply from the 2-beam dynamical theory of Chapters 8 and 10. According to the Bloch-wave formulation, a beam incident at the Bragg angle generates the two Bloch waves $\psi^{(1)}$ and $\psi^{(2)}$ for which the components of the wave vector in the direction of propagation are $k_0 \pm \xi$ where $\xi = |v_h|/2\kappa$. The two Bloch waves propagate with refractive indices $n = 1 + (\Phi_0 \pm \Phi_h)/2E$, and absorption coefficients $\mu_0 \pm \mu_h$. After passing through a thickness z of crystal the waves are out of phase by an amount $2\xi z$ which is equal to 2π for $z = 2h^2\lambda/\pi m e \Phi_h \equiv \xi_h$, the extinction distance. This progressive phase difference leads to the appearance of the sinusoidal thickness fringes for a wedge-shaped crystal. The Bloch waves may be written

$$\begin{aligned} \psi^{(1)} &= 2^{-1/2}\,[\exp\{2\pi i \boldsymbol{k}^{(1)}\cdot\boldsymbol{r}\} - \exp\{2\pi i(\boldsymbol{k}^{(1)}+\boldsymbol{h})\cdot\boldsymbol{r}\}]\;, \\ \psi^{(2)} &= 2^{-1/2}\,[\exp\{2\pi i \boldsymbol{k}^{(2)}\cdot\boldsymbol{r}\} + \exp\{2\pi i(\boldsymbol{k}^{(2)}+\boldsymbol{h})\cdot\boldsymbol{r}\}]\;. \end{aligned} \tag{16}$$

At the position of a stacking fault each of these Bloch waves of the initial part of the crystal acts as an incident plane wave and each generates two Bloch waves in the second part of the crystal:

$$\psi^{(1)} \rightarrow \psi^{11} + \psi^{12} ,$$

$$\psi^{(2)} \rightarrow \psi^{21} + \psi^{22} .$$

From the diagram, Fig. 9.1, it is clear that if $\boldsymbol{b}\cdot\boldsymbol{s} = 1/2$, the wave which had nodes on the atom planes now has nodes between them and vice versa. Hence the roles of the two Bloch waves are interchanged. The phase difference in a crystal of thickness H with a fault at z is then

$$(k_1-k_2)z + (k_2-k_1)(H-z) = (k_2-k_1)H + 2(k_1-k_2)z . \tag{17}$$

Thus the variation of the phase change with z is twice as great as with H. Hence fringes occur in the image as for a wedge crystal but with a spacing half as great.

For a thick crystal in which absorption is important we may use the argument of Hashimoto et al. [1961] based on Fig. 18.3. If the fault is at the bottom of the crystal, as on the left side, the wave $\psi^{(1)}$ is strongly attenuated in the top part of the crystal so that at the fault there is only the wave $\psi^{(2)}$,

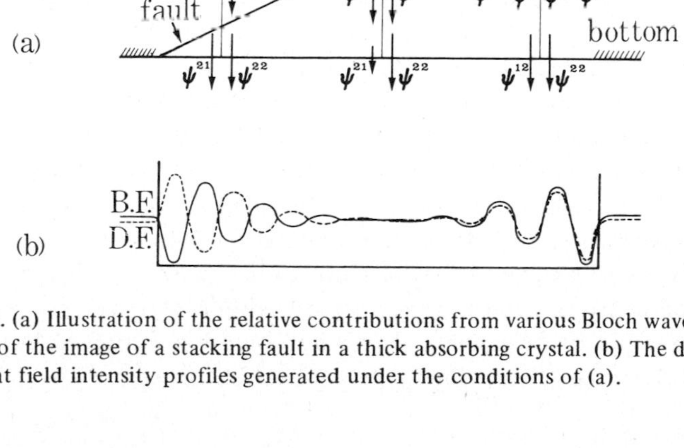

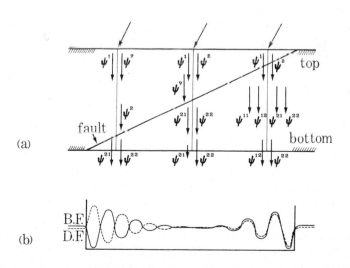

Fig. 18.3. (a) Illustration of the relative contributions from various Bloch waves to the contrast of the image of a stacking fault in a thick absorbing crystal. (b) The dark-field and bright field intensity profiles generated under the conditions of (a).

giving ψ^{21} and ψ^{22} which interfere with almost equal amplitude at the exit face to give strong fringes in bright field and dark field, as for a thin wedge-shaped crystal.

For the fault in the middle of the crystal, the Bloch wave $\psi^{(2)}$ is strongest at the fault and gives ψ^{21} and ψ^{22}, but in the second half of the crystal ψ^{21} is more strongly attenuated. Then the image is given by the interference of ψ^{22} with a much weaker ψ^{21} and the fringes have very low contrast.

For the fault at the top of the crystal all four waves ψ^{11}, ψ^{12}, ψ^{21} and ψ^{22} are generated at the fault but ψ^{11} and ψ^{21} are more strongly attenuated so that the image is given by interference of equally strong ψ^{12} and ψ^{22}, giving strong fringes. However, it can be seen from (16) that for this case the phase of the diffracted wave component relative to the incident wave component of the Bloch wave will be opposite for ψ^{12} and ψ^{22} whereas they will be the same for ψ^{21} and ψ^{22} which were important for the fault at the bottom of the crystal. Hence for the dark field image, the fringes will be out of phase with the bright-field fringes for the fault at the bottom but in phase for the fault at the top, as suggested by Fig. 18.3(b). The top and bottom of the foil may thereby be distinguished.

For deviations from the Bragg angle the difference between $k^{(1)}$ and $k^{(2)}$ increases and the initial amplitudes of the Bloch waves cease to be equal. The fringes therefore become weaker and more closely spaced.

Deviations from the ideal two beam conditions add more Bloch waves and render the fringe pattern more complicated. Interpretation then depends on complete n-beam calculations.

18.4. Dislocations

18.4.1. Diffraction effects

Since the usual diffraction and imaging methods are not sensitive to the detailed configuration of atoms around the core of a dislocation, it is usually sufficient to assume the simple classical models of dislocation strain fields based on macroscopic elastic theory. Often the considerations are limited even further by the assumption of isotropic elastic properties for the material.

For a screw dislocation the Burgers vector $\boldsymbol{b}$ is parallel to the dislocation line. The displacements of the atoms are in the direction of $\boldsymbol{b}$ and decrease in inverse proportion to the distance from the dislocation line. The spacings of the planes of atoms parallel to the dislocation line are assumed to be unaffected.

The pure edge dislocation can be envisaged as the edge of an additional half-plane of atoms. The Burgers vector $\boldsymbol{b}$ is perpendicular to the extra half-plane and so perpendicular to the dislocation line. Planes of atoms perpendicular to the dislocation line maintain their normal spacings. Within these planes, the displacements of the atoms have components R_1 parallel to $\boldsymbol{b}$ and R_2 perpendicular to $\boldsymbol{b}$ where

$$R_1 = \frac{b}{2\pi}\left[\Phi + \frac{\sin 2\Phi}{4(1-\nu)}\right]$$

$$R_2 = -\frac{b}{2\pi}\left[\frac{1-2\nu}{2(1-\nu)}\ln|r| + \frac{\cos 2\Phi}{4(1-\nu)}\right]. \tag{18}$$

Here Φ is the angle measured from the direction perpendicular to the extra half-plane, $\boldsymbol{r}$ is the distance from the dislocation line, and ν is the Poisson's ratio for the material.

The problem of the kinematical diffraction from a needle-shaped crystal having an axial screw dislocation was worked out by Wilson [1952] who showed that the reciprocal lattice points were broadened into discs perpendicular to the dislocation axis, assumed to be the c-axis. The width of these discs then increased with $|\boldsymbol{b}|l$ where $\boldsymbol{b}$ is the Burgers vector and l the appropriate index. The reciprocal lattice maxima for $l = 0$ were unaffected by the dislocation. The equivalent results for a pure edge and mixed dislocation have also been obtained (see Krivoglaz and Ryaboshapka [1963]).

The possibility of observing the diffraction effects due to individual dislocations by using X-ray diffraction is remote since the volume of material appreciably affected by the presence of one dislocation is much too small to give measurable intensities. In most materials containing large numbers of dislocations the orientations of the dislocations may be more or less random or else the dislocations may be segregated into dislocation networks forming small-angle grain boundaries. The diffraction problem then comes within the province of diffraction by a mosaic crystal or by a crystal having internal strain. In each case the effect on the intensities can be evaluated statistically (see Chapter 16). However, in some naturally occurring grain boundaries and in grain boundaries between specially-prepared, superimposed crystals where the lattices on the two sides are related by a simple rotation, periodic arrays of parallel dislocations may be formed. Electron diffraction patterns (Guan and Sass [1973]; Sass and Baluffi [1976]) and X-ray diffraction patterns (Gaudig and Sass [1979]) from such arrays show superlattice reflections corresponding to the dislocation periodicity. It is important, especially in the electron diffraction case to differentiate between extra spots owing to a dislocation network and similar spots which could be produced by double diffraction from the two superimposed crystals. The patterns from dislocation networks, however, do have distinctive features, and agreement has been found between observed patterns and those calculated on this basis (see Sass [1980]).

It now appears quite feasible to obtain electron diffraction patterns from thin needle crystals containing screw dislocations (Cowley [1954]) or from small regions of thin crystals containing individual dislocations of any type (Cockayne et al. [1967]; Cowley [1970]). However, experimental difficulties and the uncertainties of interpretation of intensities strongly affected by dynamical diffraction have so far discouraged any detailed studies by these methods.

18.4.2. The imaging of dislocations

Some indication of the form of the contrast in electron microscope images of dislocations can be derived from arguments based on the column approximation. For columns passing through the dislocation core the planes of atoms are displaced as at a stacking fault except that the displacement takes place over a distance of tens or hundreds of Angstroms. Therefore the projected dislocation line may be expected to show oscillatory contrast similar to that of the stacking fault fringes.

Away from the dislocation line the main effect of the dislocation is to tilt the lattice planes towards or away from the Bragg angle, the tilt being in opposite directions on the two sides. Therefore, except when the undistorted crystal is at the exact Bragg angle, the contrast may be expected to be asymmetric across the dislocation line image.

These deductions are in accord with experimental observations for some cases, but the observed images show a great amount of detail, in general, depending on the diffraction conditions, the Burgers vector and the elastic constants of the material. A system for the rapid computer-generation of theoretical images for various values of these parameters has been evolved by Head [1967]. The method is normally used with a 2-beam dynamical theory approximation but extension to n-beam is possible. By this means images of dislocations, stacking faults or other defects may be calculated for all possible combinations of parameters for a particular system. Then comparison with observed images allows a unique identification of the form of the defect (Humble [1970]; Head et al. [1973]).

For the simplest 2-beam theory for isotropic material the lack of distortion of the spacings of some planes gives the result that for $\boldsymbol{g} \cdot \boldsymbol{b} = 0$ the contrast of a dislocation line image will be zero, where $\boldsymbol{g}$ is the diffraction vector. This simple relationship has been used widely for identification of dislocations, but, as shown by detailed calculations, it may not be valid for anisotropic materials or under *n*-beam conditions, and so must be used with care.

18.4.3. Averaging over angles of incidence

In electron microscopy the incident beam is normally made convergent at the specimen level by the focussing action of the condenser lenses and the fore-field of the objective lens, but the angle of convergence rarely exceeds 10^{-3} radians and is usually less than the angular width for the occurrence of a strong reflection. Hence the assumption of a plane parallel incident beam does not give any serious errors.

For scanning transmission electron microscopy it is often convenient to use a wide-angle detector to collect a large proportion of the scattered radiation. By application of the Reciprocity relationship (Cowley [1969]) this is seen to be equivalent to the use of a very large angle of incidence for conventional transmission electron microscopy. The image contrast is then given by averaging over a large range of incident beam directions. The effect is a strong reduction of the contrast of stacking fault fringes. For dislocation images the overall contrast tends to be somewhat reduced, the oscillatory component tends to vanish and the dislocation image tends to be uniformly dark (in a positive print of a bright field image) (Booker et al. [1974]).

The images of dislocations in X-ray topographs represent an extreme case of averaging over incident angle. While the incident beam is reasonably well collimated by normal X-ray diffraction standards, its convergence angle tends to be several orders of magnitude greater than the angular widths of reflections from perfect crystal regions (typically 10^{-5} radians). The dislocation images are almost uniform black lines with very little indication of any oscillatory contrast or asymmetric profiles (Lang [1959]). It is possible, of course, to obtain well collimated incident X-ray beams; for example by using a beam reflected or transmitted from a thick near-perfect crystal. Then the full range of dynamical contrast effects can be observed as in electron microscopy. Fig. 18.4 is an X-ray topograph of a silicon crystal showing thickness fringes and dislocation images with some oscillatory contrast.

18.4.4. n-beam diffraction effects

If the column approximation is assumed, calculations of diffraction amplitudes or image intensities can be made by suitable modification of the computer programs used for perfect single crystals. The Howie and Whelan [1961] equations, given in Chapter 10, have been used by a number of people. The

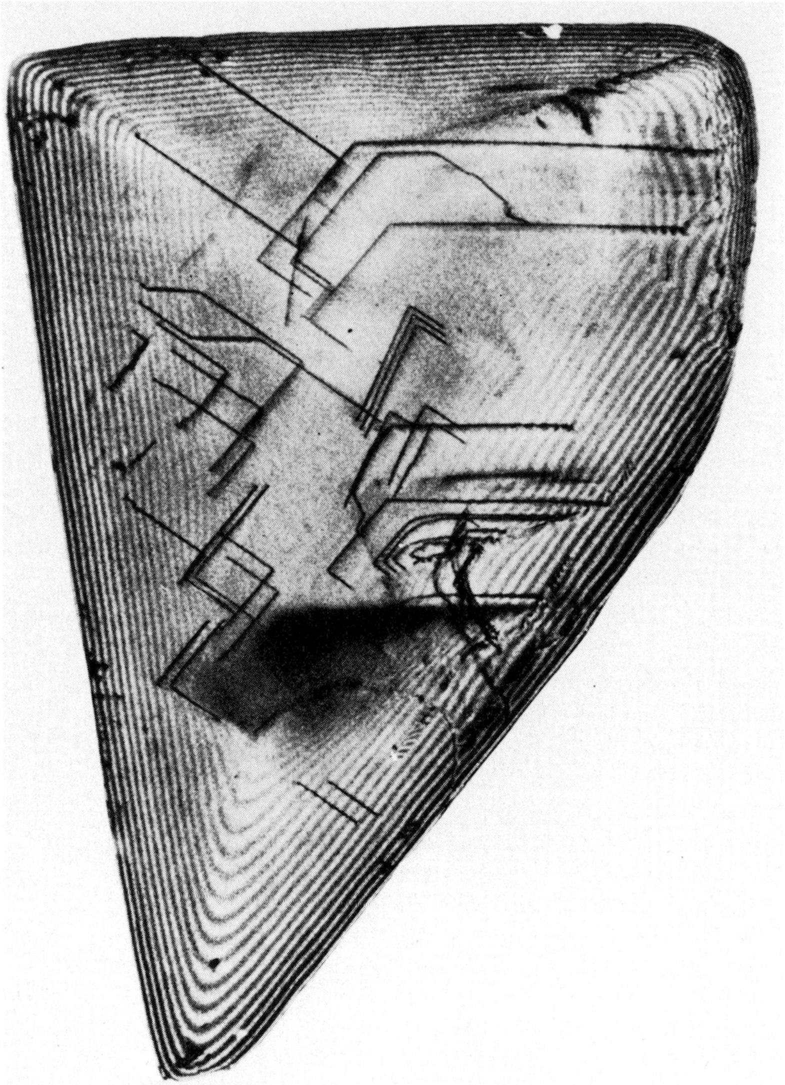

Fig. 18.4. Equal thickness fringes and dislocation images in an X-ray topograph of a pyramidal silicon crystal of maximum diameter 1.5 mm. Obtained using AgK_{α} radiation and a (111)-type reflection. (From Borrmann [1964].)

methods of Chapter 11 may be applied equally well by replacing the constant values of the Fourier coefficients of the potential distributions of the various slices by values which are functions of the depth in the crystal. The values of these Fourier coefficients for the various slices are then generated by a subroutine from the assumed form of the perturbation of the structure.

If the column approximation is not valid, as in the calculation of high resolution (2 to 3 Å) images of the distorted crystal structure, it is possible to use the technique mentioned at the end of Chapter 11 whereby the dislocation or other defect is assumed to occur periodically and the calculation is made as for a structure having a large unit cell, using a very large number of diffracted beams.

One important outcome of the *n*-beam approach is the weak-beam dark-field imaging method of Cockayne et al. [1969]. These authors showed that if, instead of a strongly excited inner reflection, a weak reflection corresponding to a reciprocal lattice point far from the Ewald sphere is used to form a dark-field image, the images of dislocations produced are much sharper. Widths of 10 to 20 Å are common, as compared with the 100 to 200 Å normally observed. Consequently a wealth of fine detail on dislocation separations and interactions has been made available.

Cockayne has pointed out that a simple pseudo-kinematic description gives reasonable agreement with the observations. It may be argued that for a reciprocal lattice point far from the Ewald sphere appreciable intensity will be generated only when the lattice planes are tilted through a relatively large angle and this occurs only near the dislocation core. However the resolution of the images is now such that a description in terms of lattice planes is becoming inadequate for the core region of a dislocation and the calculation of the intensity must, in any case, involve *n*-beam dynamical calculations.

The methods available for the calculation of high-resolution images of defects in crystals have been discussed by Rez [1978] and by Anstis and Cockayne [1979]. The column approximation is valid if the crystal is sufficiently thin. The limiting thickness depends on the accuracy of the results required but may be estimated roughly, as mentioned in Section 10.6, from the requirement that that the lateral spread of the waves owing to Fresnel diffraction should be small compared with the resolution limit. Another way of looking at the column approximation is to note that only the finite set of sharp Bragg reflections is considered and no account is taken of the fact that defects give rise to diffuse scattering around or between the Bragg peaks. Thus the assumption is made that the excitation error for the scattering by the defects is the same as that for the neighboring Bragg reflection so that the Ewald sphere is approximated by a set of planar patches perpendicular to the beam and centered on the reciprocal lattice points. This approximation will clearly be good even for large thicknesses if the diffuse scattering occurs only in the regions very close to the Bragg reflections, i.e. if the defectd involve only slow variations of the lattice spacings.

A better approximation is that introduced by Takagi [1962] and Jouffrey and Taupin [1967]. This introduces a variation of excitation error which is linear with the distance of the scattering from a Bragg reflection. Thus the Ewald sphere is, in effect, approximated by a set of planar patches, which are tangential to the sphere, around each reciprocal lattice point. This is an obvious improvement over the column approximation but fails when the diffuse scattering from the defects is far from the Bragg reflections, i.e. when the variations of atom spacings become appreciable within the distance separating the lattice planes.

The method of Howie and Basinski [1968] goes one step further in including second-order as well as the first-order terms and so gives an accurate representation of the Ewald sphere for all diffuse scattering. It can be used to represent the diffraction and imaging from defects containing quite large local distortions of the crystal structure. Calculations using this method are laborious if the scattering, far from Bragg reflections must be included, but these are feasible with reasonably localized scattering such as is given by strain fields extending quite close to dislocation cores. The method of periodic continuation mentioned in Section 11.5 and Subsection 13.4.2 may best be applied when the distortions of the crystal structure occur over only a small region or where the distortion can be divided into small distinct regions which may be calculated separately. It can deal with arbitrarily large displacements of atoms or with disorder in the occupancy of lattice sites but is not appropriate for extended strain fields. Hence it is complimentary to the above methods which treat successive levels of deviation from a perfectly periodic structure.

Recently images of dislocations, viewed end-on, have been obtained with high-resolution, electron microscopes having a resolution approaching 3 Å. (Spence and Kolar [1979]; Bourret and Desseaux [1979]). This resolution is insufficient to show the positions of atom rows directly resolved in an intuitively interpretable way for the common metals and semiconductors. In the case of silicon and germanium, the pairs of closely spaced atom rows seen in the [110] direction appear as white or black spots about 3 Å apart and the displacements of these pairs of rows near the cores of dislocations or across stacking faults are clearly visible. However, since the separations of these spots are close to the resolution limit, comparison with models of defect structure can be made only with careful, detailed calculations of image intensities for accurately known experimental conditions. Since in such images the extended strain fields of the defects do not affect the image intensity appreciably the method of periodic continuation may be applied for these calculations and is beginning to show valuable results (Bourret et al. [1979]).

As the improvement of resolution of electron microscope continues and especially with the increased application of the high-voltage, high-resolution, electron microscopes, the details of the atom configurations in extended defects will become increasingly clear. Already in one outstanding case Krivanek [1979] has shown the detailed structure of a grain boundary in germanium, interpretable in terms of individual atom rows parallel to the beam. This presages a future in which, to an ever increasing extent, questions on the structure of crystals and their defects will be answered by the determination of the position of each atom individually.

REFERENCES

Allpress, J.G., Elizabeth Hewat, A.F. Moodie and J.V. Sanders, 1972, Acta Cryst. A28, 528. [Ch. 13].

Allpress, J.G. and J.V. Sanders, 1973, J. Appl. Crystallogr. 6, 105. [Ch. 13].

Amelinckx, S., 1964, *Solid state physics*, Supplement 6 (Academic Press, New York). [Ch. 13].

Anderson, J.S., 1969, Bull. Soc. Chim. France, 2203. [Ch.17].

Anderson, J.S., 1978–1979, Chem. Scripta 14, 287. [Ch. 13].

Andersson, B., J.K. Gjønnes and J. Tafto, 1974, Acta Cryst. A30, 216. [Ch. 12].

Anstis, G.R., D.F. Lynch, A.F. Moodie and M.A. O'Keefe, 1973, Acta Cryst. A29, 138. [Ch. 13].

Anstis, G.R. and M.A. O'Keefe, 1976, In: Proc. 34th annual meet. electron microscopy, Ed. G.W. Bailey (Claitors, Baton Rouge) p. 480. [Ch. 13].

Anstis, G.R. and D.J.H. Cockayne, 1979, Acta Cryst. A35, 511. [Ch. 18].

Arsac, J., 1966, *Fourier transforms and the theory of distributions* (Prentice Hall, Englewood-Cliffs, N.J.). [Ch. 2].

Authier, A., 1970, in: *Modern diffraction and imaging techniques in materials science*, Eds. S. Amelinckx et al. (North-Holland, Amsterdam). [Ch. 15].

Authier, A., 1970, in: *Advances in structural research by diffraction methods*, Eds. R. Brill and R. Mason, Vol. 3 (pergamon, Oxford) p. 1.

Avilov, A.S., R.M. Imamov, R.K. Karakhanyan and Z.G. Pinsker, 1973, Kristallografiya 18, 49. [Ch. 6].

Bacon, G.E., 1962, *Neutron diffraction*, 2nd edition (Clarendon Press, Oxford). [Ch. 4, 6, 12].

Balter, S., R. Feldman and B. Post, 1971, Phys. Rev. Letters 27, 307. [Ch. 14].

Bardhan, P. and J.B. Cohen, 1976, Acta Cryst. A32, 597. [Ch. 17].

Bartell, L.S., 1975, J. Chem. Phys. 63, 3750. [Ch. 4].

Batterman, B.W., 1957, J. Appl. Phys. 28, 556. [Ch. 17].

Batterman, B.W., 1962, Appl. Phys. Letters 1, 68. [Ch. 9].

Batterman, B.W., 1964, Phys. Rev. 133, A759. [Ch. 9].

Batterman, B.W. and H. Cole, 1964, Rev. Mod. Phys. 36, 681. [Ch. 8, 9].

Beauvillain, J., 1970, J. de Microscopie 9, 455. [Ch. 14].

Becker, P.J. and P. Coppens, 1974, Acta. Cryst. A30, 129. [Ch. 16].

Bell, W., 1971, Proc. 29th Annual EMSA Meeting, p. 184. [Ch. 15].

Benedek, R. and P.S. Ho, 1973, J. Phys. F: Metal Physics 3, 1285. [Ch. 12].

Berry, M.V., 1971, J. Phys. C. 4, 697. [Ch. 13].

Berry, M.V. and K.E. Mount, 1972, Rep. Progr. Phys. 35, 315. [Ch. 13, 14].

Bethe, H.A., 1928, Ann. Physik 87, 55. [Ch. 1, 8, 16].

Biscoe, J. and B.E. Warren, 1942, J. Appl. Phys. 13, 364. [Ch. 7].
Blackman, M., 1939, Proc. Roy. Soc. Lond. A173, 68. [Ch. 16].
Bless, P.W., R.B. vonDreele, E. Kostiner and R.E. Hughes, 1972, J. Solid State Chem. 4, 262. [Ch. 18].
Bonse, U. and M. Hart, 1965, Appl. Phys. Letters 6, 155. [Ch. 15].
Bonse, U. and M. Hart, 1970, Physics Today, August, p. 26. [Ch. 15].
Booker, G.R., A.M.B. Shaw, M.J. Whelan and P.B. Hirsch, 1967, Phil. Mag. 16, 1185. [Ch. 14].
Booker, G.R., 1970, in: *Modern diffraction and imaging techniques in materials science,* Eds. S. Amelinckx et al. (North-Holland, Amsterdam). [Ch. 14].
Booker, G.R., D.C. Joy, J.P. Spencer, H. Graf von Harrach and M.N. Thompson, 1974, in: *Scanning electron microscopy, 1974*, Proc. 7th annual scanning electron microscopy symposium, Eds. Om Johari and Irene Corvin (I.I.T. Research Institute, Chicago) p. 225. [Ch. 18].
Borie, B., 1957, Acta Cryst. 10, 89. [Ch. 12, 17].
Borie, B., 1959, Acta Cryst. 12, 280. [Ch. 12].
Borie, B., 1961, Acta Cryst. 14, 472. [Ch. 12].
Borie, B., 1966, Acta Cryst. 21, 470. [Ch. 8].
Borie, B., 1970, Acta Cryst. A26, 533. [Ch. 12].
Borie, B. and C.J. Sparks, 1971, Acta Cryst. A27, 198. [Ch. 17].
Born, M. and R.D. Misra, 1940, Proc. Camb. Phil. Soc. 36, 466. [Ch. 12].
Born, M. and E. Wolf, 1975, *Principles of optics,* 5th edition (Pergamon Press, London). [Ch. 3].
Borrmann, G., 1936, Ann. d. Phys. 27, 669. [Ch. 14].
Borrmann, G., 1941, Phys. Zeit. 43, 157. [Ch. 9].
Borrmann, G., 1950, Zeit. f. Phys. 127, 297. [Ch. 9].
Borrmann, G., 1964, Z. f. Kristallogr. 120, 143. [Ch. 14].
Borrmann, G. and W. Hartwig, 1965, Z. f. Kristallogr. 121, 6 and 401. [Ch. 10, 14].
Brout, R., 1965, *Phase transitions* (Benjamin, New York). [Ch. 17].
Bourret, A., A. Renault and G.R. Anstis, 1978–1977, Chem. Scripta 14, 207. [Ch. 18].
Bourret, A. and Desseaux, J., 1979, Philos. Mag. 39, 405. [Ch. 18].
Buerger, M.J., 1959, *Vector space* (Wiley, New York). [Ch. 6].
Bunyan, P., 1963, Proc. Phys. Soc. 82, 1051. [Ch. 4].
Buseck, P.R. and S. Iijima, 1974, Amer. Mineral. 59, 1. [Ch. 13].
Buseck, P.R., 1980, in: *Electron microscopy and analysis 1979,* Ed. T. Mulvey (Institute of Physics, Bristol) p. 93. [Ch. 13].
Buxton, B.F., J.A. Eades, J.W. Steeds and G.M. Rackham, 1976, Phil. Trans. R. Soc. London 281, 171. [Ch. 15].

Cairns, J.A. and R.S. Nelson, 1968, Phys. Letters 27A, 15. [Ch. 14].
Castaing, R. and A. Guinier, 1951, C.R. Acad. Sci., Paris 232, 1948. [Ch. 14].
Castaing, R., 1966, in: *Electron microscopy 1966*, Sixth Internat. Conf. on Electron Microscopy, Kyoto, Vol. 1 (Maruzen Co., Tokyo) p. 63. [Ch. 12].
Chadderton, L.T., 1970, J. Appl. Cryst. 3, 429. [Ch. 14].
Chadderton, L.T., 1973, in: *Channelling: theory, observation and application*, Ed. D.V. Morgan (Wiley, London) p. 287. [Ch. 14].
Chen, H., R.J. Comstock and J.B. Cohen, 1979, Ann. Rev. Mater. Sci. 9, 51. [Ch.17].

Cherns, D.. 1974, Phil. Mag. 30, 549. [Ch. 15].
Chipman, D.R. and C.B. Walker, 1972, Phys. Rev. B5, 3823. [Ch. 17].
Clapp, P.C. and S.C. Moss, 1968, Phys. Rev. 171, 754. [Ch. 17].
Clarke, G.L. and W. Duane, 1922, Proc. Nat. Acad. Sci. 8, 90. [Ch. 14].
Coates, D.G., 1967, Phil. Mag. 16, 1179. [Ch. 14].
Cockayne, D.J.H., P. Goodman, J.C. Mills and A.F. Moodie, 1967, Rev. Sci. Inst. 38, 1093. [Ch. 9, 18].
Cockayne, D.J.H., I.L.F. Ray and M.J. Whelan, 1969, Phil. Mag. 20, 1265. [Ch. 18].
Cockayne, D.J.H., J.R. Parsons and C.W. Hoelke, 1971, Phil. Mag. 24, 139. [Ch. 13].
Cohen, J.B., 1968, *Recent developments concerning the order-disorder transformation, 1968 Seminar on phase transformations*, Detroit, Mich. [Ch. 17].
Cohen, J.B. and L.H. Schwarz, 1978, *Diffraction from materials* (Academic Press, New York) pp. 403–423; 520–530. [Ch. 17].
Colella, R., 1972, Acta Cryst. A28, 11. [Ch. 8, 10].
Colella, R. and J.F. Menadue, 1972, Acta Cryst. A28, 16. [Ch. 8, 15].
Colella, R., 1974, Acta Cryst. A30, 413. [Ch. 15].
Cooper, M.J. and K.D. Rouse, 1970, Acta Cryst. A26, 213. [Ch. 16].
Cowley, J.M. and A.L.G. Rees, 1946, Nature 158, 550. [Ch. 9].
Cowley, J.M. and A.L.G. Rees, 1947, Proc. Phys. Soc. 59, 283. [Ch. 9].
Cowley, J.M., 1950a, Phys. Rev. 77, 669. [Ch. 17].
Cowley, J.M., 1950b, J. Appl. Phys. 21, 24. [Ch. 17].
Cowley, J.M., A.L.G. Rees and J.A. Spink, 1951, Proc. Phys. Soc. 64, 609. [Ch. 16].
Cowley, J.M., 1954, J. Electrochem. Soc. 101, 277. [Ch. 18].
Cowley, J.M. and A.F. Moodie, 1957, Acta Cryst. 10, 609. [Ch. 8, 11].
Cowley, J.M. and A.F. Moodie, 1957a, Proc. Phys. Soc. B70, 486. [Ch. 1].
Cowley, J.M. and A.F. Moodie, 1957b, Proc. Phys. Soc. B70, 497. [Ch. 1].
Cowley, J.M. and A.F. Moodie, 1957c, Proc. Phys. Soc. B70, 505. [Ch. 1].
Cowley, J.M., P. Goodman and A.L.G. Rees, 1957, Acta Cryst. 10, 19. [Ch. 9].
Cowley, J.M. and A.F. Moodie, 1958, Proc. Phys. Soc. 71, 533. [Ch. 11].
Cowley, J.M., 1959, Acta Cryst. 12, 367. [Ch. 13].
Cowley, J.M. and A.F. Moodie, 1959, Acta Cryst. 12, 423. [Ch. 13].
Cowley, J.M. and A.F. Moodie, 1959, Acta Cryst. 12, 360. [Ch. 15].
Cowley, J.M., 1960, Phys. Rev. 120, 1648. [Ch. 17].
Cowley, J.M. and A.F. Moodie, 1960, Proc. Phys. Soc. 76, 378. [Ch. 1].
Cowley, J.M., A.F. Moodie, S. Miyake, S. Takagi and F. Fujimoto, 1961, Acta Cryst. 14, 87. [Ch. 15].
Cowley, J.M. and A.F. Moodie, 1962, J. Phys. Soc. Japan. 17, B-II, 86. [Ch. 11].
Cowley, J.M., 1964, Acta Cryst. 17, 33. [Ch. 14].
Cowley, J.M., 1965a, Phys. Rev. 138, A1384. [Ch. 17].
Cowley, J.M., 1965b, Proc. Int. Conf. on *Electron diffraction and the nature of defects in crystals,* Melbourne (Australian Acad. of Science) Paper J-5. [Ch. 12, 17].
Cowley, J.M., 1967, Crystal structure determination by electron diffraction, in: *Progress in materials science*, Vol. 13, No. 6 (Pergamon Press, Oxford). [Ch. 6].
Cowley, J.M., 1968, Phys. Letters 26A, 623. [Ch. 14].
Cowley, J.M., 1968, Acta Cryst. A24, 557. [Ch. 17].
Cowley, J.M. and R.J. Murray, 1968, Acta Cryst. A24, 329. [Ch. 12, 17].
Cowley, J.M. and A.P. Pogany, 1968, Acta Cryst. A24, 109. [Ch. 12, 13, 17].

Cowley, J.M. and A. Strojnik, 1968, in: *Electron microscopy, 1968*, Proc. Fourth European Regional Conf. on Electron Microscopy, Rome, Vol. 1, 71. [Ch. 13].
Cowley, J.M., 1969, Acta Cryst. A25, 129. [Ch. 15].
Cowley, J.M., 1969, Appl. Phys. Letters 15, 58. [Ch. 13].
Cowley, J.M., 1969, Z. f. Angew. Phys. 27, 149. [Ch. 16].
Cowley, J.M., 1969, Appl. Phys. Letters 15, 58. [Ch. 18].
Cowley, J.M., 1970, J. Appl. Cryst. 3, 49. [Ch. 13, 18].
Cowley, J.M., 1971, in: *Advances in high-temperature chemistry*, Vol. 3 (Academic Press, New York). [Ch. 17].
Cowley, J.M. and S. Kuwabara, 1971, Phys. Letters 34A, 135. [Ch. 12].
Cowley, J.M. and Sumio Iijima, 1972, Z. f. Naturforsch. 27a, 445. [Ch. 13].
Cowley, J.M. and S.W. Wilkins, 1972, in: *Interatomic potentials and simulation of lattice defects*, Eds. P.C. Gehlen, J.R. Beeler and R.I. Jaffee (Plenum Press, New York) p. 265. [Ch. 17].
Cowley, J.M., 1973, Acta Cryst. A29, 529. [Ch. 13].
Cowley, J.M., 1973, Acta Cryst. A29, 537. [Ch. 17].
Cowley, J.M., 1974, in: *Principles and techniques of electron microscopy*, Ed. M.A. Hayat, in press. [Ch. 13].
Cowley, J.M., 1975, in: *Microscopie electronique à haute tension,* Eds. B. Jouffrey and P. Favard (Société Française de Microscopie Electronique, Paris) p. 129. [Ch. 13].
Cowley, J.M., 1976a, Acta Cryst. A32, 83. [Ch. 18].
Cowley, J.M., 1976b, Acta Cryst. A32, 88. [Ch, 18].
Cowley, J.M., 1978a, in: *Electron microscopy 1978 Vol. III,* Ed. J.M. Sturgess (Microscopical Soc. Canada, Toronto) p. 207. [Ch. 13].
Cowley, J.M., 1978b, in: *High voltage electron microscopy 1977,* Eds. T. Imura and H. Hashimoto (Japanese Soc. Electron Microscopy, Tokyo) p. 9. [Ch. 13].
Cowley, J.M. and A.Y. Au, 1978, Acta Cryst A34, 738. [Ch. 18].
Cowley, J.M., 1978–1979, Chemica Scripta 14, 279. [Ch. 13].
Cowley, J.M. and R.E. Bridges, 1979, Ultramicroscopy 4, 419. [Ch. 13].
Cowley, J.M., J.B. Cohen, M.B. Salamon and B.J. Wuensch, Eds., 1979, *Modulated structures-1979 (Kailua Kona, Hawaii)* (American Inst. Physics, New York). [Ch. 18].
Cowley, J.M. and P.M. Fields, 1979, Acta Cryst A35, 28. [Ch. 12].
Cowley, J.M., 1980, in: *Scanning electron microscopy/1980,* Ed., O. Johari (SEM Inc., Illinois). [Ch. 13].
Craven, A.J., J.M. Gibson, A. Howie and D.R. Spalding, 1978, Philos. Mag. A38, 519. [Ch. 12].
Crewe, A.V. and J. Wall, 1970, Optik. 30, 461. [Ch. 13].
Cundy, S.L., A.J.F. Metherell and M.J. Whelan, 1966, in: *Electron Microscopy 1966*, Sixth Internat. Conf. on Electron Microscopy, Kyoto, Vol. 1 (Maruzen Co., Tokyo) p. 87. [Ch. 12].
Cundy, S.L., A. Howie and U. Valdre, 1969, Phil. Mag. 20, 147. [Ch. 12].

Darwin, C.G., 1914, Phil. Mag. 27, 315, 675. [Ch. 1, 8, 10, 16].
Datz, S., C. Erginsoy, G. Liefried and H.O. Lutz, 1967, Ann. Rev. Nucl. Sci. 17, 129.
Dawson, B., 1967, Proc. Roy. Soc. A298, 255, 379. [Ch. 15].
Dawson, B., P. Goodman, A.W.S. Johnson, D.F. Lynch and A.F. Moodie, 1974, Acta Cryst. A30, 297. [Ch. 4].

De Marco, J.J. and P. Suortti, 1971, Phys. Rev. B4, 1028. [Ch. 12].
de Rosier, D.J. and A. Klug, 1968, Nature 217, 130. [Ch. 13].
Ditchburn, R.W., 1976, *Light,* Vol. 1, 3rd edition (Blackie & Sons, London). [Ch. 1].
Doyle, P.A. and P.S. Turner, 1968, Acta Cryst. A24, 390. [Ch. 4].
Doyle, P.A., 1969, Ph.D. Thesis, University of Melbourne. [Ch. 4].
Doyle, P.A., 1969, Acta Cryst. A25, 569. [Ch. 12, 14].
Doyle, P.A., 1970, Acta Cryst. A26, 133. [Ch. 12].
Doyle, P.A., 1971, Acta Cryst. A27, 109. [Ch. 12, 14].
Duffieux, P.M., 1946, *L'integral de Fourier et ses applications à l'optique* (Privately printed, Besançon). [Ch. 1, 3].
Duncumb, P., 1962, Phil. Mag. 7, 2101. [Ch. 9].
Dupuoy, G. and J. Beauvillain, 1970, in: *Microscopie electronique, 1970,* Vol. II, 207. [Ch. 14].

Ehrhart, P., 1978, J. Nucl. Mater. 69-70, 200. [Ch. 12].
Eisenhandler, Clare B. and B.M. Siegel, 1965, J. Appl. Phys. 37, 1613. [Ch. 13].
Erdelyi, A., 1954, *Tables of integral transforms*, Vol. 1, Bateman Mathematical Project (McGraw-Hill, New York). [Ch. 2].
Erickson, H.P. and A. Klug, 1971, Phil. Trans. Roy. Soc. B261, 105 [Ch. 13].
Ewald, P.P., 1916, Ann. Physik 49, 1, 117. [Ch. 1, 8].
Ewald, P.P., 1917, Ann. Physik. 54, 519. [Ch. 8].
Ewald, P.P. and Y. Heno, 1968, Acta Cryst. A24, 5. [Ch. 10].
Ewald, P.P. and Y. Heno, 1968, Acta Cryst. A24, 16. [Ch. 14].

Fejes, P.L., 1970, M. Sc. Thesis, University of Melbourne. [Ch. 15].
Fejes, P., 1973, Ph.D. Thesis, Arizona State University. [Ch. 11].
Fejes, P.L., Sumio Iijima and J.M. Cowley, 1973, Acta Cryst. A29, 710. [Ch. 10, 15].
Fellgett, P.B. and E.H. Linfoot, 1955, Phil. Trans. Roy. Soc. A247, 369. [Ch. 3].
Ferrell, R.A., 1956, Phys. Rev. 101, 554. [Ch. 12].
Fields, P.M. and J.M. Cowley, 1978, Acta Cryst. A34, 103. [Chs. 12, 17].
Fisher, P.M.J., 1965, Proc. Int. Conf. on *Electron diffraction and the nature of defects in crystals* (Australian Acad. of Science, Melbourne) paper IH-4. [Ch. 12].
Fisher, P.M.J., 1965, M.Sc. Thesis, University of Melbourne. [Ch. 17].
Fisher, P.M.J., 1968, Jap. J. Appl. Phys. 7, 191. [Ch. 10].
Fisher, P.M.J., 1969, Ph.D. Thesis, University of Melbourne. [Ch. 15].
Fisher, P.M.J., 1972, private communication. [Ch. 15, 17].
Fisher, P.M.J., 1971, private communication. [Ch. 10].
Fisher, R.M., J.S. Lally, C.J. Humphreys and A.J.E. Metherell, in: *Microscopie Electronique, 1970,* 1, 107. [Ch. 15].
Flinn, P.A., 1956, Phys. Rev. 104, 350. [Ch. 12, 17].
Flocken, J.W. and J.R. Hardy, 1970, Phys. Rev. B1, 2447. [Ch. 12].
Fues, E., 1949, Z. f. Phys. 125, 531. [Ch. 14].
Fujimoto, F., 1959, J. Phys. Soc. Japan 14, 1558. [Ch. 8, 10, 11].
Fujimoto, F. and Y. Kainuma, 1963, J. Phys. Soc. Japan 18, 1792. [Ch. 12, 14].
Fujimoto, F. and A. Howie, 1966, Phil. Mag. 13, 1131. [Ch. 12].
Fujimoto, F. and K. Komaki, 1968, J. Phys. Soc. Japan 25, 1679. [Ch. 12].

Fujimoto, F., 1077, in: *High coltage electron microscopy 1977,* Eds. T. Imura and H. Hashimoto (Japanese Soc. Electron Microscopy, Tokyo) p. 271. [Ch. 14].
Fujiwara, K., 1957, J. Phys. Soc. Japan 12, 7. [Ch. 17].
Fujiwara, K., 1959, J. Phys. Soc. Japan 14, 1513. [Ch. 8, 10, 11].
Fujiwara, K., 1961, J. Phys. Soc. Japan 16, 2226. [Ch. 8].
Fukuhara, A., 1966, J. Phys. Soc. Japan 21, 2645. [Ch. 10].

Gaudig, W. and S.L. Saas, 1979, Philos. Mag. A39, 725. [Ch. 18].
Gehlen, P.C. and J.B. Cohen, 1965, Phys. Rev. 139, A844. [Ch. 17].
Georgopolous, P. and J.B. Cohen, 1977, J. de Physique C-7, Suppl. 12, 191. [Ch. 17].
Giardina, M.D. and A. Merlini, 1973, Z. Naturforsch 28a, 1360. [Ch. 4].
Gjønnes, J.K., 1962, Acta Cryst. 15, 703. [Ch. 8, 16].
Gjønnes, J.K., 1962, J. Phys. Soc. Japan 17, Suppl. BII, 137. [Ch. 17].
Gjønnes, J.K., 1964, Acta Cryst. 17, 1075. [Ch. 4].
Gjønnes, J.K., 1965, Proc. Int. Conf. on *Electron diffraction and the nature of defects in crystals*, Melbourne (Australian Acad. of Science) paper IH-2. [Ch. 12].
Gjønnes, J.K. and A.F. Moodie, 1965, Acta Cryst. 19, 65. [Ch. 13, 15].
Gjønnes, J.K., 1966, Acta Cryst. 20, 240. [Ch. 10, 12, 14].
Gjønnes, J.K. and D. Watanabe, 1966, Acta Cryst. 21, 297. [Ch. 12].
Gjønnes, J.K. and R. Høier, 1971, Acta Cryst. A27, 166. [Ch. 12].
Gjønnes, J.K. and R. Høier, 1971, Acta Cryst. A27, 313. [Ch. 15].
Glauber, R. and V. Schomaker, 1953, Phys. Rev. 89, 667. [Ch. 4].
Goldberger, M.L. and F. Seitz, 1947, Phys. Rev. 71, 294. [Ch. 8].
Goodman, J.W., 1968, *Introduction to fourier optics* (McGraw Hill, New York). [Ch. 1].
Goodman, P. and G. Lehmpfuhl, 1964, Z. Naturforsch. 19a, 818. [Ch. 9].
Goodman, P. and A.F. Moodie, 1965, in: Proc. Int. Conf. on *Electron diffraction and the nature of defects in crystals,* Melbourne (Australian Acad. of Science). [Ch. 11].
Goodman, P. and G. Lehmpfuhl, 1967, Acta Cryst. 22, 14. [Ch. 15].
Goodman, P., 1968, Acta Cryst. A24, 400. [Ch. 15].
Goodman, P. and G. Lehmpfuhl, 1968, Acta Cryst. A24, 339. [Ch. 15].
Goodman, P., 1971, Acta Cryst. A27, 140. [Ch. 15].
Goodman, P. and A.F. Moodie, 1974, Acta Cryst. A30, 280. [Ch. 8, 11, 15].
Goodman, P., 1975, Acta Cryst. A31, 804. [Ch. 15].
Goodman, P. and A.W.S. Johnson, 1977, Acta Cryst. A33, 997. [Ch. 15].
Goodman, P., 1978, in: *Electron diffraction 1927–77,* Eds. P.J. Dobson, J.B. Pendry and C.J. Humphreys (Institute of Physics, Bristol) 116. [Ch. 15].
Gragg, J.E., 1970, Ph.D. Thesis, Northwestern University. [Ch. 17].
Grinton, G.R. and J.M. Cowley, 1971, Optik. 34, 221. [Ch. 11, 13].
Guan, D.Y. and S.L. Sass, 1973, Philos. Mag. 26, 1211. [Ch. 18].
Guinier, A., 1963, *X-ray diffraction in crystals, imperfect crystals and amorphous bodies* (W.H. Freeman and Co., San Francisco and London). [Ch. 5, 18].

Hall, C.R., 1965, Phil. Mag. 12, 815. [Ch. 12].
Hall, C.R. and P.B. Hirsch, 1965, Proc. Roy. Soc. A286, 158. [Ch. 12].
Hall, C.R., P.B. Hirsch and G.R. Booker, 1966, Phil. Mag. 14, 979. [Ch. 12].
Hall, C.R. and P.B. Hirsch, 1968, Phil. Mag. 18, 115. [Ch. 9].

Harburn, G., C.A. Taylor and T.R. Welberry, 1975, *Atlas of optical transforms* (Cornell Univ. Press, New York). [Ch. 2].
Hart, M. and A.D. Milne, 1970, Acta Cryst. A26, 223. [Ch. 15].
Hashimoto, H., A. Howie and M.J. Whelan, 1960, Phil. Mag. 5, 967. [Ch. 18].
Hashimoto, H., 1962, Proc. Roy. Soc. A269, 80. [Ch. 18].
Hashimoto, H., M. Mannami and T. Naiki, 1961, Phil. Trans. Roy. Soc. 253, 459 and 490. [Ch. 13].
Hashimoto, H., A. Kumao, K. Hino, H. Yotsumoto and A. Ono, 1973, Jap. J. Appl. Phys. 10, 1115. [Ch. 13].
Hashimoto, S. and S. Ogawa, 1970, J. Phys. Soc. Japan 29, 710. [Ch. 17].
Haubold, H.G., 1974, Proc. Int. discussion meeting on *studies of lattice distortions and local atomic arrangements by X-ray, neutron and electron diffraction*, Jülich, April–May, 1974. [Ch. 12].
Hauptman, H.A., 1972, *Crystal structure determination: the role of the cosine invariants* (Plenum Press, New York). [Ch. 6].
Hauptman, H., 1978, Acta Cryst. A34, 525. [Ch. 6].
Hashimoto, H., H. Endoh, T. Tanji, A. Ono and E. Watanabe, 1977, J. Phys. Soc. Japan 42, 1073. [Ch. 13].
Head, A.K., 1967, Australian J. Phys. 20, 557. [Ch. 15, 18].
Head, A.K., P. Humble, L.M. Clarebrough, A.J. Morton and C.T. Forwood, 1973, *Computed electron micrographs and defect identification* (North-Holland, Amsterdam). [Ch. 18].
Heidenreich, R.D., 1942, Phys. Rev. 62, 291. [Ch. 9].
Heidenreich, R.D., 1962, J. Appl. Phys. 33, 2321. [Ch. 12].
Heidenreich, R.D., 1964, *Fundamentals of transmission electron microscopy* (Interscience Publishers, New York). [Ch. 8, 13].
Heidenreich, R.D. and R.W. Hamming, 1965, Bell System Tech. J. 44, 207. [Ch. 13]
Helmholtz, H.V., 1886, Crelles Journ. 100, 213. [Ch. 1].
Heno, Y. and P.P. Ewald, 1968, Acta Cryst. A24, 16. [Ch. 10].
Hendricks, S. and E. Teller, 1942, J. Chem. Phys. 10, 147. [Ch. 18].
Hiraga, K., M. Hirabayashi and D. Shindo, 1977, in: *High voltage electron microscopy 1977*, Eds. T. Imura and H. Hashimoto (Japanese Soc. Electron Microscopy) 309. [Ch. 13].
Hirsch, P.B., 1952, Acta Cryst. 5, 176. [Ch. 14].
Hirsch, P.B., A. Howie, R.B. Nicholson, D.W. Pashley and M.J. Whelan, 1965, *Electron microscopy of thin crystals* (Butterworth and Co., London). [Ch. 8, 13, 15, 18].
Hoerni, J., 1950, Helv. Phys. Acta. 23, 587. [Ch. 9].
Hoerni, J.A. and J.A. Ibers, 1953, Phys. Rev. 91, 1182. [Ch. 4].
Hoerni, J.A., 1956, Phys. Rev. 102, 1534. [Ch. 4].
Høier, R., 1969, Acta Cryst. A25, 516. [Ch. 14].
Høier, R., 1973, Acta Cryst. A29, 663. [Ch. 12].
Holmes, R.J., I.E. Pollard and C.J. Ryan, 1970, J. Appl. Cryst. 3, 200. [Ch. 15].
Honjo, G., 1947, J. Phys. Soc. Japan. 2, 133. [Ch. 9].
Honjo, G. and K. Mihama, 1954, J. Phys. Soc. Japan 9, 184. [Ch. 9].
Hopkins, H.H., 1950, *Wave theory of aberrations* (Oxford University Press). [Ch. 3].
Hopkins, H.H., 1953, Proc. Roy. Soc. A217, 408. [Ch. 3].
Hoppe, W., 1956, Z. f. Krist. 107, 406. [Ch. 12].

Hoppe, W., 1964, in: *Advances in structural research by diffraction methods*, Ed. Brill (Interscience Publ.) p. 90. [Ch. 12].
Horiuchi, S., 1978–1979, Chemica Scripta 14, 75. [Ch. 13].
Horstmann, M. and G. Meyer, 1962, Acta Cryst. 15, 271. [Ch. 16].
Horstmann, M. and G. Meyer, 1965, Zeit. f. Phys. 182, 380. [Ch. 16].
Hosemann, R. and R.N. Baggchi, 1962, *Direct analysis of diffraction by matter* (North-Holland, Amsterdam). [Ch. 5].
Hove, L. van, 1954, Phys. Rev. 95, 249. [Ch. 12].
Howie, A. and M.J. Whelan, 1961, Proc. Roy. Soc. A263, 217. [Ch. 8, 10, 18].
Howie, A., 1963, Proc. Roy. Soc. A271, 268. [Ch. 12].
Howie, A., 1966, Phil. Mag. 14, 223. [Ch. 14].
Howie, A. and Z.S. Basinski, 1968, Phil. Mag. 17, 1039. [Ch. 10].
Howie, A. and Z.S. Basinski, 1968, Philos. Mag. 17, 1038. [Ch. 18].
Hren, J.J., J.I. Goldstein and D.C. Joy, Eds., 1979, *Introduction to analytical electron microscopy* (Plenum Press, New York). [Ch. 13].
Huang, K., 1947, Proc. Roy. Soc. A190, 102. [Ch. 12, 17].
Humble, P., 1970, in: *Modern diffraction and imaging techniques in materials science*, Eds. S. Amelinckx et al. (North-Holland, Amsterdam) p. 99. [Ch. 18].
Humphreys, C.J. and P.B. Hirsch, 1968, Phil. Mag. 18, 115. [Ch. 12].
Humphreys, C.J. and M.J. Whelan, 1969, Phil. Mag. 20, 165. [Ch. 12].
Humphreys, C.J. and R.M. Fisher, 1971, Acta Cryst. A27, 42. [Ch. 8].
Humphreys, C.J., L.E. Thomas, J.S. Lally and R.M. Fisher, 1971, Phil. Mag. 23, 87. [Ch. 13].
Humphreys, C.J., J.P. Spencer, R.J. Woolf, D.C. Joy, J.M. Titchmarsh and G.R. Booker, 1972, in: *Scanning electron microscopy 1972* (I.I.T. Research Institute, Chicago). [Ch. 14].
Humphreys, C.J., R. Sandstrom and J.P. Spencer, 1973, in: *Scanning electron microscopy 1973* (I.I.T. Research Institute, Chicago). [Ch. 14].

Iijima, Sumio, 1971, J. Appl. Phys. 42, 5891. [Ch. 13].
Iijima, Sumio, 1973, Acta Cryst. A29, 18. [Ch. 13].
Iijima, S. and J.G. Allpress, 1974a, Acta Cryst. A30, 22. [Ch. 13].
Iijima, S. and J.G. Allpress, 1974b, Acta Cryst. A30, 29. [Ch. 13].
Iijima, S., 1978, Acta Cryst. A34, 922. [Ch. 13].
Iizumi, M., J.D. Axe, G. Shirane and K. Shimaoka, 1977, Phys. Rev. B15, 4392. [Ch. 17].
Imamov, R.M. and Z.G. Pinsker, 1965, Soviet Physics – Crystallography 10, 148. [Ch. 16].
Imamov, R.M., V. Pannkhorst, A.S. Avilov and Z.G. Pinsker, 1976, Kristallografiya 21, 364. [Ch. 16].
Imeson, D., J.R. Sellar and C.J. Humphreys, 1979, Proc. 37th ann. meet. electron microscopy Soc. Amer., Ed. G.W. Bailey (Claitor's, Baton Rouge). [Ch. 15].
Ishida, K., 1970, J. Phys. Soc. Japan 28, 450. [Ch. 12].
Ishizuka, K., 1980, Ultramicroscopy 5, 55. [Ch. 13].
Ishizuka, K. and N. Uyeda, 1977, Acta Cryst. A33, 740. [Ch. 11].
Izui, K., S. Furuno and H. Otsu, 1977, J. Electron Microscopy 26, 129. [Ch. 13].

James, R.W., 1948, *The optical principles of the diffraction on X-rays* (G. Bell.and Sons, London). [Ch. 8, 12, 14].
Jap, B.K. and R.M. Glaeser, 1978, Acta Cryst. A34, 94. [Ch. 11].
Jennison, R.C., 1961, *Fourier transforms and convolutions for the experimentalist* (Pergamon Press, Oxford). [Ch. 2].
Johnson, A.W.S., 1968, Acta Cryst. A24, 534. [Ch. 10].
Johnson, A.W.S., 1972, Acta Cryst. A28, 89. [Ch. 15].
Jouffrey, B. and D. Taupin, 1967, Phil. Mag. 15, 507. [Ch. 10].
Jouffrey, B. and D. Taupin, 1967, Philos. Mag. 16, 703. [Ch. 18].

Kainuma, Y., 1953, J. Phys. Soc. Japan 8, 685. [Ch. 14].
Kainuma, Y. and M. Kogiso, 1968, Acta Cryst. A24, 81. [Ch. 12, 14].
Kakinoki, J. and Y. Komura, 1951, J. Inst. Poly Tech., Osaka City Uni. 2, 1. [Ch. 18].
Kakinoki, J. and Y. Komura, 1952, J. Inst. Poly Tech., Osaka City Uni. 3B, 1 and 35. [Ch. 18].
Kakinoki, J. and Y. Komura, 1952, J. Phys. Soc. Japan 7, 30. [Ch. 18].
Kambe, K., 1957, J. Phys. Soc. Japan 12, 13, 25. [Ch. 14].
Kambe, K. and K. Moliere, 1970, *Advances in structure research by diffraction methods*, Eds. R. Brill and R. Mason, 3, 53. [Ch. 8].
Kambe, K., G. Lehmpfuhl and F. Fujimoto, 1974, Z.f. Naturforsch., 29a, 1034. [Ch. 13, 14].
Kamiya, Y. and R. Uyeda, 1961, J. Phys. Soc. Japan 16, 1361. [Ch. 13].
Kanzaki, H., 1957, J. Phys. Chem. Solids 2, 107. [Ch. 12].
Kato, N., 1952, J. Phys. Soc. Japan 7, 397 and 406. [Ch. 9].
Kato, N., 1955, J. Phys. Soc. Japan 10, 46. [Ch. 14].
Kato, N. and A.R. Lang, 1959, Acta Cryst. 12, 787. [Ch. 9, 15, 18].
Kato, N., 1960, Z. f. Naturforsch. 15a, 369. [Ch. 18].
Kato, N., 1961, Acta Cryst. 14, 526 and 627. [Ch. 9, 18].
Kato, N., 1969, Acta Cryst. A25, 119. [Ch. 9, 15].
Kato, N., 1973, Z. f. Naturforsch. 28a, 604. [Ch. 18].
Kato, N., 1976, Acta Cryst. A32, 543, 458. [Ch. 16].
Kato, N., 1979, Acta Cryst. A35, 9. [Ch. 16].
Kikuchi, S., 1928, Proc. Jap. Acad. Sci. 4, 271, 275, 354, 475. [Ch. 14].
Kikuta, S., T. Matsushita and K. Kohra, 1970, Phys. Letters 33A, 151. [Ch. 15].
Kikuta, S., 1971, Phys. Stat. Sol.(b) 45, 333. [Ch. 15].
Kinder, E., 1943, Naturwiss. 31, 149. [Ch. 9].
Knowles, J.W., 1956, Acta Cryst. 9, 61. [Ch. 9].
Kogiso, M. and Y. Kainuma, 1968, J. Phys. Soc. Japan 25, 498. [Ch. 16].
Kohra, K., 1954, J. Phys. Soc. Japan 9, 690. [Ch. 14, 15].
Kohra, K. and S. Kikuta, 1968, Acta Cryst. A24, 200. [Ch. 15].
Kossel, W., V. Loeck and H. Voges, 1934, Z. f. Phys. 94, 139. [Ch. 14].
Kossel, W. and G. Mollenstedt, 1939, Ann. Phys. 36, 113. [Ch. 9, 14].
Kreutle, M. and G. Meyer-Ehmsen, 1969, Phys. Stat. Sol. 35, K17. [Ch. 15].
Krivanek, O.L., 1978–1979, Chem. Scripta 14, 213. [Ch. 18].
Krivoglaz, M.A. and K.P. Ryaboshapka, 1963, Fiz. Metal. i Metalloved. 15, 18. [Ch. 18].
Krivoglaz, M.A., 1969, *Theory of X-ray and thermal neutron scattering by real crystals* (Plenum Press, New York). [Ch. 12].

Kumm, H., F. Bell, R. Sizmann, H.J. Kreiner and D. Harder, 1972, Rad. Effects 12, 52. [Ch. 14].
Kuriyama, M., 1970, Acta Cryst. A26, 56 and 667. [Ch. 8].
Kuriyama, M., 1972, Acta Cryst. A28, 588. [Ch. 8].
Kuriyama, M., 1973, Z. f. Naturforsch. 28a, 622. [Ch. 18].
Kuwabara, Shigeya, 1959, J. Phys. Soc. Japan 14, 1205. [Ch. 6].
Kuwabara, S., 1961, J. Phys. Soc. Japan 16, 2226. [Ch. 16].
Kuwabara, S., 1962, J. Phys. Soc. Japan 17, 1414. [Ch. 12].
Kuwabara, S., 1967, J. Phys. Soc. Japan 22, 1245. [Ch. 16].
Kuwabara, S. and J.M. Cowley, 1973, J. Phys. Soc. Japan 34, 1575. [Ch. 12].

Ladd, M.F.C. and R.A. Palmer, 1977, *Structure determination by X-ray crystallography* (Plenum, New York). [Ch. 6].
Lally, J.S., C.J. Humphreys, A.J.F. Metherell and R.M. Fisher, 1972, Phil. Mag. 25 321. [Ch. 15].
Lang, A., 1958, J. Appl. Phys. 29, 597. [Ch. 18].
Lang, A., 1959, Acta Cryst. 12, 249. [Ch. 9, 15, 18].
Lang, A., 1970, in: *Modern diffraction and imaging techniques in materials science*, Eds. S. Amelinckx et al. (North-Holland, Amsterdam). [Ch. 15].
Langmore, J.P., J. Wall and M.S. Isaacson, 1973, Optik. 38, 335. [Ch. 13].
Laue, M. von, 1931, Ergeb. Exakt. Natur. 10, 133. [Ch. 1, 8].
Laue, M. von, 1935, Ann. Phys. Lpz. 23, 705. [Ch. 1, 14].
Laue, M. von, 1952, Acta Cryst. 5, 619. [Ch. 9].
Laval, J., 1958, Rev. Mod. Phys. 30, 222. [Ch. 12].
Lehmpfuhl, G. and A. Reissland, 1968, Z.f. Naturforsch. 23a, 544. [Ch. 9, 15].
Lervig, P., J. Lindhard and V. Nielsen, 1967, Nucl. Phys. A96, 481. [Ch. 14].
Lighthill, M.J., 1960, *Fourier analysis and generalized functions* (Cambridge University Press, Cambridge, UK). [Ch. 2].
Lindhard, J., 1965, Mat. Fys. Medd. Dan. Vid. Selsk. 34, 1. [Ch. 14].
Linfoot, E.H., 1955, *Recent advances in optics* (Oxford University Press). [Ch. 3].
Lipson, H. and C.A. Taylor, 1958, *Fourier transforms and X-ray diffraction* (G. Bell and Sons, London). [Ch. 3].
Lipson, H. and W. Cochran, 1966, *The determination of crystal structures*, 3rd revised edition (G. Bell and Sons, London). [Ch. 6].
Lipson, S.G. and H. Lipson, 1969, *Optical physics* (Cambridge University Press, Cambridge, UK). [Ch. 1].
Lonsdale, K., 1947, Phil. Trans. Roy. Soc. A240, 219. [Ch. 14].
Lynch, D., 1971, Acta Cryst. A27, 399. [Ch. 11, 15].
Lynch, D.F. and M. O'Keefe, 1973, Acta Cryst. A28, 536. [Ch. 13].

MacGillavry, C.H., 1940, Physica 7, 329. [Ch. 9].
Mackay, K.J.H., 1966, Proc. IV Congr. *X-ray optics and microanalysis* (Hermann, Paris) p. 544. [Ch. 14].
Massover, W.H., 1978, in: *Electron microscopy 1978,* Vol II, Ed. J.M. Sturgess (Microscopical Soc, Toronto, Canada) 182. [Ch. 13].
Mazel, A. and R. Ayroles, 1968, J. de Microscopie 7, 793. [Ch. 15].

McMahon, A.G., 1969, M.Sc. Thesis, University of Melbourne. [Ch. 15].
Menadue, J.F., 1972, Acta Cryst. A28, 1. [Ch. 8].
Menter, J.W., 1956, Proc. Roy. Soc. A236, 119. [Ch. 13].
Meyer, G., 1966, Phys. Letters 20, 240. [Ch. 15].
Miyake, S. and R. Uyeda, 1950, Acta Cryst. 3, 314. [Ch. 15].
Miyake, S. and R. Uyeda, 1955, Acta Cryst. 8, 335. [Ch. 14, 15].
Miyake, S., 1959, J. Phys. Soc. Japan 14, 1347. [Ch. 8].
Miyake, S., K. Fujiwara, M. Tokonami and F. Fujimoto, 1964, Jap. J. Appl. Phys. 3, 276. [Ch. 13].
Miyake, S., K. Hayakawa and R. Miida, 1968, Acta Cryst. A24, 182. [Ch. 9].
Moliere, K. and H. Wagenfeld, 1958, Z. Krist. 110, 3. [Ch. 9].
Moodie, A.F. and C.E. Warble, 1967, Phil. Mag. 16, 891. [Ch. 13].
Moodie, A.F., 1972, Dynamical *n*-beam theory of electron diffraction, in: *Encyclopaedia dictionary of physics*. Suppl. 4, Ed. A. Thewlis. [Ch. 3, 11].
Moodie, A.F., 1978–1979, Chem. Scripta 14, 21. [Ch. 10].
Moon, A.R., 1972, Zeit. f. Naturforsch. 27b, 390. [Ch. 10].
Morse, P.M., 1930, Phys. Rev. 35, 1310. [Ch. 8].
Moser, B., D.T. Keating and S.C. Moss, 1968, Phys. Rev. 175, 868. [Ch. 17].
Moss, S.C., 1964, J. Appl. Phys. 35, 3547. [Ch. 17].
Moss, S.C., 1965, in: *Local atomic arrangements studied by X-ray diffraction*, Eds. J.B. Cohen and J.E. Hilliard (Gordon and Breach, New York). [Ch. 17].

Nagata, F. and A. Fukuhara, 1967, Jap. J. Appl. Phys. 6, 1233. [Ch. 15].
Nakayama, K., S. Kikuta and K. Kohra, 1971, Phys. Letters 37A, 29. [Ch. 15].
Niehrs, H., 1959a, Z. f. Phys. 156, 446. [Ch. 8].
Niehrs, H., 1959b, Z. f. Naturforsch. 149, 504. [Ch. 8].
Niehrs, H., 1962, J. Phys. Soc. Japan 17, Suppl. B-II, 104. [Ch. 13].
Nussbaum, A. and R.A. Phillips, 1976, *Contemporary optics for scientists and engineers* (Prentice Hall, New Jersey). [Ch. 1].

O'Connor, D.A., 1967, Proc. Phys. Soc. 91, 917. [Ch. 12].
Ogawa, S., 1962, J. Phys. Soc. Japan 17, Suppl. B-II, 253. [Ch. 17].
Ohtsuki, Y.H. and S. Yanagawa, 1965, Phys. Letters 14, 186. [Ch. 8].
Ohtsuki, Y.H. and S. Yanagawa, 1966, J. Phys. Soc. Japan 21, 326. [Ch. 8].
Ohtsuki, Y.H., 1967, Phys. Letters A24, 691. [Ch. 12].
Okamoto, K., T. Ichinokawa and Y. Ohtsuki, 1971, J. Phys. Soc. Japan 30, 1690. [Ch. 14].
O'Keefe, M., 1973, Acta Cryst. A29, 389. [Ch. 13].
O'Keefe, M.A. and S. Iijima, 1978, in: Electron Microscopy 1978 Vol. I, Ed. J.M. Sturgess (Microscopical Soc., Toronto, Canada) p. 282. [Ch. 11, 13].

Parsons, J.R. and C.W. Hoelke, 1969, J. Appl. Phys. 40, 866. [Ch. 13].
Parsons, J.R. and C.W. Hoelke, 1970, Phil. Mag. 22, 1071. [Ch. 13].
Paskin, A., 1958, Acta Cryst. 11, 165. [Ch. 12].
Paskin, A., 1959, Acta Cryst. 12, 290. [Ch. 12].
Pendry, J.B., 1974, Low Energy Electron Diffraction (Academic Press, London). [Ch. 8].
Pfister, H., 1953, Ann. Phys. 11, 239. [Ch. 14].
Phillips, V.A. and J.A. Hugo, 1968, Appl. Phys. Letters 13, 67. [Ch. 13].

Phillips, V.A. and J.A. Hugo, 1970, Acta Met. 18, 123. [Ch. 13].
Pines, D., 1955, in: *Solid state physics* (Academic Press, New York). [Ch. 12].
Pines, D., 1956, Rev. Mod. Phys. 28, 184. [Ch. 12].
Pines, D., 1964, *Elementary excitations in solids* (Benjamin, New York). [Ch. 12].
Pinsker, Z.G., 1964, *Structure analysis by electron diffraction*, Translated from the Russian (Butterworths, London). [Ch. 6, 8].
Pinsker, Z.G., 1978, *Dynamical theory of X-ray scattering in ideal Crystals* (Springer-Verlag, Berlin) (Russian Edition, 1974). [Ch. 8, 9].
Pogany, A.P., 1968, Ph.D. Thesis, University of Melbourne. [Ch. 12].
Pogany, A.P. and P.S. Turner, 1968, Acta Cryst. A24, 103. [Ch. 1].
Pollard, I.E., 1970, Ph.D. Thesis, University of Melbourne. [Ch. 15].
Pollard, I.E., 1972, private communication. [Ch. 15].
Post, B., 1979, Acta Cryst. A35, 17. [Ch. 15].

Raether, H.A., 1980, *Excitation of plasmons and interband transitions by electrons* (Springer-Verlag, Berlin). [Ch. 12].
Rayleigh, Lord, 1881, Phil. Mag. 11, 196. [Ch. 1].
Rees, A.L.G. and J.A. Spink, 1950, Acta Cryst. 3, 316. [Ch. 6].
Rez, P., 1977, Ph.D. Thesis (Oxford Univ., U.K.). [Ch. 8, 10].
Rez, P., 1978, in: *Electron Diffraction 1927–77,* Eds. P.J. Dobson, J.B. Pendry and C.J. Humphreys (Institute of Physics, Bristol) p. 61. [Ch. 18].
Ritchie, R.H., 1957, Phys. Rev. 106, 874. [Ch. 12].

Sass, S.L. and R.W. Baluffi, 1976, Phil. Mag. 33, 103. [Ch. 18].
Sass, S.L., 1980, J. Appl. Cryst. 13, 109. [Ch. 18].
Sato, H. and R.S. Toth, 1963, in: *Alloying behaviour and effects in concentrated solid solutions*, Ed. T.B. Massalski (Gordon and Breach, New York). [Ch. 17].
Scherzer, O., 1949, J. Appl. Phys. 20, 20. [Ch. 13].
Schomaker, V. and R. Glauber, 1952, Nature 170, 290. [Ch. 4].
Schwarzenberger, D.R., 1959, Phil. Mag. 47, 1242. [Ch. 14].
Seeman, H., 1919, Phys. Zeit. 20, 169. [Ch. 14].
Seeman, H., 1926, Naturwiss. 23, 735. [Ch. 14].
Sellar, J.R. and J.M. Cowley, 1973, in: *Scanning electron microscopy, 1973*, Ed. Om Johari (I.I.T. Research Institute, Chicago) p. 243. [Ch. 13].
Shinohara, K., 1932, Sci. Pap. Inst. Phys. Chem. Res. Tokyo 21, 21. [Ch. 14].
Shirley, C.G., 1970, M.Sc. Thesis, University of Melbourne. [Ch. 17].
Shirley, C.G. and Stephen Wilkins, 1972, Phys. Rev. B6, 1252. [Ch. 17].
Shull, C., 1968, Phys. Rev. Letters 21, 1585. [Ch. 9].
Skarnulis, A.J., S. Iijima and J.M. Cowley, 1976, Acta Cryst. A32, 799. [Ch. 13].
Smith, David J. and J.M. Cowley, 1971, J. Appl. Cryst. 4, 482. [Ch. 14].
Smith, P.J. and G. Lehmpfuhl, 1975, Acta Cryst. A31, S220. [Ch. 15].
Sneddon, I.N., 1951, *Fourier transforms,* Appendix C (McGraw-Hill, New York). [Ch. 2].
Spence, J.C.H. and A.E.C. Spargo, 1970, Phys. Letters 33A, 116. [Ch. 12].
Spence, J.C.H., M.A. O'Keefe and S. Iijima, 1978, Phil. Mag. 38, 463. [Ch. 13].
Spence, J.C.H., J.M. Cowley and R. Gronsky, 1979, Ultramicroscopy 4, 429. [Ch. 13].
Spence, J.C.H. and H. Kolar, 1979, Phil. Mag. A39, 59. [Ch. 13, 18].
Steeds, J.W., 1970, Phys. Stat. Sol. 38, 203. [Ch. 15].

Sturkey, L., 1957, Acta Cryst. 10, 858. [Ch. 8].
Sturkey, L., 1962, Proc. Phys. Soc. 80, 321. [Ch. 8, 10].
Sumida, N., Y. Uchida, F. Fujimoto and H. Fujita, 1977, in: *High voltage electron microscopy 1977,* Eds. T. Imura and H. Hashimoto (Japanese Soc. Electron Microscopy, Tokyo) p. 281. [Ch. 14].

Tachiki, M. and K. Teramoto, 1966, J. Phys. Chem. Solids 27, 335. [Ch. 17].
Takagi, S., 1958, J. Phys. Soc. Japan 13, 278, 287. [Ch. 12].
Takagi, S., 1962, Acta Cryst. 15, 1311. [Ch. 18].
Talbot, F., 1836, Phil. Mag. 9, 401. [Ch. 1].
Tanaka, M. and G. Honjo, 1964, J. Phys. Soc. Japan 19, 549. [Ch. 15].
Tanaka, M. and G. Lehmpfuhl, 1972, Acta Cryst. A28, 5202. [Ch. 15].
Taylor, C.A. and H. Lipson, 1964, *Optical transforms* (G. Bell and Sons, London). [Ch. 3, 6].
Terasaki, O., D. Watanabe and J. Gjønnes, 1979, Acta Cryst. A35, 895. [Ch. 15].
Tewordt, L., 1958, Phys. Rev. 109, 61. [Ch. 12].
Tibballs, J.E., 1975, J. Appl. Cryst. 8, 111. [Ch. 17].
Tixier, R. and C. Wache, 1970, J. Appl. Cryst. 3, 466. [Ch. 14].
Tournarie, M., 1960, Bull. Soc. Franc. Miner. Crist. 83, 179. [Ch. 8].
Tournarie, M., 1961, C.R. Acad. Sci. 252, 2862. [Ch. 8].
Turner, P.S. and J.M. Cowley, 1969, Acta Cryst. A25, 475. [Ch. 16].

Unwin, P.N.T., 1971, Phil. Trans. Roy. Soc. London B261, 95. [Ch. 13].
Uyeda, N., T. Kobayashi, E. Suito, Y. Harada and M. Watanabe, 1970, in: *Microscopie électronique, 1970*, Proc. 7th Int. Conf. on Electron Microscopy, Grenoble, Vol. I, p. 23; 1972, J. Appl. Phys. 43, 5181. [Ch. 13].
Uyeda, R. and M. Nonoyama, 1965, Jap. J. Appl. Phys. 4, 498. [Ch. 9, 15].
Uyeda, R., 1968, Acta Cryst. A24, 175. [Ch. 10, 15].
Uyeda, N., T. Kobayashi, K. Ishizuka and Y. Fujiyoshi, 1978–1979, Chemica Scripta 14, 47. [Ch. 13].

Vainshtein, B.K., 1964, *Structure analysis by electron diffraction*, Translated from the Russian (Pergamon Press, Oxford). [Ch. 6, 16].
Van Dyck, D., 1978, in: *Electron microscopy 1978,* Vol. I, Ed. J.M. Sturgess (Microscopical Soc., Toronto, Canada) p. 196. [Ch. 13].
van Hove, L., 1954, Phys. Rev. 93, 268; 95, 249 and 1374. [Ch. 5].
Voges, H., 1936, Ann. Phys. 27, 694. [Ch. 14].

Wadsley, A.D., 1961, Acta Cryst. 14, 660. [Ch. 13].
Warren, B.E. and B.L. Averbach, 1950, J. Appl. Piys. 21, 595. [Ch. 16].
Warren, B.E., B.L. Averbach and B.W. Roberts, 1951, J. Appl. Cryst. 22, 1493. [Ch. 17].
Warren, B.E. and B.L. Averbach, 1952, J. Appl. Phys. 23, 497. [Ch. 16].
Warren, B.E., 1969, *X-ray diffraction* (Addison–Wesley, Reading, Mass.). [Ch. 5, 8, 12, 16, 18].
Watanabe, D. and P.M.J. Fisher, 1965, J. Phys. Soc. Japan 20, 2170. [Ch. 17].
Watanabe, D., R. Uyeda and M. Kogiso, 1968, Acta Cryst. A24, 249. [Ch. 15].
Watanabe, D., R. Uyeda and A. Fukuhara, 1969, Acta Cryst. A25, 138. [Ch. 15].

Watanabe, H., 1966, in: *Electron microscopy, 1966*, 6th Int. Conf. on Electron Microscopy, Kyoto, Vol. 1 (Maruzen Co., Tokyo) p. 63. [Ch. 12].

Weisel, H., 1910, Ann. Phys. Lpz. 33, 995. [Ch. 1].

Whelan, M.J., 1965, J. Appl. Phys. 36, 2099. [Ch. 12].

Willis, B.T.M. and A.W. Pryor, 1975, *Thermal vibrations in crystallography* (Cambridge Univ. Press, Oxford). [Ch. 12].

Wilson, A.J.C., 1942, Proc. Roy. Soc. A180, 277. [Ch. 18].

Wilson, A.J.C., 1952, Acta Cryst. 5, 318. [Ch. 18].

Wolfke, M., 1913, Ann. Phys., Lpz. 40, 194. [Ch. 1].

Woolfson, M.M., 1961, *Direct methods in crystallography* (Oxford Univ. Press, New York). [Ch. 6].

Woolfson, M.M., 1970, *An introduction to X-ray crystallography* (Cambridge Univ. Press, Oxford). [Ch. 6].

Wu, T. and T. Ohmura, 1962, *Quantum theory of scattering* (Prentice Hall, New York). [Ch. 1, 4].

Yoshioka, H., 1957, J. Phys. Soc. Japan 12, 618. [Ch. 12].

Young, R.A., R.J. Gerdes and A.J.C. Wilson, 1967, Acta Cryst. 22, 155. [Ch. 16].

Zachariasen, W.H., 1945, *Theory of X-ray diffractions in crystals* (Wiley, New York). [Ch. 8, 14].

Zachariasen, W.H., 1968, Acta Cryst. A24, 212, 324, 421, 425. [Ch. 16].

Zachariasen, W.H., 1969, Acta Cryst. A25, 102. [Ch. 16].

Index